Lecture Notes in Mathematics

Edited by A. Dold and B. Eckmann

1352

L. L. Avramov K. B. Tchakerian (Eds.)

Algebra
Some Current Trends

Proceedings of the 5th National School in Algebra held in Varna, Bulgaria, Sept. 24 – Oct. 4, 1986

Springer-Verlag
Berlin Heidelberg New York London Paris Tokyo

Editors

Luchezar L. Avramov
Kerope B. Tchakerian
University of Sofia, Institute of Mathematics
ul. "Akad. G. Bončev" bl. 8, 1113 Sofia, Bulgaria

Mathematics Subject Classification (1980): 13B25, 16-02, 16A27, 14C30, 16A39, 16A62, 15A66, 16A38, 22-02, 68P15, 16A38, 13D03, 11L40, 20G40, 20F17, 17-02

ISBN 3-540-50371-4 Springer-Verlag Berlin Heidelberg New York
ISBN 0-387-50371-4 Springer-Verlag New York Berlin Heidelberg

Printed in Germany

Printing and binding: Druckhaus Beltz, Hemsbach/Bergstr.
2146/3140-543210

FOREWORD

The present volume is based on the lectures of invited speakers at the Fifth National School in Algebra, held at the Black Sea coast, near the city of Varna, from September 24 to October 4, 1986.

The preceding National Algebraic Schools in Bulgaria were held biannually starting in 1975, with the primary aim of introducing young algebraists to some specific areas. Accordingly, small numbers of lecturers were invited to give comprehensive accounts of their particular fields.

At the Fifth National School, the number of invited speakers was increased considerably, as was the range of topics covered. The lecturers were requested to give broad surveys, at an advanced level, on topics of current research. We are glad to express the gratitude of the participants and the organizers to the speakers for sharing their insight and enthusiasm for many beautiful algebraic problems.

A second objective of this School was to host an international conference in Algebra, which was organized in five Special Sessions, held in the afternoons. Thanks are due to all participants who gave short communications on their results.

The Editors

M A I N   L E C T U R E S

L.Bokut' - Some new combinatorial results on rings and groups
A.Bovdi - The multiplicative group of a group ring
R.-O.Buchweitz - Maximal Cohen-Macaulay modules and Tate cohomology over Gorenstein rings
K.Buzási - On representations of infinite groups
J.Carlson - Extensions of mixed Hodge structures
P.M.Cohn - Valuations on skew fields
E.Formanek - Invariants of nxn matrices
J.Herzog - Matrix factorizations of homogeneous polynomials
V.Iskovskih - On the rationality problem for conic bundles
T.Józefiak - Characters of projective representations of symmetric groups
H.Koch - Unimodular lattices and self-dual codes
G.Margulis - Lie groups and ergodic theory
B.Plotkin - Algebraic models of data bases
A.Regev - PI-algebras and representation theory
J.-E.Roos - Decomposition of injective modules, von Neumann regular rings and factors
A.Šmel'kin - The Specht property of some varieties of representations of groups and Lie algebras
A.Tietäväinen - Incomplete sums and two applications of Deligne's result
M.Vaughan-Lee - The restricted Burnside problem
W.Vogel - Castelnuovo bounds for algebraic sets in n-space
R.A.Wilson - Maximalsubgroups of sporadic simple groups
A.Zalesskii - Recognition problems of linear groups and the theory of group representations
G.Zappa - Normal Fitting classes of groups and generalizations

SPECIAL SESSIONS

I. RING THEORY

I.Chipchakov - On the normality of locally finite division algebras
P.M.Cohn - The specialization lemma for skew fields
D.Dikranjan - Closure operators in algebras and topology
V.Drensky - A combinatorial approach to PI-algebras
E.Formanek - A conjecture of Regev on the Capelli polynomials
T.Gateva-Ivanova - Recognizable properties of associative algebras
P.Koshlukov - Polynomial identities for a family of Jordan algebras
K.Murata - On arithmetical ideal theory of a non-maximal order of a ring
A.Orsatti - The basic ring of a locally Artinian category
A.Popov - Varieties of algebras and their lattices of subvarieties
A.Regev - On Kemer's algebras
L.Vladimirova - The codimension sequences of some T-ideals

II. GROUP THEORY

J.Alonso - Growth functions of amalgams
Ts.Gentchev - Products of finite simple groups
A. Kondrat'ev - Finite linear groups of small degree
T. Krausz - Crossed group algebras over finite fields
J.Krempa - On Gel'fand-Kirillov dimension of tensor products
O. Macedońska - On lifting automorphisms
S.Mihovski - Simple crossed products of groups and rings
T.Mollov, N.Nachev - Semi-simple crossed group algebras of cyclic p-groups of odd order
N.Petrova - Some results in the theory of right-ordered groups
E.Szabó - Abelian group-pairs
S.Todorinov - Semi-group conditions in the theory of partially ordered groups
G.Zappa - Remarks on Fitting classes of finite groups

III. COMBINATORICS

E.Ademaj - Projective planes of order 13 with a Frobenius group of order 39

B.Brindza - On minimal distance of (p,p) codes

G.Fazekas - Permutation source coding methods in digital signal processing

P.Lakatos - Wreath products of groups and their applications in coding theory

V.Tonchev - Symmetric designs without ovals and extremal self-dual codes

R.A.Wilson - New families of ovoids on O_8^+

V.Yorgov - A (72, 36, 16) doubly even code does not have an automorphism of order 11

IV. COMMUTATIVE ALGEBRA

R.Achilles - Stückrad-Vogel intersection numbers

A.Aramova - Constants of a derivation in positive characteristic

L.Avramov - Periodicity of modules over local rings

J.Backelin - Now many semigroup rings are there?

R.Fröberg - Gorenstein rings as maximal subrings of $K[[x]]$ with fixed conductor

C.Jacobsson, C.Löfwall - The radical of graded Lie algebras

P.Schenzel - Noetherian symbolic blow-up rings

A.Tyc - Differential basis, p-basis and smoothness in characteristic $p > 0$

V. ALGEBRAIC GEOMETRY

J.Carlson - Maximality theorems for variations of mixed Hodge structures

R.-O.Buchweitz - Algebraic cycles on smooth hypersurfaces

S.Dimiev - Hypocomplex structures on some Lie groups

V.Kanev - Universal properties of Prym varieties of singular curves

A.Kasparian - Deformations of complete intersections of symmetric spaces

A.Todorov - Deformations of Calabi-Yau manifolds

G.Tomanov - The congruence problem for some anisotropic algebraic groups

W.Vogel - Castelnuovo bounds for locally Cohen-Macaulay schemes

LIST OF PARTICIPANTS

R.Achilles (Halle)
E.Ademaj (Prishtina)
J.Alonso (Stockholm)
L.Apostolova (Sofia)
A.Aramova (Sofia)
E.Arnaudova (Sofia)
L.Avramov (Sofia)
J.Backelin (Stockholm)
P.Balyuchev (Plovdiv)
L.Bokut' (Novosibirsk)
A.Bovdi (Uzhgorod)
Yu.Bozhkov (Sofia)
M.Brakalova (Sofia)
G.Brattström (Stockholm)
B.Brindza (Debrecen)
R.-O.Buchweitz (Hannover)
K.Buzási (Debrecen)
J.A.Carlson (Salt Lake City)
P.Chekova (Sofia)
I.Chipchakov (Moscow)
P.M.Cohn (London)
L.Davidov (Sofia)
D.Dikranjan (Sofia)
S.Dimiev (Sofia)
Z.Dimitrov (Plovdiv)
S.Dodunekov (Sofia)
M.Dossev (Sofia)
V.Drensky (Sofia)
G.Eneva (Sofia)
G.Fazekas (Debrecen)
E.Formanek (University Park)
R.Fröberg (Stockholm)
V.Gasharov (Sofia)
T.Gateva-Ivanova (Sofia)
M.Gavrilov (Sofia)
Ts.Gentchev (Shoumen)
P.Georgieva (Plovdiv)
S.Grozdev (Sofia)
J.Herzog (Essen)
S.Hineva (Sofia)
I.Ilarionov (Sofia)
V.Iliev (Sofia)
V.Iskovskih (Moscow)
C.Jacobsson (Stockholm)
T.Józefiak (Toruń)
I.Kalev (Sofia)
V.Kanev (Sofia)
S.Kapralov (Gabrovo)
A.Kasparian (Sofia)
P.Kitanov (Sofia)
H.Koch (Berlin)
K.Kolikov (Plovdiv)
A.Kondrat'ev (Sverdlovsk)
P.Koshlukov (Sofia)
T.Krausz (Debrecen)
J.Krempa (Warsaw)
P.Lakatos (Debrecen)
D.Levchenko (Sofia)
C.Löfwall (Stockholm)
O.Macedońska (Gliwice)
N.Manev (Sofia)
G.Margulis (Moscow)
V.Markova (Sofia)
S.Mihovski (Plovdiv)
H.Minchev (Sofia)
T.Mollov (Plovdiv)
K.Murata (Yamaguchi City)
N.Nachev (Plovdiv)
N.Nenov (Sofia)
R.Nikolaev (Varna)
R.Nikolov (Sofia)
D.Nikolova (Sofia)

A.Orsatti (Padova)
T.Pantev (Sofia)
I.Peeva (Sofia)
K.Petrov (Sofia)
N.Petrov (Shoumen)
N.Petrova (Plovdiv)
B. Plotkin (Riga)
A.Popov (Sofia)
P.Radnev (Plovdiv)
A.Rahnev (Plovdiv)
V.Raichev (Sofia)
Ts.Rashkova (Russe)
A.Regev (University Park)
J.-E.Roos (Stockholm)
P.Schenzel (Halle)
A.Šmel'kin (Moscow)
A.Shterev (Sofia)
P.Siderov (Sofia)
A.Stoyanova (Plovdiv)
E.Szabó (Debrecen)

K.Tchakerian (Sofia)
S.Teneva (Plovdiv)
A.Tietäväinen (Turku)
S.Todorinov (Plovdiv)
A.Todorov (Sofia)
K.Todorov (Sofia)
G.Tomanov (Sofia)
V.Tonchev (Sofia)
T.Tonkov (Sofia)
I.Tonov (Sofia)
A.Tyc (Toruń)
M.Vaughan-Lee (Oxford)
L.Vladimirova (Sofia)
W.Vogel (Halle)
A.R.Wilson (Cambridge)
V.Yorgov (Shoumen)
A.Zalesskii (Minsk)
G.Zappa (Firenze)
N.Zyapkov (Shoumen)

TABLE OF CONTENTS

MATRIX FACTORIZATIONS OF HOMOGENEOUS POLYNOMIALS

JÖRGEN BACKELIN
Matematiska institutionen
Stockholms Universitet
Box 6701
S-113 85 Stockholm
Sweden

JÜRGEN HERZOG and HERBERT SANDERS
Universität - Gesamthochschule - Essen
FB 6 Mathematik
Universitätsstraße 3
D-4300 Essen 1
Federal Republic of Germany

Contents.

Introduction

This paper originates in the aim to describe the linear maximal Cohen-Macaulay modules (*MCM-modules*) over a graded hypersurface ring $R = S/(f)$, where $S = k[X_1, \ldots, X_n]$ is the polynomial ring over a field k and f is a homogeneous form of degree d. If M is a graded MCM-module, then it has a free homogeneous S-resolution

$$0 \longrightarrow S^m \xrightarrow{\alpha} S^m \longrightarrow M \longrightarrow 0\,.$$

With respect to homogeneous bases, α can be described as an $m \times m$-matrix with homogeneous polynomials as entries. M is called a *linear* MCM-module if all the entries of α are linear forms. Quite generally, if M is an MCM-module over R admitting a rank, then $\mu(M) \leq e(R) \cdot \operatorname{rank} M$, where μ denotes the minimal number of generators of M and $e(R)$ the multiplicity of R. Ulrich calls an MCM-module *maximally generated* if $\mu(M) = e(R) \cdot \operatorname{rank} M$, and asks in his paper [U] if any CM-ring admits a maximally generated MCM-module. It turns out that the maximally generated MCM-modules (which are sometimes called *Ulrich modules*) are just the linear MCM-modules if R is a graded hypersurface ring. The existence of Ulrich-modules is confirmed for one-

and normal two-dimensional CM-rings, as well as for rings of minimal multiplicity, see [HK] and [BHU]. We show in the STOP PRESS at the end of the paper that e.g. all hypersurface rings of characteristic zero admit Ulrich-modules.

The study of MCM-modules over a hypersurface ring leads to matrix factorizations as introduced by Eisenbud in [E]. Given a homogeneous polynomial $f \in S$, an equation $fE = \alpha\beta$, where α and β are square matrices with homogeneous polynomials as entries, is called a *matrix factorization* of f. (Here E denotes the unit matrix; for simplicity we write f for fE in the sequel.)

As an example of a matrix factorization, consider

$$X_1^2 + X_2^2 = \begin{pmatrix} X_1 & X_2 \\ X_2 & -X_1 \end{pmatrix} \begin{pmatrix} X_1 & X_2 \\ X_2 & -X_1 \end{pmatrix} .$$

According to Eisenbud's theory the MCM-modules over $R = S/(f)$ correspond to matrix factorizations $f = \alpha\beta$, and the linear MCM-modules to such factorizations for which α is a matrix of linear forms. The precise statement is given in section 4.

For a non-singular quadratic form f it is shown in the paper of Buchweitz-Eisenbud-Herzog [BEH], that there exists essentially just one matrix factorization $f = \alpha\beta$, where α and β are matrices of linear forms, and consequently there exist up to isomorphisms at most two indecomposable linear MCM-modules (one corresponding to α, the other to β). This result is obtained by considering the Clifford algebra C of f, and showing that there is as well a correspondence between the matrix factorizations of f and the $\mathbf{Z}/2\mathbf{Z}$-graded modules over C.

In this paper we imitate this method in order to obtain similar but somewhat weaker results for homogeneous polynomials $f \neq 0$ of degree $d > 2$.

In section 1 we introduce the notion of a *generalized Clifford algebra for* f. Let

$$V = \bigoplus_{i=1}^{n} k e_i$$

be an n-dimensional k-vectorspace with basis $e_1, \ldots, e_n$, then f defines a map $f : V \longrightarrow k$, by $f(x_1e_1 + \ldots + x_ne_n) := f(x_1, \ldots, x_n)$ for all $x_i \in k$. A $\mathbf{Z}/d\mathbf{Z}$-graded algebra C together with a monomorphism $V \to C_1$ is called a generalized Clifford algebra, if C is generated by V, and $f(x) = x^d$ for all $x \in V$. It is not at all clear that such an object exists for f. Of course, the most natural thing to do is to form the tensor algebra $T(V)$ of V and to divide by the relations $x \otimes \cdots \otimes x - f(x)$. We call

$$C(f) := T(V)/(\{x \otimes \cdots \otimes x - f(x) | x \in V\})$$

the *universal Clifford algebra for* f.

This algebra was first introduced by N. Roby ([R]) in this generality. The special case of binary cubic forms was already studied by N. Heerema ([H]) in 1954.

Using Gröbner basis arguments we show in theorem 1.8 that the natural map $V \to C_1(f)$ is an inclusion.

Of course any other generalized Clifford algebra for f is a quotient of $C(f)$ by a homogeneous two-sided ideal a in $C(f)$ with $a \cap V = 0$.

The essential observation in this paper is that the $\mathbf{Z}/d\mathbf{Z}$-graded modules over a generalized Clifford algebra correspond to *linear* matrix factorizations

$$f = \alpha_0 \cdot \ldots \cdot \alpha_{d-1} ,$$

where $\alpha_0, \ldots, \alpha_{d-1}$ are square matrices of linear forms, and moreover that if $f = \alpha_0 \cdot \ldots \cdot \alpha_{d-1}$ corresponds to $M = \bigoplus_{i=0}^{d-1} M_i$, then the size of α_i equals $\dim_k M_i$, which is independent of i (theorem 1.3).

A similar observation was made by Roby ([R]). Not taking into account the $\mathbf{Z}/d\mathbf{Z}$-grading of $C(f)$ he assigns to each $C(f)$-module a factorization $f = \alpha^d$ of f as a pure power of a linear square matrix. Such factorizations have been studied thoroughly by L. N. Childs ([C]).

Unfortunately $\dim_k C(f)$ is infinite if and only if $n > 1$ and $d > 2$, as is shown in theorem 1.8. (In the case that Char $k > d$ this theorem has been shown by Childs ([C, theorem 3]), using the early results by Heerema ([H]).) As a consequence, it is not clear whether there exists a finite dimensional $\mathbf{Z}/d\mathbf{Z}$-graded module over $C(f)$ for an arbitrary form f of degree at least three in more than two variables. Thus we don't know whether all homogeneous forms have linear matrix factorizations in the above sense with finite matrices.

On the other hand, if we weaken the conditions on the factorization slightly, we easily obtain factorizations of f with finite matrices:

— If we do *not* demand the factorization to be *linear*, we may just choose a non-trivial MCM-module over R, for instance the $(n-1)$ th syzygy module Ω_R^{n-1} of k. To this module corresponds according to Eisenbud's theory a matrix factorization $f = \alpha\beta$. (Both matrices cannot have linear forms as entries, unless f is a quadratic form.)

— If we allow *non-square* matrices, we may by quite elementary means decompose the 1×1-matrix (f) into factors, all of whose entries are linear:
For $i = 1, \ldots, d-1$, let B_i be a $1\times\binom{n+i-1}{i}$ row matrix, whose entries are all the different monomials of degree i in the variables $X_1, \ldots, X_n$. For $i \in \{1, \ldots, d-2\}$ there is a matrix α_i', whose non-zero entries all are variables, such that

$$B_i \alpha_i' = B_{i+1} .$$

Finally, there is an $\binom{n+d-2}{d-1} \times 1$ column matrix β, such that $(f) = B_{d-1}\beta$. If we let $\alpha_0 = B_1$, $\alpha_i = \alpha_i'$ for $i = 1, \ldots, d-2$, and $\alpha_{d-1} = \beta$, then indeed

$$(f) = \alpha_0 \cdot \ldots \cdot \alpha_{d-1} .$$

Even though such factorizations may be useful in other contexts, they do not provide us with linear MCM-modules over the hypersurface ring $S/(f)$.

In section 2 we extend a result of Atiyah-Bott-Shapiro [ABS] to generalized Clifford algebras. It essentially says that the category of $\mathbf{Z}/d\mathbf{Z}$-graded modules over a generalized Clifford algebra C is equivalent to the category of C_0-modules. This result considerably simplifies the further considerations, and applied to matrix factorizations it gives a deeper insight into the relations among the factors of a factorization $fE = \alpha_0 \cdot \ldots \cdot \alpha_{d-1}$. For instance we are able to say which of the factors α_i that are equivalent to each other (corollary 2.7), or under which circumstances fE is a power of a single matrix (corollary 2.4).

Section 3 is devoted to the study of diagonal forms. Most of its results may be found (explicitly or implicitly) in [L] or in [C]; we however give an essentially self-contained presentation of the results and proofs.

For the diagonal forms, finite-dimensional generalized Clifford algebras may be constructed. Just as for quadratic forms one obtains these Clifford algebras as tensor products of cyclic algebras. More generally, suppose f_1 and f_2 are forms of degree d in disjoint sets of variables, and let C_i be a generalized Clifford algebra for f_i $(i = 1, 2)$. We assume that k contains a d-th primitive root ξ of unity. Then we define the $\mathbf{Z}/d\mathbf{Z}$-graded tensor product $C_1 \widehat{\otimes} C_2$ as the ordinary tensor product equipped with the multiplication defined by $(a \otimes b)(c \otimes d) = \xi^{(\deg b)(\deg c)} ac \otimes bd$ for homogeneous elements $b \in C_2$ and $c \in C_1$. It turns out (theorem 3.1) that $C_1 \widehat{\otimes} C_2$ is a generalized Clifford algebra for $f_1 + f_2$. Now if $f = a_1 X_1^d + \ldots a_n X_n^d$ is a diagonal form with $a_i \in k$, $a_i \neq 0$ for $i = 1, \ldots, n$, then $C_i = k[e_i]/(e_i^d - a_i)$ is a generalized Clifford algebra for $a_i X_i^d$, whence $C(f, \xi) = C_1 \widehat{\otimes} \ldots \widehat{\otimes} C_n$ is a generalized Clifford algebra for f, whose dimension over k is d^n. The structure of this algebra can be described quite easily. In theorem 3.6 it is shown that $C_0(f, \xi)$ is simple if n is odd and semisimple if n is even. The consequences for matrix factorizations of diagonal forms are formulated in theorem 3.9. At the end of this section we work out explicit factorizations of $\sum_{i=1}^n X_i^d$ over $\mathbf{C}$.

Finally, in section 4 we show that a linear matrix factorization $f = \alpha_0 \cdot \ldots \cdot \alpha_{d-1}$ corresponds to a free module F over the hypersurface ring $R = S/(f)$ together with a filtration of F, whose quotients are linear MCM-modules over R. In particular, together with the results of section 3, it follows that a hypersurface ring of a diagonal form admits linear MCM-modules.

Many questions remain open [but see the STOP PRESS!]. We list a few of them:

1) Does every (homogeneous) form admit a finite-dimensional generalized Clifford algebra?
2) Do the linear MCM-modules together with R generate the Grothendieck group of R?
3) Can the periodicity theorem of Knörrer [K] be generalized to forms of higher degree?
4) Which forms can be transformed into diagonal forms?

We wish to thank T. G. Ivanova with whom we had many stimulating and helpful discussions, and P. M. Cohn for his valuable comments and suggestions. We also thank Bokut who informed us that L'vov and Nesterenko (answering a question of Krendelev) reported on the solution of question 1 at the 17:th All Union Algebra Conference in Minsk 1983, and announced this and related results (without proofs) in the Proceedings of that conference (pp 118 and 137, in Russian).

In particular, we thank the referee for putting our attention to the extensive work already done concerning generalized Clifford algebras (e.g. in [C], [H], [L], and [R]).

Finally we would like to express our gratitude to the organizers of the Fifth National School in Algebra in Varna, who brought together two of the authors of this paper and made possible many fruitful discussions with other participants of this conference that were indispensable for writing this paper.

1. Matrix factorizations and Clifford algebras

Let $f \neq 0$ be a homogeneous polynomial of degree $d \geq 2$ in the indeterminates $X_1, \ldots, X_n$ with coefficients in a field k.

DEFINITION 1.1. A (*linear*) *matrix factorization* of f (of size m) is an equation $f = \alpha_0 \cdot \ldots \cdot \alpha_{d-1}$, where the α_i are square matrices (of size m), whose entries are linear forms in the indeterminates $X_1, \ldots, X_n$ with coefficients in k, and f simply stands for f times the unit matrix E of size m.

We allow m to be infinite. In that case, however, we require that each row of the matrices has only finitely many nonzero entries, whence their products are defined, and that the product of any cyclic permutation of the matrices α_i is f again.

Given a matrix factorization $f = \alpha_0 \cdot \ldots \cdot \alpha_{d-1}$ and a $j \in \mathbf{Z}$, we set $\alpha_j := \alpha_i$, where $0 \leq i \leq d-1$ and $i \equiv j \pmod d$.Then, since any cyclic permutation of the factors again yields f as their product, it follows that $f = \alpha_i \cdot \alpha_{i+1} \cdot \ldots \cdot \alpha_{i+d-1}$ also is a matrix factorization for all $i \in \mathbf{Z}$.

Two matrix factorizations $f = \alpha_0 \cdot \ldots \cdot \alpha_{d-1}$, $f = \beta_0 \cdot \ldots \cdot \beta_{d-1}$ of the same size are called *equivalent* if there exists matrices $S_j \in \mathrm{Gl}(m,k)$ such that $\beta_j = S_j \alpha_j S_{j+1}^{-1}$ for all j.

The *sum* of the matrix factorizations $f = \alpha_0 \cdot \ldots \cdot \alpha_{d-1}$ and $f = \beta_0 \cdot \ldots \cdot \beta_{d-1}$ is the matrix factorization $f = \gamma_0 \cdot \ldots \cdot \gamma_{d-1}$, where

$$\gamma_i = \begin{pmatrix} \alpha_i & 0 \\ 0 & \beta_i \end{pmatrix}$$

for all i.

The matrix factorization $f = \alpha_0 \cdot \ldots \cdot \alpha_{d-1}$ is called *indecomposable* if it is not equivalent to a sum of matrix factorizations of f.

We consider the k-vector space $\bigoplus_{i=1}^{n} kX_i$ as the dual space of $V := \bigoplus_{i=1}^{n} ke_i$, where the basis $X_1, \ldots, X_n$ is dual to the basis $e_1, \ldots, e_n$.

Recall that a matrix α of size m with linear forms in n variables may be interpreted as a k-linear map $\phi \colon V \to \mathrm{Hom}_k(V_1, V_2)$, where V_1 and V_2 are m-dimensional k-vectorspaces (with specified bases): given a matrix α of linear forms and $x \in V$, we let $\alpha(x)$ be the matrix with coefficients in k, which is obtained from α by evaluating the entries of α at x. With respect to the given bases of V_1 and V_2 $\alpha(x)$ defines a linear map $\phi(x) \colon V_1 \to V_2$. We therefore may define

$$\begin{array}{ccc} V & \longrightarrow & \mathrm{Hom}_k(V_1, V_2) \\ x & \longmapsto & \phi(x) \end{array}.$$

Similarly one associates with $\phi \colon V \to \mathrm{Hom}_k(V_1, V_2)$ a matrix α of linear forms.

Therefore, given a matrix factorization $f = \alpha_0 \cdot \ldots \cdot \alpha_{d-1}$, there exist k-vector spaces $V_0 = V_d, V_1, \ldots, V_{d-1}$ such that α_i yields a linear map

$$\begin{array}{ccc} V & \longrightarrow & \mathrm{Hom}_k(V_i, V_{i+1}) \\ x & \longmapsto & \phi_i(x) \end{array}.$$

If we set $f(x) := f(x_1, \ldots, x_n)$ for $x = \sum x_i e_i \in V$, we obtain $f(x) \cdot \mathrm{id}_{V_0} = \phi_{d-1}(x) \circ \ldots \circ \phi_0(x)$ for all $x \in V$.

In this paper we are often dealing with $\mathbf{Z}/d\mathbf{Z}$-graded modules over $\mathbf{Z}/d\mathbf{Z}$-graded rings. If

$$M = \bigoplus_{i=0}^{d-1} M_i$$

is a $\mathbf{Z}/d\mathbf{Z}$-graded module and $j \in \mathbf{Z}$, we set $M_j = M_i$, where $0 \leq i \leq d-1$ and $i \equiv j \pmod d$. Then if we use the convention that $M(a)$ denotes the module shifted by a (so that $M(a)_i = M_{a+i}$ for all i), it follows that $M(a)$ is obtained from M by a cyclic permutation of the homogeneous components of M.

Given a matrix factorization $f = \alpha_0 \cdot \ldots \cdot \alpha_{d-1}$ of size m, we assign to it a $\mathbf{Z}/d\mathbf{Z}$-graded module M over the tensor algebra $T := T(V)$. We first choose a $\mathbf{Z}/d\mathbf{Z}$-graded k-vector space $M = \bigoplus_{i=0}^{d-1} M_i$, where $\dim_k M_i = m$ for all i.

Let $\phi_i \colon V \to \operatorname{Hom}_k(M_i, M_{i+1})$ be the k-linear maps associated with the matrices α_i, as described above. The T-module structure of M is then defined by the equation $x \cdot m = \phi_i(x)(m)$ for all $x \in V$, $m \in M_i$, and $i = 0, \ldots, d-1$.

Let $m \in M_i$ and $x \in V$. Then

$$(x^{\otimes d})m = (\phi_{i+d-1}(x) \circ \ldots \circ \phi_{i+1}(x) \circ \phi_i(x))m = f(x)m$$

(where $x^{\otimes d} = \overbrace{x \otimes \ldots \otimes x}^{d \text{ copies}}$). It follows that the two-sided ideal $I(f) = \langle \{x^{\otimes d} - f(x) \mid x \in V\} \rangle$ is contained in the annihilator of M, so that M is a module over the $\mathbf{Z}/d\mathbf{Z}$-graded algebra

$$C(f) := T/I(f).$$

We call $C(f)$ the *universal (generalized) Clifford algebra* of f.

More generally we define (for $f \neq 0$)

DEFINITION 1.2. A *generalized Clifford algebra* for f is a $\mathbf{Z}/d\mathbf{Z}$-graded k-algebra C together with a monomorphism $V \hookrightarrow C_1$ of vector spaces such that

1) C is generated by V, and 2) $x^d = f(x)$ for all $x \in V$.

We shall see later (in theorem 1.8) that the universal Clifford algebra of f is indeed a generalized Clifford algebra for f. Then clearly $C(f)$ is universal in the sence that for any generalized Clifford algebra C for f there is a unique $\mathbf{Z}/d\mathbf{Z}$-graded epimorphism $\epsilon \colon C(f) \longrightarrow C$ such that

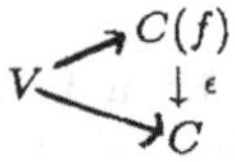

commutes.

If f is a quadratic form, then $C(f)$ is the usual Clifford algebra.

If k is finite and $d \gg 0$, then we may pick an $f \neq 0$ of degree d, such that $f(x) = 0$ for *all* $x \in V$. In this 'pathological' case, $C(f)$ is a $\mathbf{Z}$-graded ring in the natural manner. On the other hand, if there is a $u \in V$ such that $f(u) = y \in k$, $y \neq 0$, then u is a *unit of degree* 1 in $C(f)$ (since $y^{-1}u^{d-1} \cdot u = y^{-1}f(u) = 1$). It is well-known that if k is infinite and $f \neq 0$ then f cannot act trivially on V; therefore we sometimes will demand k to be infinite, in order to ensure the existence of such a unit. (This is not a serious restriction, as remark 1.10 below shows.)

THEOREM 1.3. *Assume that k is infinite. Let $f \neq 0$ be a homogeneous polynomial of degree d.*

i. *The equivalence classes of matrix factorizations of f correspond bijectively to the isomorphism classes of $\mathbf{Z}/d\mathbf{Z}$-graded modules over the universal Clifford algebra of f.*

ii. *Let $M = \bigoplus_{i=0}^{d-1} M_i$ correspond to the matrix factorization $f = \alpha_0 \cdot \ldots \cdot \alpha_{d-1}$. Then*

1) $\dim_k M_i$ is equal to the size of the matrices α_i for all i.

2) If $j \in \mathbf{Z}$, then the shifted module $M(j)$ corresponds to the matrix factorization $f = \alpha_j \cdot \alpha_{j+1} \cdot \ldots \cdot \alpha_{j+d-1}$.

3) This matrix factorization is decomposable if and only if M is decomposable.

Proof. We just indicate how a $\mathbf{Z}/d\mathbf{Z}$-graded module $M = \bigoplus_{i=0}^{d-1} M_i$ defines a matrix factorization of f. Choose a $u \in V$ such that $f(u) \neq 0$; u is a unit in $C(f)$. Since $u \in C_1(f)$, the multiplication by u induces k-isomorphisms u: $M_i \xrightarrow{\sim} M_{i+1}$ for $i = 0, \cdots, d-1$, whence all M_i have the same k-vectorspace dimension. This implies that the k-linear maps ϕ_i: $V \to \mathrm{Hom}_k(M_i, M_{i+1})$ for $i = 0, \cdots, d-1$ define square matrices α_i of linear forms (with respect to some bases of the M_i). Clearly $f = \alpha_0 \cdot \ldots \cdot \alpha_{d-1}$. □

We now describe the algebra $C(f)$ more precisely: For any $\ell \geq 0$, let N_ℓ be the set of n-tuples $\nu = (\nu_1, \ldots, \nu_n)$ with $\nu_i \geq 0$ for $i = 1, \ldots, n$ and $\sum_{i=1}^n \nu_i = \ell$. Let $N = \bigcup_{\ell \geq 0} N_\ell$. As usual, we set $x^\nu := x_1^{\nu_1} \cdot \ldots \cdot x_n^{\nu_n}$. Then (for some $a_\nu \in k$) we have

$$f = \sum_{\nu \in N_d} a_\nu x^\nu .$$

Let $\nu \in N$; a monomial in the generators $e_1, \ldots, e_n$ is said to have *multidegree* ν, if e_i occurs exactly ν_i times as a factor in this monomial for $i = 1, \ldots, n$. For example, the monomials of multidegree $(2,1)$ are $e_1^2 e_2$, $e_1 e_2 e_1$, and $e_2 e_1^2$.

We let g_ν be the sum of all monomials of multidegree ν, so that for instance $g_{(2,1)} = e_1^2 e_2 + e_1 e_2 e_1 + e_2 e_1^2$. For convenience we put $g_\nu = 0$ if ν is an n-tuple *not in* N, so that for instance $g_{(4,-1)} = 0$.

Let $J(f)$ be the two-sided ideal of T generated by the elements $g_\nu - a_\nu$, $\nu \in N_d$, and let $S(f) = T/J(f)$. Then we have

LEMMA 1.4.

i. $I(f) \subseteq J(f)$

ii. $I(f) = J(f)$, *if k is infinite.*

Proof. If $x = \sum_{i=1}^n x_i e_i \in V$, then

$$x^{\otimes d} - f(x_1, \ldots, x_n) = \sum_{\nu \in N_d} (g_\nu - a_\nu) x^\nu .$$ □

In other words, there is a natural epimorphism $C(f) \twoheadrightarrow S(f)$, which is an isomorphism if k is infinite.

Next we shall employ the "Diamond lemma" techniques (cf [Be]), in order to study the ideal $J(f)$.

If we set $e_1 < e_2 < \ldots < e_n$, we can order the monomials of T in the e_i in the standard way: first by length, then (for monomials of the same length) lexicographically.

Let $g \in T$ be an arbitrary non-zero element in the tensor algebra. g uniquely is a linear combination of monomials with non-zero coefficients. We denote by g^* (the *leading monomial* of g) the highest monomial occurring in this linear combination. If $I \subset T$ is a two-sided ideal, then we let I^* (the *associated monomial ideal* to I) be the two-sided ideal which is generated by all g^*, $g \in I$.

A subset $S \subset I$ is called a *standard basis* (or a *Gröbner basis*) of I, if g^*, $g \in S$, generates I^*. Any standard basis of I is a basis of I as well. (In the terminology of [Be], a given basis S of I is standard iff the corresponding system of reductions has no unresolvable ambiguities; c.f. e.g. [Be, 5.3].)

The importance of these notions results from the following well-known

LEMMA 1.5. *Let B be the set of all monomials of T not belonging to I^*. Then the residue classes of the elements of B form a k-vector space basis of T/I.* □

Hence given a standard basis of I one can easily describe a k-vector space basis of T/I.

THEOREM 1.6. *The basis $\{g_\nu - a_\nu \mid \nu \in N_d\}$ of $J(f)$ described in lemma 1.4 is a standard basis of $J(f)$.*

Proof. For $\nu = (\nu_1, \ldots, \nu_n) \in N$, let $m_\nu = g_\nu^* = e_n^{\nu_n} \cdots e_1^{\nu_1}$ and let $h_\nu = g_\nu - m_\nu$. Then what we want to prove is that

$$J(f)^* = I^* := T(m_\nu)_{\nu \in N_d} T \tag{1}$$

Also note that m_ν, $\nu \in N$, are the *non-increasing* monomials in $e_1, \ldots, e_n$, i.e. the 'words' $e_{i_1} e_{i_2} \cdots e_{i_r}$ such that $j < l \implies i_j \geq i_l$.

In the sequel we adopt the terminology of [Be, **1**].

The *system of reductions* corresponding to the alleged standard basis is $S = \{\sigma_\nu \mid \nu \in N_d\}$, where $\sigma_\nu = (m_\nu, a_\nu - h_\nu)$.

The *ambiguities* all are on the form

$$a = (\sigma_\mu, \sigma_\nu, A, B, C)\ , \quad \mu, \nu \in N_d\ ,$$

where A, B, and C are non-trivial monomials not in I^* such that $m_\mu = AB$ and $m_\nu = BC$. Since m_μ and m_ν are non-increasing, so are A, B, C, and their product, whence $ABC = m_\lambda$ for some $\lambda \in N_\ell$, $\ell = d + \operatorname{length} A$. Clearly λ is determined by ℓ and by μ and ν, since m_μ and m_ν are a right factor and a left factor, respectively, of m_λ. Conversely, the whole ambiguity a above is determined by λ, whence we put $a_\lambda :=$ this a.

Note that not all (μ, ν, ℓ) give rise to ambiguities, but that there is an ambiguity a_λ for any $\lambda \in N_\ell$, $d + 1 \leq \ell \leq 2d - 1$. However, also note that if this $\ell > d + 1$, then there are non-trivial monomials D and E, and a $\rho \in N_d$, such that $m_\lambda = D m_\rho E$. Then (as an easy and well-known argument shows) the ambiguity m_λ indeed is resolvable. Thus the only remaining ambiguities to check are

$$a_\lambda = (\sigma_\mu, \sigma_\nu, e_i, m_\kappa, e_j)\ , \quad \lambda \in N_{d+1}\ ,$$

where i and j are the highest and the lowest non-vanishing index, repectively, of the n-tuple λ, and where $\mu \in N_d$, $\nu \in N_d$, or $\kappa \in N_{d-1}$ is obtained by subtracting 1 from the j-component, from the i-component, or from both the i-component and the j-component of λ, respectively.

By inspection it is clear that if exactly s of the variables $e_1, \ldots, e_n$ occur in the 'word' m_λ, then no other variables can occur in any image of m_λ under any finite sequence of reductions. Hence we may forget the other variables in our analysis, and thus actually assume that $\lambda_l \geq 1$ *for* $l = 1, \ldots, n$. In particular we get

$$m_\lambda = m_\mu e_1 = e_n m_\nu = e_n m_\kappa e_1 \ .$$

We must show that there is some common 'image under reduction' of the two 'branches' $b_1 = r_{\sigma_\mu e_1}(m_\lambda)$ and $b_2 = r_{e_n \sigma_\nu}(m_\lambda)$. Recall that by definition b_1 is obtained by replacing m_μ by $a_\mu - h_\mu$ in m_λ, and similarly for b_2. Thus

$$b_1 = a_\mu e_1 - h_\mu e_1$$

and

$$b_2 = a_\nu e_n - e_n h_\nu \quad .$$

As a starter, let us note that in the case $n = 1$ we have $\mu = \nu = (d)$, and $b_1 = a_{(d)} e_1 = b_2$, and we are through.

Next, assume that $n > 1$. For any $i, j = 1, \ldots, n$ and any $\rho = (\rho_1, \ldots, \rho_n) \in N$, let

$$\rho(i) = (\rho_1, \ldots, \rho_i - 1, \ldots, \rho_n)$$

and

$$\rho(i,j) = (\rho(i))(j) \ ;$$

thus e.g. $\mu = \lambda(1)$, $\nu = \lambda(n)$, and $\kappa = \mu(n) = \nu(1) = \lambda(1,n)$. Let us write $\lambda = (\lambda_1, \ldots, \lambda_n)$. We distinguish four cases, depending on whether $\lambda_1 = 1$ or $\lambda_1 > 1$, and on whether $\lambda_n = 1$ or $\lambda_n > 1$.

The case $\lambda_1 > 1$, $\lambda_n > 1$: In this case we have

$$g_\mu = \sum_{i=1}^{n} e_i g_{\mu(i)} = \sum_{i=1}^{n} e_i g_{\lambda(1,i)}$$

whence

$$b_1 = a_{\lambda(1)} e_1 - e_n h_{\lambda(1,n)} e_1 - \sum_{i=1}^{n-1} e_i g_{\lambda(1,i)} e_1 \ .$$

We may reduce the leading terms $m_{\lambda(i)}$ in $g_{\lambda(1,i)} e_1$ for $i = 1, \ldots, n-1$:

$$r_{e_1 \sigma_{\lambda(1)}} \cdots r_{e_{n-1} \sigma_{\lambda(n-1)}}(b_1) = a_{\lambda(1)} e_1 - e_n h_{\lambda(1,n)} e_1 - \sum_{i=1}^{n-1} e_i (a_{\lambda(i)} - \sum_{j=2}^{n} g_{\lambda(i,j)} e_j)$$

$$= - \sum_{i=2}^{n-1} a_{\lambda(i)} e_i - e_n h_\kappa e_1 + \sum_{\substack{1 \leq i \leq n-1 \\ 2 \leq j \leq n}} e_i g_{\lambda(i,j)} e_j = \alpha \ ,$$

say. Similarly we get

$$b_2 = a_{\lambda(n)} e_n - e_n h_\kappa e_1 - \sum_{j=2}^{n} e_n g_{\lambda(j,n)} e_j$$

and

$$r_{\sigma_{\lambda(2)} e_2} \cdots r_{\sigma_{\lambda(n)} e_n}(b_2) = \alpha \ ,$$

indeed.

The case $\lambda_1 = 1 < \lambda_n$: We get (proceeding as above):

$$r_{e_2\sigma_{\lambda(2)}}\cdots r_{e_{n-1}\sigma_{\lambda(n-1)}}(b_1) = a_{\lambda(1)}e_1 - e_n h_\kappa e_1 - \sum_{i=2}^{n-1} a_{\lambda(i)}e_i + \sum_{\substack{2\le i\le n-1\\ 2\le j\le n}} e_i g_{\lambda(i,j)} e_j = \beta\,,$$

say, and

$$r_{\sigma_{\lambda(2)}e_2}\cdots r_{\sigma_{\lambda(n)}e_n}(b_2) = a_{\lambda(n)}e_n - e_n h_\kappa e_1 - \sum_{j=2}^{n} a_{\lambda(j)}e_j + \sum_{\substack{1\le i\le n-1\\ 2\le j\le n}} e_i g_{\lambda(i,j)} e_j = \gamma\,,$$

say. Now, since $\lambda(1)_1 = 0$, $e_1 m_{\lambda(1)}$ appears in $e_1 g_{\lambda(1,2)} e_2$ and may be reduced:

$$r_{e_1\sigma_{\lambda(1)}}(\gamma) = a_{\lambda(n)}e_n - e_n h_\kappa e_1 - \sum_{j=2}^{n} a_{\lambda(j)}e_j + \sum_{\substack{2\le i\le n-1\\ 2\le j\le n}} e_i g_{\lambda(i,j)} e_j + a_{\lambda(1)}e_1 = \beta\,,$$

indeed.

The case $\lambda_n = 1 < \lambda_1$ is symmetrical to the latter case above, and in *the case* $\lambda_1 = \lambda_n = 1$ both b_1 and b_2 may be reduced to

$$a_{\lambda(1)}e_1 + a_{\lambda(n)}e_n - \sum_{i=2}^{n-1} a_{\lambda(i)}e_i - e_1 g_{\lambda(1,n)} e_n + \sum_{\substack{2\le i\le n-1\\ 2\le j\le n-1}} e_i g_{\lambda(i,j)} e_j$$

by similar means. □

For any subset $B \subset \{1, \dots, n\}$, we let $X_B = (X_i : i \in B)$ and we get a form $f_B(X_B) \in k[X_B]_d$ as the "f-part with support in B"; in other words, f_B is obtained from f by putting $X_i = 0$ for all $i \in B$. We may identify the dual space to $\bigoplus_{i\in B} kX_i$ with $V_B := \bigoplus_{i\in B} ke_i \subset V$. From theorem 1.6 and by the "inspection" argument in its proof we then may deduce:

COROLLARY 1.7. *For any subset B of $\{1, \dots, n\}$, the inclusion $T(V_B) \hookrightarrow T(V)$ induces an injection $S(f_B) \hookrightarrow S(f)$ of generalized Clifford algebras. In particular, if k is infinite, then the natural morphism $C(f_B) \to C(f)$ is injective.* □

Theorem 1.6 showed that the associated monomial ideal $J(f)^*$ *only* depends on d and on n; indeed, we may write

$$J(f)^* = I(d,n) := T\,(m_\nu)_{\nu\in N_d}\,T$$

Now we may easily derive the following fundamental results (generalizing [C, theorem 3]):

THEOREM 1.8. *Let f be an homogeneous polynomial of degree $d \geq 2$. Then we have*

i. The natural map $V \to C(f)$ is injective. In particular, $C(f) \neq 0$.

ii. $\dim_k C(f) < \infty$, if and only if $n \leq 1$ or $d = 2$.

Proof. It suffices to prove both assertions for $S(f)$ in place of $C(f)$. Then i follows from the commutative diagram

$$\begin{array}{ccc} V & \longrightarrow & C(f) \\ & \searrow & \downarrow \\ & & S(f) \end{array}$$

and *ii* follows, since $C(f) \to S(f)$ is always surjective and since it becomes an isomorphism if $n = 1$ or $d = 2$.

Now we apply lemma 1.5 and theorem 1.6 in order to find a monomial k-basis of $S(f)$. All monomials of total-degree $< d$ are basis elements; in particular the images of $e_1, \ldots,$ and e_n, are linearly independent, and *i* follows. As for *ii*, it is enough to calculate the 'Hilbert series' of the 'associated monomial rings' $T/J(f)^*$, e.g. with the techique from [Ba]. Actually, only one Hilbert series has to be calculated: the cases $n \leq 1$ and $d = 2$ are well-known, whence we may assume that $n \geq 2$ and $d \geq 3$ and try to prove that $S(f)$ is infinite-dimensional. Furthermore, by corollary 1.7 we may assume that $n = 2$, and since $I(2,3) \supset I(2,4) \supset \ldots$, we may assume that $d = 3$. Finally, by concrete calculations we get the multiple Hilbert series

$$\mathrm{Hilb}_{T(ke_1 \oplus ke_2)/I(2,3)}(t_1, t_2) = \frac{1 + t_1 + t_2 + t_1^2 + t_1t_2 + t_2^2 + t_1^2t_2 + t_1t_2^2 + t_1^2t_2^2}{1 - t_1t_2}$$

(whence in particular $\mathrm{Hilb}_{T(ke_1 \oplus ke_2)/I(2,3)}(t) = (1+t+t^2)^2/(1-t^2)$ has all its coefficients positive). □

COROLLARY 1.9. *Every homogeneous polynomial has a matrix factorization (possibly of infinite size).*

Proof. Let ℓ be an infinite extension field of k. Then we have natural algebra homomorphisms

$$C(f) \hookrightarrow \ell \otimes_k C(f) \twoheadrightarrow C(\tilde{f}),$$

where $\tilde{f}$ is the polynomial f considered as an element in $\ell[X_1, \ldots, X_n]$. Thus any $\mathbf{Z}/d\mathbf{Z}$-graded $C(\tilde{f})$-module M (e.g., $M = C(\tilde{f})$ itself) is a $\mathbf{Z}/d\mathbf{Z}$-graded $C(f)$-module as well. Furthermore, there is a $u \in \ell \otimes_k V$ such that $\tilde{f}(u) \neq 0$, whence multiplication by u acts as ℓ-vector space isomorphisms $M_i \xrightarrow{\sim} M_{i+1}$, as we saw in the proof of theorem 1.3. These isomorphisms are k-vector space isomorphisms as well, and actually this was all we ever needed in order to deduce that M *regarded as a* $C(f)$*-module corresponds to a matrix factorization of* f. □

REMARK 1.10. The arguments above work as well for any field extension ℓ/k, such that $\tilde{f}(\ell \otimes_k V) \neq 0$. In particular we have:

If there is an extension field of k, over which f has a finite matrix factorization, then f has a finite matrix factorization over k.

Indeed, if ℓ' is such an extension field, then the subfield ℓ'' of ℓ', generated by k and by the coefficients of the variables in the entries of the matrix factors of f, is a *finite* extension of k, and furthermore there is a finite field extension ℓ/ℓ'' such that $\tilde{f}(\ell \otimes_k V) \neq 0$. Thus the finite matrix factorization of f over ℓ' (and over ℓ) corresponds to a $\mathbf{Z}/d\mathbf{Z}$-graded $C(\tilde{f})$-module M of finite ℓ-dimension, and then M has finite k-dimension as well. (In fact $\dim_k M = [\ell : k] \cdot \dim_\ell M$.)

By the same argument, *any homogeneous form in two variables has a matrix factorization of finite size,* since over the algebraic closure of k this form decomposes into linear factors.

We call a homogeneous polynomial $f \in k[X_1, \ldots, X_n]$ *degenerated* if, after a suitable linear transformation, f becomes a form in less than n variables. Otherwise f is called *non-degenerated.*

COROLLARY 1.11. *Let f be a non-degenerated polynomial. Then the following conditions are equivalent:*

(a) *There exists a matrix factorization of finite size.*

(b) *There exists a finite-dimensional $\mathbf{Z}/d\mathbf{Z}$-graded $C(f)$-module.*

(c) *There exists a homogeneous left ideal a in $C(f)$, such that $0 < \dim_k C(f)/a < \infty$.*

(d) *There exists a finite-dimensional generalized Clifford algebra for f.*

Proof. The equivalence of (a) and (b) follows from theorem 1.3, while the implications (d) $\Longrightarrow$ (c) $\Longrightarrow$ (b) are trivial. It remains to show that (b) $\Longrightarrow$ (d).

Let a be the annihilator of the finite-dimensional $\mathbf{Z}/d\mathbf{Z}$-graded $C(f)$-module M. Since a is a two-sided homogeneous ideal, the quotient algebra $C = C(f)/a$ is a generalized Clifford algebra for f, unless $V \cap a \neq 0$.

Suppose there exists an element $e \in V \cap a$, $e \neq 0$; then we may choose a basis $e_1, \ldots, e_n$ of V with $e_n = e$ and a linear transformation of $k[X_1, \ldots, X_n]$ such that after this transformation

$$f(x_1, \ldots, x_n) \cdot \mathrm{id}_{M_0} = \phi_{d-1}(x) \circ \ldots \circ \phi_0(x)$$

for all $x = \sum_{i=1}^{n} x_i e_i \in V$, where $\phi_i(x)$ is the linear map from M_i to M_{i+1} given by the multiplication with x. Since e_n is in the annihilator of M it follows that $\phi_i(e_n) = 0$ for all $i = 0, \cdots, d-1$. Therefore the entries of the corresponding matrices α_i are linear forms in the variables $X_1, \ldots, X_{n-1}$ only. This in turn implies that f is a form in $X_1, \ldots, X_{n-1}$, contradicting the assumption that f is non-degenerated.

Finally, in order to see that $\dim_k C < \infty$, notice that M is a *faithful* module over C, whence the left-multiplication induces a monomorphism of k-vector spaces $C \hookrightarrow \mathrm{Hom}_k(M, M)$, but $\dim_k \mathrm{Hom}_k(M, M) < \infty$, since $\dim_k M < \infty$. □

We call a matrix factorization of f a *C-matrix factorization* if the corresponding $\mathbf{Z}/d\mathbf{Z}$-graded module is defined over the generalized Clifford algebra C.

In section 3 we will give explicit presentations of finite dimensional generalized Clifford algebras.

We conclude this section by considering transposes of matrix factorizations. Let α^t denote the transpose of a matrix α. If $f = \alpha_0 \cdot \ldots \cdot \alpha_{d-1}$ is a matrix factorization of f, then $f = \alpha_{d-1}^t \cdot \ldots \cdot \alpha_0^t$ also is a matrix factorization of f.

Let M be the $\mathbf{Z}/d\mathbf{Z}$-graded module over $C(f)$ corresponding to the matrix factorization $f = \alpha_0 \cdot \ldots \cdot \alpha_{d-1}$ of f. Then the graded dual $M^* = \mathrm{Hom}_k(M, k)$ is a $\mathbf{Z}/d\mathbf{Z}$-graded module over the opposite algebra $C(f)^{\mathrm{op}}$ of the universal Clifford algebra for f.

Since the natural isomorphism $T \xrightarrow{\sim} T^{\mathrm{op}}$, which assigns to each monomial in $e_1, \ldots, e_n$ the monomial in the reverse order, obviously maps $I(f)$ into $I(f)$, we get an induced isomorphism $C(f) \xrightarrow{\sim} C(f)^{\mathrm{op}}$. Considering M^* as a $C(f)$-module by means of this isomorphism, M^* corresponds to the matrix factorization $f = \alpha_{d-1}^t \cdot \ldots \cdot \alpha_0^t$ of f.

2. The Atiyah-Bott-Shapiro equivalence

Throughout this section we assume that k is an *infinite* field, and (as before) that $f \neq 0$ is a homogeneous polynomial of degree $d \geq 2$. Thus we may fix a $u \in V$ and a $y \in k$ such that $f(u) = y \neq 0$. Furthermore, let C be a generalized Clifford algebra for f. Note that u is a unit in C (since it is a unit in $C(f)$). We denote by $\mathcal{M}_0$ the category of C_0-modules, and by $\mathcal{M}$ the category of $\mathbf{Z}/d\mathbf{Z}$-graded C-modules with homogeneous homomorphisms (of degree 0) as morphisms.

There are functors in both ways:

$$F: \begin{array}{ccc} \mathcal{M}_0 & \longrightarrow & \mathcal{M} \\ N & \mapsto & C\otimes_{C_0} N \end{array} \quad \text{and} \quad G: \begin{array}{ccc} \mathcal{M} & \longrightarrow & \mathcal{M}_0 \\ M & \mapsto & M_0 \end{array} .$$

THEOREM 2.1. (The Atiyah-Bott-Shapiro equivalence.) *The functors F and G are adjoint to each other. In other words, for all N in $\mathcal{M}_0$ and all M in $\mathcal{M}$ there exists a natural isomorphism*

$$\alpha: \operatorname{Hom}_C(F(N), M) \xrightarrow{\sim} \operatorname{Hom}_{C_0}(N, G(M)) .$$

Moreover, the functorial homomorphisms $\alpha(\mathrm{id}_{F(N)})\colon N \to GF(N)$ and $\alpha^{-1}(\mathrm{id}_{G(M)})\colon FG(M) \to M$ are isomorphisms for all N in $\mathcal{M}_0$ and M in $\mathcal{M}$. In particular, the categories $\mathcal{M}_0$ and $\mathcal{M}$ are equivalent.

Proof. Notice first that there is a functorial isomorphism

$$\eta: \begin{array}{ccc} N & \longrightarrow & C_0\otimes_{C_0} N \\ n & \mapsto & 1\otimes n \end{array} = F(N)_0 = GF(N) .$$

Let $\phi \in \operatorname{Hom}_C(F(N), M)$ with 0-th homogeneous component ϕ_0. We define α by $\alpha(\phi) = \phi_0 \circ \eta$. It is straightforward to check that α is natural.

In order to see that α is an isomorphism, we define the functorial inverse:

$$\beta: \operatorname{Hom}_{C_0}(N, G(M)) \to \operatorname{Hom}_C(F(N), M) .$$

Given $\psi \in \operatorname{Hom}_{C_0}(N, G(M))$ we define $\beta(\psi)$ by its homogeneous components:

$$\beta(\psi)_i: F(N)_i = \begin{array}{ccc} C_i\otimes_{C_0} N & \longrightarrow & M_i \\ c\otimes n & \mapsto & c\cdot\psi(n) \end{array}$$

It is easily checked that α and β are inverse to each other, and that $\alpha(\mathrm{id}_{F(N)})$ is an isomorphism for all N in $\mathcal{M}_0$. We only verify here that $\beta(\mathrm{id}_{G(M)})$ is an isomorphism for all M in $\mathcal{M}$; that is, that

$$\mu_i: \begin{array}{ccc} C_i\otimes_{C_0} M_0 & \longrightarrow & M_i \\ c\otimes m & \mapsto & c\cdot m \end{array}$$

is an isomorphism for all M in $\mathcal{M}$ and all i.

μ_i is surjective, since for for any $m \in M_i$ one has $\mu_i(u^i \otimes u^{-i}m) = m$.

Let $\sum_j c_j \otimes m_j \in C_i\otimes_{C_0} M_0$, and assume that $0 = \mu_i(\sum c_j \otimes m_j) = \sum c_j m_j$, then $\sum c_j \otimes m_j = \sum u^i \otimes (u^{-i}c_j)m_j = u^i \otimes u^{-i} \sum c_j m_j = 0$, which shows that μ_i is injective as well. □

COROLLARY 2.2. *Equivalence classes of C-matrix factorizations of f correspond bijectively to isomorphism classes of C_0-modules.* □

This bijection can be made explicit: Given a C_0-module N we define k-linear maps $\phi_i\colon V \to \mathrm{End}_k(N)$ for $i = 0,\dots,d-1$. We set

$$\phi_i(x)(n) = \begin{cases} (u^{-i-1}xu^i)n & \text{for } i = 0,\dots,d-2 \\ (xu^{d-1})n & \text{for } i = d-1 \end{cases} \tag{2}$$

for all $x \in V$ and all $n \in N$.

It follows that

$$(\phi_{d-1}(x)\circ\ldots\circ\phi_1(x)\circ\phi_0(x))(n) = (xu^{d-1})\cdot(u^{-d+1}xu^{d-2})\cdot\ldots\cdot(u^{-2}xu)\cdot(u^{-1}x)(n) = x^d\cdot n = f(x)\cdot n\,.$$

Thus N determines a matrix factorization of f, which, in fact, belongs to the $\mathbf{Z}/d\mathbf{Z}$-graded module $F(N)$, as can be seen from the following commutative diagram

$$\begin{array}{ccc} N & \xrightarrow{\phi_i(x)} & N \\ u^i\downarrow & & \downarrow u^{i+1} \\ F(N)_i & \xrightarrow[x]{} & F(N)_{i+1} \end{array}$$

In order to obtain the matrices α_i corresponding to the C_0-module N explicitly, we choose a k-basis of N. With respect to this basis the $\phi_i(x)$ are matrices with entries in k, and we have

$$\alpha_i = \sum_{r=1}^{n} \phi_i(e_r)X_r\,. \tag{3}$$

COROLLARY 2.3. *Suppose C is a finite dimensional generalized Clifford algebra for f. If C_0 is simple and if $0 \neq y \in f(V)$, then there exists a square matrix α of linear forms and a square matrix S with coefficients in k, such that*

(i) Up to a factor in k^*, S^d is the unit matrix E.

(ii) If $\alpha_i = S^{-i}\alpha S^i$ for $i = 0,\dots,d-2$ and if $\alpha_{d-1} = y\cdot S^{-d+1}\alpha S^{d-1}$, then $f = \alpha_0\cdot\ldots\cdot\alpha_{d-1}$ is an indecomposable C-matrix factorization of f.

(iii) Any other indecomposable C-matrix factorization of f is equivalent to the one given in (ii).

Proof. $f(u) = y$ where $u \in C_1$ is a unit in C.

Since C_0 is simple there exists up to isomorphisms exactly one indecomposable C_0-module N. Thus any indecomposable C-matrix factorization of f is equivalent to the matrix factorization described in (3).

We consider the C_0-automorphism

$$\begin{array}{cccc} \sigma: & C_0 & \longrightarrow & C_0 \\ & c & \longmapsto & u^{-1}cu \end{array}$$

of order dividing d. Since by the Skolem-Noether Theorem all automorphisms of C_0 are inner, there exists a $v \in C_0$ such that $u^{-1}cu = \sigma(c) = v^{-1}cv$ for all $c \in C_0$. It follows that for all $x \in V$ we have

$$u^{-i-1}xu^i = u^{-i}(u^{-1}x)u^i = v^{-i}(u^{-1}x)v^i \quad \text{for } i = 0,\dots,d-2$$

and

$$xu^{d-1} = yu^{-d} \cdot xu^{d-1} = y \cdot v^{-d+1}(u^{-1}x)v^{d-1} .$$

Hence we get

$$\phi_i(x) = v^{-i}\phi_0(x)v^i \quad \text{for } i = 0, \ldots, d-2$$

and

$$\phi_{d-1}(x) = y \cdot v^{-d+1}\phi_0(x)v^{d-1} .$$

If we let S be the matrix corresponding to the left multiplication with v and α be the matrix corresponding to $\phi_0(x)$ with respect to a given basis of N, then all assertions follow. □

COROLLARY 2.4. *In addition to the assumptions of corollary 2.3, let m be the size of an indecomposable C-matrix factorization of f. Then there exists a matrix β of size m of linear forms, and an element $t \in k^*$, such that $t \cdot f = \beta^d$.*

Proof. $f = (\alpha S^{-1})^d y S^d$, and $y \cdot S^d = t^{-1} \cdot E$, for some $t \in k^*$. The assertion follows if we set $\beta = \alpha S^{-1}$. □

We conclude this section by a few remarks concerning cyclic permutations in a matrix factorization of f.

Let C be a generalized Clifford algebra for f, and choose $u \in V$ such that $f(u) \neq 0$. Consider, as in the proof of corollary 2.3, the automorphism

$$\begin{array}{rccc} \sigma : & C_0 & \longrightarrow & C_0 \\ & c & \mapsto & u^{-1}cu \end{array}$$

of order dividing d.

Any C_0-module M may be given a new C_0-module structure by means of σ:

$$c \bullet m = \sigma(c)m$$

for all $c \in C_0$, $m \in M$. The module M, with this new C_0-module structure, will be denoted by $\sigma_* M$.

PROPOSITION 2.5. *$F(\sigma_* M) \simeq F(M)(1)$ for any C_0-module M.*

Proof. Because of the Atiyah-Bott-Shapiro equivalence it suffices to show that $\sigma_* M \simeq C_1 \bigotimes_{C_0} M$. Consider the k-vector space isomorphism

$$\begin{array}{rccc} \psi : & M & \xrightarrow{\sim} & C_1 \bigotimes_{C_0} M \\ & m & \mapsto & u \otimes m \end{array} .$$

We transfer the C_0-module structure of $C_1 \bigotimes_{C_0} M$ to M by means of ψ. Then, for $c \in C_0$ and $m \in M$ we get

$$c \bullet_\psi m = \psi^{-1}(c \cdot \psi(m)) = \psi^{-1}(c(u \otimes m)) = \psi^{-1}(c \cdot u \otimes m) = u^{-1}cum = \sigma(c) \cdot m = c \bullet_\sigma m .$$

Hence indeed $C_1 \bigotimes M \simeq \sigma_* M$. □

COROLLARY 2.6. *If the C_0-module M corresponds to the matrix factorization $f = \alpha_0 \cdot \ldots \cdot \alpha_{d-1}$, then $\sigma_*^i M$ corresponds to the matrix factorization $f = \alpha_i \cdot \alpha_{i+1} \cdot \ldots \cdot \alpha_{i+d-1}$.*

Now let $\mathcal{O}$ denote the set of isomorphism classes of indecomposable C_0-modules. Clearly σ_* operates on $\mathcal{O}$, and therefore $\mathcal{O}$ decomposes into disjoint orbits $\{\mathcal{O}_i\}_{i \in I}$ and $r_i := |\{\mathcal{O}_i\}|$ divides d for all $i \in I$. As an immediate consequence of proposition 2.5 we get:

COROLLARY 2.7.

i. *Up to cyclic permutations of the factors there are as many pairwise non-equivalent indecomposable C-matrix factorizations of f, as there are different orbits of the operation of σ_* on $\mathcal{O}$.*

ii. *If $f = \alpha_0 \cdot \ldots \cdot \alpha_{d-1}$ corresponds to the orbit $\mathcal{O}_i$, then α_r is equivalent to α_s if and only if $r \equiv s \pmod{r_i}$.* □

3. Diagonal forms

The results and examples in this section in general may be found in the works [C] by L. N. Childs or [L] by F. W. Long, also when we give no explicit references. Our proofs are also mainly based on the same ideas as theirs. (Childs considers a slightly less general situation than ours: he assumes that Char $k > d$. However, in general his proofs work as well in our situation.)

A *diagonal form of degree* d is a homogeneous polynomial $f = \sum_{i=1}^{n} a_i X_i^d$ with coefficients $a_i \neq 0$ in a field k. Throughout this section we will assume that $d \geq 2$, that (Char $k, d) = 1$, that k is infinite, and that k contains a d-th primitive root ξ of unity. Under these assumptions we will construct a finite dimensional generalized Clifford algebra for f, which for $d = 2$ is the usual Clifford algebra .

More generally, assume that we are given two homogeneous polynomials f_1 and f_2 in disjoint sets of variables. Suppose that there is given a generalized Clifford algebra C_i for f_i $(i = 1, 2)$. We will construct a generalized Clifford algebra for $f_1 + f_2$ and will use this construction later to build the Clifford algebra for a diagonal form.

Given two $\mathbf{Z}/d\mathbf{Z}$-graded k-algebras A and B, we define their $\mathbf{Z}/d\mathbf{Z}$-graded tensor product $A \widehat{\otimes} B$ as the ordinary tensor product of $\mathbf{Z}/d\mathbf{Z}$-graded k-vector spaces with the multiplication given by

$$(a \otimes b)(c \otimes d) = \xi^{(\deg b)(\deg c)}(ac \otimes bd)$$

for homogeneous elements $a, c \in A$ and $b, d \in B$. Clearly $A \widehat{\otimes} B$ has the usual universal property. Notice, however, that different choices of the primitive root of unity ξ may yield non-isomorphic tensor products.

The k-subalgebra $\{a \otimes 1 \mid a \in A\}$ of $A \widehat{\otimes} B$ is isomorphic to A. Identifying this subalgebra with A we simply write a instead of $a \otimes 1$, for all $a \in A$. Similarly we identify $1 \otimes b$ with b for all $b \in B$. Then we get the multiplication rule

$$ba = \xi^{(\deg a)(\deg b)} ab$$

for all homogeneous $a \in A$, $b \in B$.

The following result has been shown by Childs ([C, theorem 2]) for the tensor products of universal Clifford algebras. The proof in the general case is essentially the same.

THEOREM 3.1. *Let f_1 and f_2 be homogeneous polynomials of degree d in disjoint sets of variables. If C_i is a generalized Clifford algebra for f_i $(i = 1,2)$, then $C := C_1 \widehat{\otimes} C_2$ is a generalized Clifford algebra for $f := f_1 + f_2$.*

Proof. Let $V_i \subseteq (C_i)_1$ be the generating k-vector spaces of C_i, so that $f_i(x_i) = x_i^d$ for all $x_i \in V_i$. Then f is naturally defined on $V := V_1 \oplus V_2$ by setting $f(x_1+x_2) := f_1(x_1)+f_2(x_2)$ for all $x_1 \in V_1$, $x_2 \in V_2$. Since

$$(C_1 \otimes C_2)_1 = \bigoplus_{i+j \equiv 1 \pmod d} (C_1)_i \otimes (C_2)_j$$

we can naturally identify V with the subvectorspace $(V_1 \otimes 1) \bigoplus (1 \otimes V_2)$ in $(C_1 \otimes C_2)_1$, and this subspace generates $C_1 \widehat{\otimes} C_2$. Hence it remains to prove that $f(x) = x^d$ for all $x \in V$. Thus let $x = x_1 + x_2$ with $x_i \in V_i$. Then, since $x_2 x_1 = \xi x_1 x_2$, we get

$$x^d = x_1^d + x_2^d + \sum_{i=1}^{d-1} g_i x_1^i x_2^{d-i} ,$$

where

$$g_i = \sum_{\substack{u_j \geq 0 \\ \sum_{j=0}^{i} u_j = d-i}} \left(\xi^{\sum_{j=0}^{i} j u_j} \right)$$

for $i = 1, \ldots, d-1$. We will show that all $g_i = 0$, so that $x^d = x_1^d + x_2^d = f_1(x_1) + f_2(x_2) = f(x)$.

The identity $g_i = 0$ is defined in $P[\xi]$, where P is the prime field in k. Hence if we prove the identity in $\mathbf{Z}[\xi]$ or in $\mathbf{C}$ rather than in k it will follow in k as well. But in $\mathbf{C}$ we can define rational functions G_i, such that $g_i = \mathrm{Res}_0\, G_i$ (the residue at zero), and the assertion follows from the next lemma, where we show that indeed $\mathrm{Res}_0\, G_i = 0$ for these G_i.

LEMMA 3.2. *Let $\xi \in \mathbf{C}$ be a d-th root of unity. For $i = 1, \ldots, d-1$ we define the rational function*

$$G_i(z) = \frac{1}{z^{d-i+1}} \prod_{j=0}^{i} \frac{1}{1 - \xi^j z} .$$

Then $\mathrm{Res}_0\, G_i = 0$ for $i = 0, \ldots, d-1$.

Proof. For each $i \in \{1, \ldots, d-1\}$ we define two more rational functions

$$H_i(z) := \frac{1}{z^2} G_i(\tfrac{1}{z}) = z^d \prod_{j=0}^{i} \frac{1}{z - \xi^j} ,$$

and

$$J_i(z) = \prod_{j=0}^{i} \frac{1}{z - \xi^j} .$$

Then

$$\operatorname{Res}_0 G_i = - \sum_{a\in(\mathbb{C}\setminus\{0\})\cup\{\infty\}} \operatorname{Res}_a G_i = \sum_{a\in\mathbb{C}} \operatorname{Res}_a H_i$$

$$= \sum_{j=0}^{i} \Big(\prod_{\substack{\ell=0\\ \ell\neq j}}^{i} \frac{1}{\xi^j - \xi^\ell} \Big) = \sum_{a\in\mathbb{C}} \operatorname{Res}_a J_i = - \operatorname{Res}_\infty J_i = 0 \, .$$

□

Before we apply theorem 3.1 to diagonal forms, we give another simple application of this theorem.

PROPOSITION 3.3. *Let $f \in k[X_1, \ldots, X_n]$ be a homogeneous polynomial of degree d, and set $g = X_0^d + f \in k[X_0, \ldots, X_n]$. Then g has a matrix factorization $g = \alpha_0 \cdot \ldots \cdot \alpha_{d-1}$, with $\alpha_i = X_0 + \xi^i \alpha$, where α is a square matrix of linear forms in the variables $X_1, \ldots, X_n$. In particular $f = (-1)^{d+1}\alpha^d$.*

Moreover, if there is a finite dimensional generalized Clifford algebra for f, then α can be chosen to be finite.

Proof. Let C be a generalized Clifford algebra for f and let $V = \bigoplus_{i=1}^{n} ke_i \subseteq C_1$ be the generating k-vector space of C such that $f(x_1, \ldots, x_n) = x^d$ for all $x = \sum_{i=1}^{n} x_i e_i \in V$.

The universal Clifford algebra of X_0^d is isomorphic to $k[e_0]/(e_0^d - 1)$. By theorem 3.1

$$\widetilde{C} := C(X_0^d) \widehat{\otimes} C$$

is a generalized Clifford algebra for g.

Let N be a $\widetilde{C}_0$-module. We apply (2) and (3) (in section 2) to N and $u = e_0$. Then we obtain the matrix factorization $g = \alpha_0 \cdot \ldots \cdot \alpha_{d-1}$ with $\alpha_i = \sum_{r=0}^{n} \phi_i(e_r) X_r$, where $\phi_i(e_r)$ is the multiplication on N by

$$e_0^{-i-1} e_r e_0^i = \begin{cases} 1 & \text{for } r = 0 \\ \xi^i e_0^{-1} e_r & \text{for } r > 0 \end{cases}$$

It follows that

$$\phi_i(e_r) = \begin{cases} \mathrm{id}_N & \text{for } r = 0 \\ \xi^i \phi_0(e_r) & \text{for } r > 0 \end{cases}$$

whence $\alpha_i = X_0 + \xi^i \alpha$, with

$$\alpha = \sum_{r=1}^{n} \phi_0(e_r) X_r \, .$$

Clearly, if C is finite-dimensional, then so is $\widetilde{C}$, and α can be chosen to be finite. □

Let $a \in k^*$. The universal Clifford algebra $C(ax^d)$ is isomorphic to

$$(a) := k[e]/(e^d - a) \, ,$$

and $\dim_k(a) = d$.

For a non-degenerate diagonal form $f = \sum_{i=1}^{n} a_i x_i^d$ we define

$$C(f, \xi) := (a_1) \widehat{\otimes} \ldots \widehat{\otimes} (a_n) \, .$$

Using theorem 3.1 we see that $C(f,\xi)$ is a generalized Clifford algebra for f and of course $\dim_k C(f,\xi) = d^n$. Thus every diagonal form has a matrix factorization of finite size.

If $d = 2$, then $C(f,-1)$ coincides with the universal Clifford algebra of f.

REMARK 3.4. Even if we skip the basic assumptions of this section, that the field k should be infinite and contain a primitive d-th root of unity, we obtain matrix factorizations of finite size for diagonal forms. As remark 1.10 shows, it is sufficient to note that there exists an *extension field* of k (e.g. the algebraic closure!) fulfilling these assumptions. (The question is open, however, for diagonal forms over a field whose characteristic divides the degree of the forms.)

We may even strengthen this result, using the fact that homogeneous forms in two variables have matrix factorizations of finite size, in order to obtain the following proposition (which essentially is a special case of [C, theorem 11]):

PROPOSITION 3.5. *Let f be a homogeneous form of degree d over a (not necessarily infinite) field k, such that (Char $k, d) = 1$ and that $f = \sum_{i=1}^{r} f_i$ for some forms $f_1, \ldots, f_r$ in disjoint sets of variables, where each f_i is a form in at most two variables. Then f has a matrix factorization of finite size.* □

For the study of explicit matrix factorizations of a non-degenerate diagonal form $f = \sum_{i=1}^{n} a_i X_i^d$, let

$$\tilde{\delta} := a_1^{-1} \cdot a_2 \cdot a_3^{-1} \cdot a_4 \cdot \ldots$$

and define the *discriminant* of f as

$$\delta := (-1)^{\ell(d+1)} \tilde{\delta} \, ,$$

where $\ell = [\frac{n}{2}]$ (i.e., the integer part of $\frac{n}{2}$).

The following structure theorem for $C_0(f,\xi)$ essentially is proved in [L]:

THEOREM 3.6.

i. *If n is odd, then $C_0(f,\xi)$ is a central simple algebra over k.*

ii. *If n is even, then $C_0(f,\xi) = A \bigotimes_k k[X]/(X^d - \delta)$, where A is a central simple algebra over k.*

For the proof of this theorem we adopt some notation from [Sch]:

For $a, b \in k^*$ we define the *multernion algebra* $((a,b)) := (a)\widehat{\otimes}(b)$. If we consider this k-algebra as a trivially graded k-algebra we simply denote it by (a,b).

Let $e_1 \in (a)_1$ and $e_2 \in (b)_1$ with $e_1^d = a$ and $e_2^d = b$. Then $e_2 e_1 = \xi e_1 e_2$, and $\{e_1^{-i} e_2^i \mid 0 \le i < d\}$ is a k-basis for $((a,b))_0$.

Thus $((a,b))_0 = k\langle e_1^{-1} e_2 \rangle$.

The following proposition is due to Long ([L, proposition 2.2]):

PROPOSITION 3.7. *For $a, b, c \in k^*$ there is a $\mathbf{Z}/d\mathbf{Z}$-graded isomorphism*

$$(a)\widehat{\otimes}(b)\widehat{\otimes}(c) \simeq ((-1)^{d+1}a^{-1}c, (-1)^{d+1}b^{-1}c)\widehat{\otimes}((-1)^{d+1}ab^{-1}c) \, .$$

Proof. First note that both sides of the equivalence have the k-dimension d^3.

Choose $e_1 \in (a)_1$ and $e_2 \in (b)_1$ as above and $e_3 \in (c)_1$ with $e_3^d = c$, and let $A := (a)\widehat{\otimes}(b)\widehat{\otimes}(c)$. Then $\{(e_1^{-1}e_3)^i(e_2^{-1}e_3)^j \mid 0 \le i,j < d\}$ is a basis of A_0, and $e_1^{-1}e_3$ and $e_2^{-1}e_3 \in A_0$ satisfy the relations $(e_2^{-1}e_3)(e_1^{-1}e_3) = \xi(e_1^{-1}e_3)(e_2^{-1}e_3)$, $(e_1^{-1}e_3)^d = (-1)^{d+1}a^{-1}c$, and $(e_2^{-1}e_3)^d = (-1)^{d+1}b^{-1}c$. Thus

$$A_0 \simeq ((-1)^{d+1}a^{-1}c, (-1)^{d+1}b^{-1}c) \, .$$

Let $g := e_1e_2^{-1}e_3 \in A_1$. Now g generates a subring B of A, and g commutes with $e_1^{-1}e_3$ and with $e_2^{-1}e_3$ and thus with all elements in A_0. Thus $A_0\widehat{\otimes}B = A_0 \otimes B$ is a subalgebra of A.

Furthermore $g^d = (-1)^{d+1}ab^{-1}c$, whence $B \simeq (g) = ((-1)^{d+1}ab^{-1}c)$. Comparing dimensions we indeed get $A_0\widehat{\otimes}((-1)^{d+1}ab^{-1}c) \simeq A_0\widehat{\otimes}B = A$. □

The next result may be found e.g. in [M, theorem 15.1] or in [L, proposition 2.1]:

LEMMA 3.8. *Let $a, b \in k^*$. Then (a, b) is a central simple k-algebra.*

Proof. 1) *(a,b) is simple*: Pick $x \in (a,b)$, $x \neq 0$. We show that the two-sided ideal generated by x is equal to (a,b). Again we choose $e_1 \in (a)_1$ and $e_2 \in (b)_1$ as above. Then we may write

$$x = \sum_{0 \le p,q \le d-1} x_{pq} e_1^p e_2^q \, .$$

Without loss of generality we may assume that $x_{00} \neq 0$ (possibly after having multiplied with a suitable $e_1^i e_2^j$). Consider

$$e_1^i e_2^j x e_2^{-j} e_1^{-i} = \sum_{p,q} x_{pq} e_1^i e_2^j e_1^p e_2^q e_2^{-j} e_1^{-i} = \sum_{p,q} x_{pq} \xi^{pj-iq} e_1^p e_1^q \, . \tag{4}$$

Using that $\sum_{j=0}^{d-1} \xi^{pj} = 0$ for $p \neq 0$ and that $\sum_{i=0}^{d-1} \xi^{-iq} = 0$ for $q \neq 0$, we find that $\sum_{i,j} e_1^i e_2^j x e_2^{-j} e_1^{-i} = d^2 x_{00} \in k^*$, whence indeed $1 = (d^2 x_{00})^{-1} \sum_{i,j} e_1^i e_2^j x e_2^{-j} e_1^{-i}$ belongs to the two-sided ideal generated by x.

2) *The centre of (a,b) is k*: If x is in the centre, then $x = e_1^i e_2^j x e_2^{-j} e_1^{-i}$ for all i and j, whence by (4) $x_{pq} = x_{pq}\xi^{pj-iq}$ for all i,j,p,q. For $(p,q) \neq (0,0)$ we may choose i,j such that $\xi^{pj-iq} \neq 1$, whence $x_{pq} = 0$. Thus $x = x_{00} \in k$. □

Now we are ready for the

Proof of theorem 3.6. One computes $C(f,\xi)$ step by step using proposition 3.7. The first step gives

$$C(f,\xi) = ((-1)^{d+1}a_1^{-1}a_3, (-1)^{d+1}a_2^{-1}a_3)\widehat{\otimes}((-1)^{d+1}a_1a_2^{-1}a_3)\widehat{\otimes}(a_4)\widehat{\otimes}\dots \, .$$

A trivially graded multernion algebra occurs while the number of the d-dimensional factors is reduced by two.

For $n = 2\ell + 1$, we get after ℓ steps a product of ℓ trivially graded multernion algebras and one last factor (δ^{-1}). The product of the multernion algebras is $C_0(f,\xi)$; this therefore is a central simple algebra over k, by lemma 3.8.

For $n = 2\ell$ we get in the same way, after $\ell - 1$ steps, a trivially graded product A of $\ell - 1$ multernion algebras with the multernion algebra $(a)\widehat{\otimes}(a_n) = ((a, a_n))$, where $a := (-1)^{(\ell-1)(d+1)}a_1 \cdot$

$a_2^{-1} \cdot \ldots \cdot a_{n-2}^{-1} \cdot a_{n-1}$. Since $C_0(f,\xi) \simeq A \bigotimes_k ((a, a_n))_0$, we have to compute $((a,a_n))_0$. Let $e_1 \in (a)_1$ and $e_2 \in (a_n)_1$ with $e_1^d = a$ and $e_2^d = a_n$. Then $((a,a_n))_0 \simeq k\langle e_1^{-1}e_2\rangle$. Since $(e_1^{-1}e_2)^d = \xi^{-\binom{d}{2}} e_1^{-1} e_2^d = (-1)^{d+1}a^{-1}a_n = \delta$, we get $((a,a_n)) \simeq (\delta)$, so that all assertions follow.□

To formulate our main result on matrix factorizations of diagonal forms we introduce the following notations: If α is a matrix of linear forms in $X_1, \ldots, X_n$, then we let $\alpha(X_i)$ denote the matrix which is obtained from α by setting $X_j = 0$ for $j \neq i$.

Now let f be a non-degenerate diagonal form of degree d with discriminant δ. Let ρ be a root of $X^d - \delta$, and set $s := [k(\rho) : k]$ and $r := d/s$.

THEOREM 3.9.

i. *Up to equivalence and cyclic permutation of the factors, there exists a unique indecomposable matrix factorization $f = \alpha_0 \cdot \ldots \cdot \alpha_{d-1}$, such that*

$$\alpha_t(X_i) \cdot \alpha_{t+1}(X_j) = \xi \alpha_t(X_j) \cdot \alpha_{t+1}(X_i)$$

for all t and all $1 \le i < j \le n$.

ii. *If n is odd, then all α_i are pairwise equivalent.*

iii. *If n is even, then α_i is equivalent to α_j if and only if $i \equiv j \pmod r$.*

Proof. Since the $C(f,\xi)$-matrix factorizations correspond to $\mathbf{Z}/d\mathbf{Z}$-graded modules over $C(f,\xi)$, the relation $e_je_i = \xi e_ie_j$ for $i < j$ implies that $\phi_{t+1}(e_j) \circ \phi_t(e_i) = \xi\phi_{t+1}(e_i) \circ \phi_t(e_j)$. This implies the identities in i..

Hence almost all statements of this theorem follow from theorem 3.6 and the corollaries 2.3 and 2.7.

For n even, it only remains to show that (in the terminology of section 2) the operation of σ_* on the set of isomorphism classes of indecomposable $C_0(f,\xi)$-modules is transitive and that the only orbit consists of r elements. By the structure theorem 3.6 we know that $C_0(f,\xi) \simeq A \bigotimes_k k\langle a^{-1}e_n\rangle$, where A is a central simple k-algebra, $a \in \bigoplus_{i=1}^{n-1} ke_i$, and $k\langle a^{-1}e_n\rangle \simeq k[x]/(x^d - \delta)$.

Furthermore $X^d - \delta = \prod_{j=1}^r f_j$, where $f_j := X^s - (\xi^s)^j \rho^s \in k[X]$ is irreducible. It follows that $k\langle a^{-1}e_n\rangle \simeq \bigoplus_{j=1}^r k[X]/f_j$, and each $k[X]/f_j$ is isomorphic to $k(\rho)$, whence $C_0(f,\xi) = \bigoplus_{j=1}^r A_j$, where each A_j is a central simple $k(\rho)$-algebra.

Hence we see that the set $\mathcal{O}$ of isomorphism classes of indecomposable $C_0(f,\xi)$-modules has exactly r elements, represented by the unique indecomposable A_j-modules with A-annihilators $(a^{-1}e_n)^s - (\xi^j\rho)^s$ for $j = 1, \ldots, r$.

The automorphism

$$\begin{array}{rccc} \sigma : & C_0(f,\xi) & \rightrightarrows & C_0(f,\xi) \\ & c & \mapsto & e_n^{-1}ce_n \end{array}$$

applied to these annihilators yields

$$\begin{aligned} \sigma\big((a^{-1}e_n)^s - (\xi^j\rho)^s\big) &= \xi^s \cdot \sigma\big((e_na^{-1})^s - (\xi^{j-1}\rho)^s\big) = \\ &= \xi^s\big(e_n^{-1}(e_na^{-1})^s e_n - (\xi^{j-1}\rho)^s\big) = \\ &= \xi^s\big((a^{-1}e_n)^s - (\xi^{j-1}\rho)^s\big) . \end{aligned}$$

It follows that σ_* operates transitively on O, since σ does it on the annihilators of the modules. □

In order to express the size of the matrices in the unique indecomposable $C(f,\xi)$-matrix factorization of f, we summarize the results of the structure theorem 3.6, using the fact that every simple algebra over k is isomorphic to $M_j(D) \simeq M_j(k)\otimes_k D$, where D is a suitable division algebra over k (i.e., skew field over k), and that up to isomorphisms there exists only one indecomposable $M_j(D)$-module. This module has the k-dimension $j \cdot \dim_k D$. Recall that the dimension of a division algebra over its centre is a square of an integer, called the *degree* of the algebra ($= \deg D$).

Now let the same data δ, ρ, r, and s as before be given for f. The structure theorem implies that $C_0(f,\xi) \simeq (M_j(k) \times \ldots \times M_j(k))\otimes_k D(f,\xi)$, where the number of the factors is equal to one if n is odd, and equal to r if n is even, and where $D(f,\xi)$ is a central division algebra over k, if n is odd, and over $k(\rho)$, if n is even.

COROLLARY 3.10. *Let m be the size of the unique indecomposable $C(f,\xi)$-matrix factorization of f. Then*

$$m = \begin{cases} d^{\frac{n-1}{2}} \cdot \deg D(f,\xi), & \text{if } n \text{ is odd,} \\ s \cdot d^{\frac{n}{2}-1} \cdot \deg D(f,\xi), & \text{if } n \text{ is even.} \end{cases}$$

In particular, m divides d^{n-1}.

Proof. Since m is equal to the k-vector space dimension of an indecomposable $C_0(f,\xi)$-module, we get

$$m = \begin{cases} j \cdot (\deg D(f,\xi))^2, & \text{if } n \text{ is odd} \\ s \cdot j \cdot (\deg D(f,\xi))^2, & \text{if } n \text{ is even} \end{cases}$$

and

$$d^{n-1} = \dim_k C_0(f,\xi) = \begin{cases} j^2 \cdot (\deg D(f,\xi))^2, & \text{if } n \text{ is odd} \\ s \cdot r \cdot j^2 \cdot (\deg D(f,\xi))^2, & \text{if } n \text{ is even.} \end{cases}$$

The formula follows. □

EXAMPLE 3.11. Let $f = X_1^d + aX_2^d \in \mathbf{Q}[X_1,X_2]$, $a \neq 0$. Then the discriminant of f is $(-1)^{d+1}a$. Assume that the order of δ in $\mathbf{Q}^*/\mathbf{Q}^{*d}$ is d. Furthermore, let $k := \mathbf{Q}(\xi)$, where $\xi \in \mathbf{C}$ is a d-th primitive root of unity. We will determine

a) The unique indecomposable $C_0(f,\xi)$-matrix factorization of f.

b) A matrix β with linear forms in X_1 and X_2 with coefficients in $\mathbf{C}$ for entries, such that $\beta^d = (-1)^{d+1}f$.

a): $C_0(f,\xi) = ((1,a))_0 = k\langle e_1^{-1}e_2\rangle \simeq k[X]/(X^d-\delta)$, whence $C_0(f,\xi) \simeq k(\sqrt[d]{\delta})$, since $X^d-\delta$ is irreducible. Thus $C_0(f,\xi)$ itself is the only indecomposable $C_0(f,\xi)$-module. Let $f = \alpha_0 \cdot \ldots \cdot \alpha_{d-1}$ be the corresponding matrix factorization of f

By proposition 3.3 we know that $\alpha_i = X_1 + \xi^i\alpha$ with $\alpha = \phi_0(e_2)X_2$, where $\phi_0(e_2)$ is the matrix which is given by the multiplication with $e_1^{-1}e_2$ on $C_0(f,\xi)$ with respect to a basis B of $C_0(f,\xi)$. Choose $B = \{e_1^{-i}e_2^i \mid 0 \le i < d\}$. Since

$$e_1^{-1}e_2 \cdot e_1^{-i}e_2^i = \begin{cases} \xi^{-i}e_1^{-i-1}e_2^{i+1} & \text{for } i < d-1 \\ a \cdot \xi & \text{for } i = d-1 \end{cases}$$

we get

$$\alpha = \begin{pmatrix} 0 & X_2 & 0 & . & . & . & 0 \\ 0 & 0 & \xi^{-1}X_2 & . & & & . \\ . & & & . & . & . & . \\ . & & & . & . & . & . \\ . & & & & & . & 0 \\ 0 & & & & & . & \xi^2 X_2 \\ a\xi X_2 & 0 & & . & . & . & 0 \end{pmatrix}$$

b): Consider $g = X_0^d + f$. Using proposition 3.3, we know that g has a matrix factorization $g = \beta_0 \cdot \ldots \cdot \beta_{d-1}$, such that $\beta_j = X_0 + \xi^j \beta$ and $\beta^d = (-1)^{d+1} f$, where β is a square matrix of linear forms in the variables X_1 and X_2.

In order to compute β we consider $C_0(g, \xi) = ((-1)^{d+1}a, (-1)^{d+1}a) \simeq \mathbf{C}\langle e_0^{-1}e_2, e_1^{-1}e_2\rangle$ (where the isomorphism follows from the proof of proposition 3.7).

In order to describe the operation of $C_0(g, \xi)$ on its unique indecomposable module, we identify $C_0(g, \xi)$ with a full matrix ring. Given the multernion algebra $((-1)^{d+1}, 1)$ over any field k that contains a d-th primitive root of unity ξ, we define an algebra isomorphism

$$\Psi\colon ((-1)^{d+1}, 1) \longrightarrow M_d(k)$$

as follows:

Let $\epsilon_{i,j} \in M_d(k)$, $1 \le i, j \le d$, be the canonical basis of $M_d(k)$, and for $p, q \in \mathbf{Z}$ let $\epsilon_{p,q} := \epsilon_{i,j}$, where $1 \le i, j \le d$, $i \equiv p \pmod{d}$ and $j \equiv q \pmod{d}$. Then $\epsilon_1 := \sum_{i=1}^{d} \xi^{i-1}\epsilon_{i,i+1}$ and $\epsilon_2 := \sum_{i=1}^{d} \xi^{i-1}\epsilon_{i,i}$ generate $M_d(k)$, as follows from the equations $\sum_{j=0}^{d-1} \xi^{(1-i)j}\epsilon_2^j = d \cdot \epsilon_{i,i}$ and $\epsilon_{i,i} \cdot \epsilon_3^r = \epsilon_{i,i+r}$, where $\epsilon_3 := \epsilon_2^{-1} \cdot \epsilon_1 = \sum_{i=1}^{d} \epsilon_{i,i+1}$.

Choose $e_1 \in ((-1)^{d+1})_1$ and $e_2 \in (1)_1$ with $e_1^d = (-1)^{d+1}$ and $e_2^d = 1$ and set $\Psi(e_i) = \epsilon_i$ for $i = 1, 2$. Then Ψ is a well-defined algebra homomorphism, since it respects the defining relations of $((-1)^{d+1}, 1)$, as can easily be seen. It is clearly an epimorphism, and since both algebras have the same dimension, Ψ must be an isomorphism.

In particular, if k is an algebraically closed field, Ψ defines an explicit identification of (a, b) with the full matrix ring $M_d(k)$.

Returning to our particular case, the identification is given by $e_0^{-1}e_2 \mapsto x\epsilon_1$ and $e_1^{-1}e_2 \mapsto x\rho\epsilon_2$, where ρ is a root of $X^2 - \xi$, if d is even, and 1, if d is odd, and where x is a root of $X^d - a$.

We must determine the matrix $\beta = \phi_0(e_1)X_1 + \phi_0(e_2)X_2$, where $\phi_0(e_i)$ are the matrices which are given by the multiplication with $e_0^{-1}e_i$ with respect to a basis $\mathcal{B}$ of the unique indecomposable $C_0(g, \xi)$-module, which is isomorphic to $\mathbf{C}^d$, with the natural operation of $M_d(\mathbf{C})$. Since $e_0^{-1}e_1 = e_0^{-1}e_2 \cdot (e_1^{-1}e_2)^{-1}$ it follows that $\phi_0(e_1) = \rho^{-1}\epsilon_2^{-1}\epsilon_1 = \rho^{-1}\epsilon_3$ and $\phi_0(e_2) = x \cdot \epsilon_1$. Thus we get $\beta = \rho^{-1}X_1\epsilon_3 + xX_2\epsilon_1$.

For $d = 2$ we have

$$\beta = \begin{pmatrix} 0 & iX_1 + xX_2 \\ iX_1 - xX_2 & 0 \end{pmatrix}$$

and for $d = 3$ we have

$$\beta = \begin{pmatrix} 0 & X_1 + xX_2 & 0 \\ 0 & 0 & X_1 + x\xi X_2 \\ X_1 + x\xi^{-1}X_2 & 0 & 0 \end{pmatrix}$$

where $\xi = \frac{1}{2}(1 + i\sqrt{3})$.

EXAMPLE 3.12. *Matrix factorization of* $f = \sum_{i=1}^{n} X_i^d$ *over* $\mathbf{C}$.

Let $\xi \in \mathbf{C}$ be a d-th primitive root of unity. We will determine the unique indecomposable $C_0(f, \xi)$-matrix factorization $f = \alpha_0 \cdot \ldots \cdot \alpha_{d-1}$ of $f = \sum_{i=1}^{n} X_i^d$ over $\mathbf{C}$. By proposition 3.3 we know that $\alpha_i = X_1 + \xi^i \alpha_{d,n}$ with $\alpha_{d,n} = \sum_{r=2}^{d} \phi_0(e_r)X_r$, where $\phi_0(e_r)$ is the multiplication by $e_1^{-1}e_r$ on an indecomposable $C_0(f, \xi)$-module.

In order to describe $\alpha_{d,n}$ we introduce some more notations: Set $\ell := [\frac{n-1}{2}]$ and $G := (\mathbf{Z}/d\mathbf{Z})^{\ell}$, and let $\bar{\imath}$ denote the residue class of an element $i \in \mathbf{Z}$ in $\mathbf{Z}/d\mathbf{Z}$. For $g = (\bar{\imath}_1, \ldots, \bar{\imath}_\ell) \in G$ and $t \in \{1, \ldots, \ell\}$ we set

$$g(t) = \xi^{\sum_{j=1}^{t}(i_j - 1)} \in \mathbf{C}, \quad \text{and}$$
$$g\langle t\rangle = (\bar{\imath}_1, \ldots, \bar{\imath}_{t-1}, \bar{\imath}_t + 1, \bar{\imath}_{t+1}, \ldots, \bar{\imath}_\ell) \in G.$$

After having ordered the elements of G lexicographically with respect to the smallest positive remainder we define square matrices ϵ_{g_1,g_2} of size d^ℓ for all $g_1, g_2 \in G$ by setting

$$\left(\epsilon_{g_1,g_2}\right)_{g_1',g_2'} := \begin{cases} 1 & \text{if } g_1 = g_1' \text{ and } g_2 = g_2' \\ 0 & \text{otherwise.} \end{cases}$$

Then

$$\begin{array}{cccccccc} \Phi: & M_d(\mathbf{C}) & \otimes & \ldots & \otimes & M_d(\mathbf{C}) & \to & M_{d^\ell}(\mathbf{C}) \\ & \epsilon_{i_1,j_1} & \otimes & \ldots & \otimes & \epsilon_{i_\ell,j_\ell} & \mapsto & \epsilon_{g_1,g_2} \end{array},$$

where $g_1 = (\bar{\imath}_1, \ldots, \bar{\imath}_\ell)$ and $g_2 = (\bar{\jmath}_1, \ldots, \bar{\jmath}_\ell)$, is an isomorphism.

Finally, let ρ be a root of $X^2 - \xi$ if d is even and 1 if d is odd, and set

$$\mu_n := \mu := \begin{cases} \rho^{-1}, & \text{if } \ell \text{ is odd} \\ \rho, & \text{if } \ell \text{ is even} \end{cases} \quad \text{and}$$

$$\mu(i) := \begin{cases} \mu, & \text{if } i \equiv 2 \pmod 4 \\ \mu^{-1}, & \text{if } i \equiv 0 \pmod 4 \\ 1, & \text{if } i \equiv 1 \pmod 2 \end{cases}.$$

Then

$$\alpha_{d,n} = \sum_{r=2}^{n} \sum_{g \in G} g([\tfrac{r-1}{2}]) \cdot \mu(r) \cdot X_r \cdot \epsilon_{g,g\langle[\frac{r}{2}]\rangle}. \tag{5}$$

Next we want to give an inductive description of $\alpha_{d,n}$. It follows immediately from (5) that if n is odd, then $\alpha_{d,n+1}$ is obtained from $\alpha_{d,n}$ by setting $X_{n+1} = 0$.

Furthermore, if we set $\tilde{\alpha}_{d,n} := \alpha_{d,n}(X_4, \dots, X_{n+2})$, the formula (5) yields

$$\alpha_{d,n+2} = \begin{pmatrix} \tilde{\alpha}_{d,n} & \mu_{n+2}X_2 + X_3 & 0 & \cdot & \cdot & \cdot & 0 \\ 0 & \xi\tilde{\alpha}_{d,n} & \mu_{n+2}X_2 + \xi X_3 & \cdot & & & \cdot \\ \cdot & & \xi^2\tilde{\alpha}_{d,n} & & \cdot & \cdot & \cdot \\ \cdot & & & & \cdot & \cdot & \cdot \\ \cdot & & & & & \cdot & 0 \\ 0 & & & & & \cdot & \mu_{n+2}X_2 + \xi^{d-2}X_3 \\ \mu_{n+2}X^2 + \xi^{d-1}X_3 & 0 & & \cdot & \cdot & 0 & \xi^{d-1}\tilde{\alpha}_{d,n} \end{pmatrix}$$

This inductive formula is easier to remember in the form

$$\alpha_{d,n+2} = \epsilon_2 \otimes \tilde{\alpha}_{d,n} + \epsilon_3 \otimes (\mu_{n+2}X_2) + \epsilon_1 \otimes X_3 \,. \tag{6}$$

Now we prove the formula (5):

By the results of this section we know that

$$C_0(f,\xi) \simeq \begin{cases} \bigotimes_{i=1}^{\ell} \mathbf{C}\langle g_{i,1}, g_{i,2}\rangle & \text{, if } n \text{ is odd} \\ (\bigotimes_{i=1}^{\ell} \mathbf{C}\langle g_{i,1}, g_{i,2}\rangle) \otimes \mathbf{C}\langle g_{\ell+1,1}\rangle & \text{, if } n \text{ is even} \end{cases}$$

where $g^d_{i,1} = g^d_{i,2} = 1$, if d is odd, and $g^d_{i,1} = (-1)^i$ and $g^d_{i,2} = -1$, if d is even. Recall that $\bigotimes_{i=1}^{\ell} \mathbf{C}\langle g_{i,1}, g_{i,2}\rangle \simeq A$, where A is a central simple $\mathbf{C}$-algebra, and that $\mathbf{C}\langle g_{\ell+1,1}\rangle \simeq \mathbf{C}^d$, so that

$$C_0(f,\xi) \simeq \begin{cases} A & \text{if } n \text{ is odd} \\ A \otimes_{\mathbf{C}} \mathbf{C}^d \simeq \bigoplus_{i=1}^{d} A_i & \text{if } n \text{ is even} \end{cases},$$

where $A_i \simeq A$ as $\mathbf{C}$-algebras. Identifying A resp. one of the A_i (it does not matter which A_i, see theorem 3.9) with a full matrix ring we may describe the coefficients $\phi_0(e_r)$ of $\alpha_{d,n} = \sum_{r=2}^{d} \phi_0(e_r)X_r$ as matrices with entries in $\mathbf{C}$, and thereby obtain the matrix $\alpha_{d,n}$ explicitly.

The g's are expressed by the e's as follows: $g_{i,2} = e_{2i}^{-1}e_{2i+1}$, and $g_{i,1} = e_{2i-1}^{-1}e_{2i-2} \cdot \ldots \cdot e_3^{-1}e_2e_1^{-1}e_{2i+1}$ (where $e_{n+1} := e_n$). Hence we get $e_1^{-1}e_{2i+1} = g_{1,2} \cdot g_{2,2} \cdot \ldots \cdot g_{i-1,2} \cdot g_{i,1}$ and $e_1^{-1}e_{2i} = g_{1,2} \cdot g_{2,2} \cdot \ldots \cdot g_{i-1,2} \cdot g_{i,1} \cdot g_{i,2}^{-1}$ (where $g_{\ell+1,2} := 1$). Now we identify the factors $\mathbf{C}\langle g_{i,1}, g_{i,2}\rangle$ of $C_0(f,\xi)$ with full matrix rings over $\mathbf{C}$.

Since $\epsilon_1^d = (-1)^{d+1}$ and $\epsilon_2^d = \epsilon_3^d = 1$, we identify $g_{i,1}$ with $\begin{cases} e_1 & \text{, if } i \text{ is odd} \\ \mu e_1 & \text{, if } i \text{ is even} \end{cases}$ and we identify $g_{i,2}$ with $\begin{cases} \mu^{-1}e_2 & \text{, if } i \text{ is odd} \\ \mu e_2 & \text{, if } i \text{ is even} \end{cases}$.

Finally, if n is even we identify $g_{\ell+1,1}$ with $\begin{cases} \epsilon_3 & \text{if } n \equiv 0 \pmod 4 \\ \mu\epsilon_3 & \text{if } n \equiv 2 \pmod 4 \end{cases}$.

Now we get

$$\begin{aligned} \phi_0(e_{2i+1}) &= \Phi(\epsilon_2 \otimes \ldots \otimes \epsilon_2 \otimes \epsilon_1 \otimes 1 \otimes \ldots \otimes 1) \\ \phi_0(e_{2i}) &= \mu(2i)\Phi(\epsilon_2 \otimes \ldots \otimes \epsilon_2 \otimes \epsilon_3 \otimes 1 \otimes \ldots \otimes 1) \end{aligned}$$

where e_1 or e_3, respectively, occurs as the i-th factor. (If n is even, then $\phi_0(e_n) = \mu(n)\Phi(\epsilon_2 \otimes \ldots \otimes \epsilon_2)$.) This yields the matrix $\alpha_{d,n}$.

Using the induction formula (6), it is easy to construct the matrices $\alpha_{d,n}$. Here are some examples:

a) $\alpha_{3,2} = X_2$. Set $\xi := \frac{1}{2}(1 + i\sqrt{3})$. Then

$$\alpha_{3,4} = \begin{pmatrix} X_4 & X_2 + X_3 & 0 \\ 0 & \xi X_4 & X_2 + \xi X_3 \\ X_2 + \xi^2 X_3 & 0 & \xi^2 X_4 \end{pmatrix}$$

and $\alpha_{3,6} =$

$$\left(\begin{array}{ccc|ccc|ccc} X_6 & X_4 + X_5 & 0 & X_2 + X_3 & 0 & 0 & 0 & 0 & 0 \\ 0 & \xi X_6 & X_4 + \xi X_5 & 0 & X_2 + X_3 & 0 & 0 & 0 & 0 \\ X_4 + \xi^2 X_5 & 0 & \xi^2 X_6 & 0 & 0 & X_2 + X_3 & 0 & 0 & 0 \\ \hline 0 & 0 & 0 & \xi X_6 & \xi X_4 + \xi X_5 & 0 & X_2 + \xi X_3 & 0 & 0 \\ 0 & 0 & 0 & 0 & \xi^2 X_6 & \xi X_4 + \xi^2 X_5 & 0 & X_2 + \xi X_3 & 0 \\ 0 & 0 & 0 & \xi X_4 + X_5 & 0 & X_6 & 0 & 0 & X_2 + \xi X_3 \\ \hline X_2 + \xi^2 X_3 & 0 & 0 & 0 & 0 & 0 & \xi^2 X_6 & \xi^2 X_4 + \xi^2 X_5 & 0 \\ 0 & X_2 + \xi^2 X_3 & 0 & 0 & 0 & 0 & 0 & X_6 & \xi^2 X_4 + X_5 \\ 0 & 0 & X_2 + \xi^2 X_3 & 0 & 0 & 0 & \xi^2 X_4 + \xi X_5 & 0 & \xi X_6 \end{array}\right)$$

b) $\alpha_{4,2} = \rho \cdot X_2$, where $\rho = \frac{1}{\sqrt{2}}(1 + i)$

$$\alpha_{4,4} = \begin{pmatrix} \rho X_4 & \rho^{-1} X_2 + X_3 & 0 & 0 \\ 0 & i\rho X_4 & \rho^{-1} X_2 + iX_3 & 0 \\ 0 & 0 & -\rho X_4 & \rho^{-1} X_2 - X_3 \\ \rho^{-1} X_2 - iX_3 & 0 & 0 & -i\rho X_4 \end{pmatrix}$$

4. The associated linear maximal Cohen-Macaulay modules over the hypersurface ring

Let $f \in S = k[X_1, \ldots, X_n]$, $f \neq 0$, be a homogeneous polynomial of degree d. We consider graded modules over the hypersurface ring $R = S/(f)$. Such a module M is called a *maximal Cohen-Macaulay module* (an MCM-module) if $\operatorname{depth} M = \dim R = n - 1$, or equivalently, if M has an S-resolution

$$0 \longrightarrow S^m \xrightarrow{\phi} S^m \longrightarrow M \longrightarrow 0,$$

for some m. Since we assume that M is graded, the resolution can be chosen to be homogeneous, in which case all entries of the matrix corresponding to ϕ (with respect to a homogeneous basis) are homogeneous polynomials. We call M a *linear* MCM-module if all the entries are linear forms.

The following result of Eisenbud ([E]) is basic for the theory of MCM-modules over hypersurface rings.

THEOREM. (Eisenbud) *Equivalence classes of (two-factor) matrix factorizations $f = \alpha \cdot \beta$, where α and β are square matrices of homogeneous polynomials, correspond bijectively to isomorphism classes of graded MCM-modules over R.*

Under this correspondence one associates with $f = \alpha\beta$ the module $M = \operatorname{Coker}\phi$, where ϕ: $R^m \to R^m$ is the map corresponding to α (with respect to the canonical basis of R^m). Moreover, if ψ: $R^m \to R^m$ is the map corresponding to β, then the periodic sequence

$$\ldots \xrightarrow{\phi} R^m \xrightarrow{\psi} R^m \xrightarrow{\phi} R^m \longrightarrow M \longrightarrow 0$$

is exact.

Conversely, if α is a square matrix of homogeneous polynomials, such that $\operatorname{Coker}\phi$ is MCM for the associated linear map ϕ: $R^m \to R^m$, then there exists a square matrix β, such that $f = \alpha\beta$.

The main result of this section is

THEOREM 4.1. *The following conditions are equivalent:*

(a) *f has a (linear) matrix factorization of size m.*

(b) *There exists a free R-module F of rank m, and a chain of submodules*

$$\ldots \subseteq U_{i+1} \subseteq U_i \subseteq \ldots \subseteq U_1 \subseteq U_0 = F$$

such that for all i

(i) $U_{i+1} \subseteq (X_1, \ldots, X_n)U_i$

(ii) U_i/U_{i+1} is a linear MCM-module, or $U_i = 0$.

In fact, if the equivalent conditions hold, then $U_i \neq 0$ for $i < d$, and $U_d = 0$.

Proof. (a) $\Rightarrow$ (b): Let $f = \alpha_1 \cdot \ldots \cdot \alpha_d$ be a matrix factorization of size m, and let $F = R^m$. Each of the α_i defines an endomorphism ϕ_i: $F \to F$, and since $\alpha_1 \cdot \ldots \cdot \alpha_d \equiv 0 \pmod f$, we have $\phi_d \circ \ldots \circ \phi_1 = 0$.

We set $\tau_0 := \mathrm{id}_F$ and $\tau_i := \phi_i \circ \ldots \circ \phi_1$ $(1 \le i \le d)$. If we let $U_i := \operatorname{Im} \tau_i$, then all generators of U_i have degree i, and clearly

$$0 = U_d \subset U_{d-1} \subset \ldots \subset U_{i+1} \subset U_i \subset \ldots \subset U_1 \subset U_0 = F\,,$$

so that (i) is satisfied.

We will prove (ii) by showing that $U_{i-1}/U_i \xrightarrow{\sim} \operatorname{Coker} \phi_i$ for $i = 1, \ldots, d$. By the theorem of Eisenbud $\operatorname{Coker} \phi_i$ is a (linear) MCM-module, since $f = \alpha_i(\alpha_{i+1} \cdot \ldots \cdot \alpha_{i+d-1})$ is a matrix factorization of f into two factors, and ϕ_i corresponds to one of them.

We have the following commutative diagram with exact rows and columns.

$$\begin{array}{ccccccccc}
 & & 0 & & 0 & & & & \\
 & & \downarrow & & \downarrow & & & & \\
0 & \longrightarrow & \operatorname{Im} \phi_i \cap \operatorname{Ker} \tau_{i-1} & \longrightarrow & \operatorname{Ker} \tau_{i-1} & & & & \\
 & & \downarrow & & \downarrow & & & & \\
0 & \longrightarrow & \operatorname{Im} \phi_i & \longrightarrow & F & \longrightarrow & \operatorname{Coker} \phi_i & \longrightarrow & 0 \\
 & & {\scriptstyle \tau_{i-1}|\operatorname{Im}\phi_i}\downarrow & & \downarrow{\scriptstyle \tau_{i-1}} & & & & \\
0 & \longrightarrow & U_i & \longrightarrow & U_{i-1} & \longrightarrow & U_{i-1}/U_i & \longrightarrow & 0 \\
 & & \downarrow & & \downarrow & & & & \\
 & & 0 & & 0 & & & &
\end{array}$$

The asserted isomorphism $U_{i-1}/U_i \xrightarrow{\sim} \operatorname{Coker} \phi_i$ thus follows from the snake lemma once we have shown that $\operatorname{Ker} \tau_{i-1} \subseteq \operatorname{Im} \phi_i$.

Since $f = (\alpha_1 \cdot \ldots \cdot \alpha_{i-1})(\alpha_i \cdot \ldots \cdot \alpha_d)$ is a matrix factorization of f into two factors, it follows from Eisenbud's theorem that $\operatorname{Ker} \tau_{i-1} = \operatorname{Im} \sigma_{i-1}$, where $\sigma_{i-1} := \phi_d \circ \ldots \circ \phi_i$. This completes the proof, since obviously $\operatorname{Im} \sigma_{i-1} \subseteq \operatorname{Im} \phi_i$.

$(b) \Rightarrow (a)$: Since all the generators of the linear module F/U_1 have the same degree, and since $U_1 \subseteq (X_1, \ldots, X_n)F$, we may assume that all the generators of F have degree 0.

For $i = 0, \ldots, d-1$ we show by induction on i that U_i is an MCM-module with pure resolution

$$0 \longrightarrow S^m(-d) \xrightarrow{\ \phi\ } S^m(-i) \longrightarrow U_i \longrightarrow 0\,.$$

For $U_0 = F$ this is clear. Now suppose $i > 0$ and consider the exact sequence

$$0 \longrightarrow U_i \longrightarrow U_{i-1} \longrightarrow M \longrightarrow 0$$

where M is a linear MCM-module. Since $U_i \subseteq (X_1, \ldots, X_n)U_{i-1}$, and since all the generators of U_{i-1} are of degree $i-1$, the same is true for M by the induction hypothesis, and we get a commutative diagram

$$\begin{array}{ccccccc}
 & & & 0 & & 0 & \\
 & & & \downarrow & & \downarrow & \\
 & & & S^m(-d) & \xrightarrow{\phi} & S^m(-i) & \\
 & & & \downarrow & & \downarrow & \\
 & & & S^m(-i+1) & \simeq & S^m(-i+1) & \\
 & & & \downarrow & & \downarrow & \\
0 & \longrightarrow & U_i \longrightarrow & U_{i-1} & \longrightarrow & M & \longrightarrow 0 \\
 & & & \downarrow & & \downarrow & \\
 & & & 0 & & 0 &
\end{array}$$

Again from the snake lemma it follows that $U_i \simeq \operatorname{Coker}\phi$, proving our assertion.

Now for each U_i ($i = 0, \dots, d-1$) fix a minimal set of homogeneous generators $\mathcal{B}_i$ and let $\overline{\alpha}_i$ be the $m \times m$ matrix of 1-forms of R that describes the elements of $\mathcal{B}_i$ as linear combinations of the elements of $\mathcal{B}_{i-1}$. Then the inclusion $U_{d-1} \subseteq F$ is given by the matrix $\overline{\alpha}_1 \cdot \ldots \cdot \overline{\alpha}_{d-1}$. Hence we get an exact sequence

$$F \xrightarrow{\Psi} F \longrightarrow F/U_{d-1} \longrightarrow 0\,,$$

where Ψ is the R-linear map associated with $\overline{\alpha}_1 \cdot \ldots \cdot \overline{\alpha}_{d-1}$. The exact sequence

$$0 \longrightarrow U_i/U_{i+1} \longrightarrow F/U_{i+1} \longrightarrow F/U_i \longrightarrow 0$$

implies, by induction on i, that F/U_i, and in particular F/U_{d-1}, is an MCM. Thus Eisenbud's theorem yields the existence of an $m \times m$ matrix α_d (with homogeneous entries) such that $f = (\alpha_1 \cdot \ldots \cdot \alpha_{d-1}) \cdot \alpha_d$, where α_i is the canonical lifting of $\overline{\alpha}_i$ to a matrix of linear form of S, for $i = 1, \dots, d-1$. Comparing degrees on both sides it follows that the entries of α_d are linear forms. □

COROLLARY 4.2. *Let $0 = U_d \subset U_{d-1} \subset \ldots \subset U_1 \subset U_0 = F$ be the filtration of a free R-module F corresponding to the matrix factorization $f = \alpha_1 \cdot \ldots \cdot \alpha_d$. Let $V_i = (F/U_{d-i})^*$ be the R-dual of F/U_{d-i} for $i = 0, \dots, d$. Then we obtain the filtration*

(7) $$0 = V_d \subset V_{d-1} \subset \ldots \subset V_1 \subset V_0 \subset F^*$$

which corresponds to the dual matrix factorization

$$f = \alpha_d^t \cdot \alpha_{d-1}^t \cdot \ldots \cdot \alpha_1^t\,.$$

Proof. The sequence of epimorphisms

$$F \longrightarrow F/U_{d-1} \longrightarrow F/U_{d-2} \longrightarrow \ldots$$

by the left-exactness of the dualizing functor $\operatorname{Hom}_R(*, R)$ yields a sequence of monomorphisms

$$\ldots \hookrightarrow V_2 \hookrightarrow V_1 \hookrightarrow V_0 = F^* .$$

Identifying each V_i with its image in F^* we get the filtration (7). Using the notation in the proof of theorem 4.1, F/U_i has the periodic resolution

$$\ldots \longrightarrow F \xrightarrow{\sigma_i} F \xrightarrow{\tau_i} F \longrightarrow F/U_i \longrightarrow 0 .$$

Hence $F/U_{d-i} \simeq \operatorname{Im}\sigma_{d-i}$ whence $V_i \simeq (\operatorname{Im}\sigma_{d-i})^*$.

Furthermore, since R is Gorenstein, the dualizing functor maps short exact sequences of MCM-modules to short exact sequences. Applying this fact on the exact sequence

$$0 \longrightarrow \operatorname{Im}\sigma_{d-i} \longrightarrow F \xrightarrow{\tau_{d-i}} \operatorname{Im}\tau_{d-i} \longrightarrow 0$$

yields that $(\operatorname{Im}\sigma_{d-i})^* \simeq \operatorname{Im}\sigma^*_{d-i}$. It follows that V_i is minimally generated by elements of degree i. In particular $V_{i+1} \subseteq (X_1, \ldots, X_n)V_i$ for all i.

As in the proof of theorem 4.1, (a) $\Rightarrow$ (b), we get

$$V_{i-1}/V_i \simeq \operatorname{Im}\sigma^*_{d-i+1}/\operatorname{Im}\sigma^*_{d-i} \simeq \operatorname{Coker}\phi^*_{d-1+i} .$$

(Here $\sigma^*_d = \mathrm{id}_{F^*}$.) Since ϕ^*_i corresponds to the transpose matrix α^t_i of α_i, we are through. □

Let $G(R)$ denote the Grothendieck group of R, and let $\widetilde{G}(R)$ denote the reduced Grothendieck group (that is, $G(R)$ modulo the class of R). If M is an R-module we let $[M]$ denote the image of M in $\widetilde{G}(R)$.

COROLLARY 4.3. *Let $f = X_1^d + \ldots + X_n^d$, where $d \geq 2$ and n is odd, and let $f = \alpha_1 \cdot \ldots \cdot \alpha_d$ be the matrix factorization of f over $\mathbf{C}$ as described in theorem 3.9. Let M be the linear MCM-module $\operatorname{Coker}\phi_1$, where the endomorphism ϕ_1: $R^m \to R^m$ is given by α_1. Then $d[M] = 0$ in $\widetilde{G}(R)$. If $n = 3$, then $[M]$ defines an element of order d in the divisor class group of R.*

Proof. Consider the filtration $0 = U_d \subset U_{d-1} \subset \ldots \subset U_1 \subset U_0 = F$ given by the factorization $f = \alpha_1 \cdot \ldots \cdot \alpha_d$. We know by theorem 3.9 that all α_i are pairwise equivalent. Therefore all linear MCM-modules U_{i-1}/U_i are isomorphic to M, which implies that $d[M] = 0$.

Let as in the examples 3.11 and 3.12 $\xi \in \mathbf{C}$ be a d-th primitive root of unity, and let ρ be a root of $x^2 - \xi$ if d is even, and 1 if d is odd. Then, if $n = 3$, M is a rank 1 module given by the image of β: $R^d \to R^d$, where $\beta = \prod_{i=1}^{d-1}(X_1 + \xi^i \alpha)$ with

$$\alpha = \begin{pmatrix} 0 & a_0 & 0 & . & . & . & 0 \\ 0 & 0 & a_1 & . & & & . \\ . & & & . & . & & . \\ . & & & & . & . & . \\ . & & & & & . & 0 \\ 0 & & & & & . & a_{d-2} \\ a_{d-1} & 0 & . & . & . & . & 0 \end{pmatrix}$$

and $a_j = \rho X_2 + \xi^j X_3$ for $j = 0, \ldots, d-1$.

M is isomorphic to the ideal I generated by the first row of β. Using the fact that $\beta = \sum_{i=0}^{d-1}(-1)^i X_1^{d-1-i}\alpha^i$, the ideal I easily can be computed to be $(x_1^{d-1}, x_1^{d-2}\bar{a}_0, x_1^{d-3}\bar{a}_0\bar{a}_1, \ldots, \bar{a}_0 \cdot \ldots \cdot \bar{a}_{d-2})$, where x_1 and $\bar{a}_i$ are the residue classes of X_1 and a_i in R, respectively.

For an ideal J in R we denote its class in $\mathcal{C}\ell(R)$ by $[J]$. In the given situation we have $\widetilde{G}(R) \simeq \mathcal{C}\ell(R)$; see for instance [HS]. Thus

$$[M] = [I = [(x_1, \bar{a}_0 \cdot \ldots \cdot \bar{a}_{d-2})] = \sum_{i=1}^{d-2}[(x_1, \bar{a}_i)] = (d-1)[P_0]\,.$$

Here we used the fact that the prime ideals $P_i := (x_1, \bar{a}_i)$ all are isomorphic as R-modules.

To conclude the proof it suffices to show that $i \cdot [P_0] \neq 0$ in $\mathcal{C}\ell(R)$ for $i = 1, \ldots, d-1$. But $i[P_0] = \sum_{j=0}^{i-1}[P_j] = [(x_1, \bar{a}_0 \cdot \ldots \cdot \bar{a}_{i-1})]$. Since $f \in (X_1, a_0 \cdot \ldots \cdot a_{i-1})$, and since X_1, $a_0 \cdot \ldots \cdot a_{i-1}$ is a regular sequence, it follows that $R/(x_1, \bar{a}_0 \cdot \ldots \cdot \bar{a}_{i-1}) \simeq \mathbf{C}[X_1, X_2, X_3]/(X_1, a_0 \cdot \ldots \cdot a_{i-1})$ is CM of $\dim R - 1$. Hence $(x_1, \bar{a}_0 \cdot \ldots \cdot \bar{a}_{i-1})$ is a non-principal CM-ideal, which implies that $[(x_1, \bar{a}_0 \cdot \ldots \cdot \bar{a}_{i-1})] \neq 0$. □

REMARK. For $d = 2$ the divisor class group $\mathcal{C}\ell(R)$ of $R = \mathbf{C}[X_1, X_2, X_3]/(X_1^d + X_2^d + X_3^d)$ equals $\mathbf{Z}[M]$ and is isomorphic to $\mathbf{Z}/2\mathbf{Z}$. However, for $d > 2$, $\mathbf{Z} \cdot [M]$ only generates the discrete part of $\mathcal{C}\ell(R)$, as we were told by Storch (see [St]).

– STOP PRESS –

Using Childs's results and an argument of 'genericity', we may get a fast affirmative answer to our "open question 1" in almost all cases:

THEOREM. *Let $f \in k[X_1, \ldots, X_n]$ be a form of degree d, where $(\operatorname{Char} k, d) = 1$. Then f has a linear matrix factorization of finite size, and (equivalently) a generalized Clifford algebra of finite dimension. Thus $k[X_1, \ldots, X_n]/(f)$ has an Ulrich module.*

Proof. We may write f as a sum of monomials; $f = \sum_{i=1}^{s} m_i$, say. Let $g = \sum_{i=1}^{s} g_i$ be the "generic s-term" d-form in $k[X_1, X_2, \ldots, X_{sd}]$ defined by $g_i = X_{(i-1)d+1} \cdot \ldots \cdot X_{id}$. There is an obvious homogeneous algebra homomorphism $\varphi\colon k[X_1, \ldots, X_{sd}] \to k[X_1, \ldots, X_n]$, such that $\varphi(g) = f$.

(Essentially) by [C, theorem 11], g has a linear matrix factorization

$$g = \beta_0 \cdot \ldots \cdot \beta_{d-1}\,,$$

say, of finite size. (In fact, each g_i has a linear factorization and thus a linear matrix factorization with matrices of size 1×1, and the g_i are monomials in disjoint sets of variables. Now apply theorem 3.1.) Now, for $j = 0, \ldots, d-1$ we let α_j be the matrix formed by β_j by applying φ to each entry. Then indeed

$$f = \alpha_0 \cdot \ldots \cdot \alpha_{d-1}$$

is a linear matrix factorization of finite size.

The other statements of the theorem follow immediately from this. □

REFERENCES

[ABS] M. F. Atiyah, R. Bott, A. Shapiro, *Clifford modules*, Topology **3**, Suppl. (1964), 3–38.

[Ba] J. Backelin, *La série de Poincaré-Betti d'une algèbre graduée de type fini à une relation est rationnelle*, C. R. Acad. Sc. Paris **287** (1978), 846–849.

[Be] G. M. Bergman, *The diamond lemma for ring theory*, Advances in Math. **29** (1978), 178–218.

[BEH] R. O. Buchweitz, D. Eisenbud, J. Herzog, *Cohen-Macaulay modules on quadrics*, to appear in the Proceedings of the Lambrecht Conference 'Representations of algebras, singularities and vector bundels' (S.L.N.).

[BHU] J. P. Brennan, J. Herzog, B. Ulrich, *Maximally generated Cohen-Macaulay modules*, to appear in Math. Scand..

[C] L. N. Childs, *Linearizing of n-ic forms and generalized Clifford algebras*, Linear and Multilinear Algebra **5** (1978), 267–278.

[E] D. Eisenbud, *Homological algebra on a complete intersection with an application to group representations*, Trans. AMS **260** (1980), 35–64.

[H] N. Heerema, *An algebra determined by a binary qubic form*, Duke math J. **21** (1954), 423–444.

[HK] J. Herzog, M. Kühl, *Maximal Cohen-Macaulay modules over Gorenstein rings and Bourbaki-sequences*, Advanced Studies in Pure Mathematics 11, 1987, Commutative Algebra and Combinatorics.

[HS] J. Herzog, H. Sanders, *The Grothendieck group of invariant rings and of simple hypersurface singularities*, to appear in the Proceedings of the Lambrecht Conference 'Representations of algebras, singularities and vector bundels' (S.L.N.).

[K] H. Knörrer, *Maximal Cohen-Macaulay modules on hypersurface singularities I*, Invent. Math. **88** (1985), 153–164.

[L] F. W. LONG, *Generalized Clifford algebras and dimodule algebras*, J. London Math. Soc. (2) **13** (1976), 438–442.

[M] J. MILNOR, *Introduction to algebraic K-theory*, (Annals of Math. Studies No. 72, Princeton, 1971).

[R] N. ROBY, *Algèbres de Clifford des formes polynomes*, C. R. Acad. Sc. Paris **268** (1969), 484–486.

[Sch] W. SCHARLAU, *Quadratic and Hermitian Forms*, Grundlehren **270**, Springer-Verlag, Berlin Heidelberg, 1985.

[St] U. STORCH, *Die Picard-Zahlen der Singularitäten* $t_1^{r_1}+t_2^{r_2}+t_3^{r_3}+t_4^{r_4}=0$, J. Reine Angew. Math. **350** (1984), 188–202.

[U] B. ULRICH, *Gorenstein rings and modules with high numbers of generators*, Math. Z. **188** (1984), 23–32.

"This paper is in final form and no version of it will be submitted for publication elsewhere"

SOME NEW RESULTS
IN THE COMBINATORIAL THEORY OF RINGS AND GROUPS

L.A. Bokut'
Institute of Mathematics, Siberian Branch,
Academy of Sciences of the USSR,
630090 Novosibirsk, USSR

Contents

In our survey we formulate (with indications of proofs) some new results in ring theory and in group theory obtained in Novosibirsk and Barnaul during the last few years.

§1. Almost finitely presentable varieties of rings

G. Bergman (1974) constructed a finite ring B (namely, $B = \mathrm{End}(\mathbb{Z}_p \oplus \mathbb{Z}_{p^2})$) which is not presentable by matrices over a commutative ring. Yu.N. Mal'tsev (1985) initiated the study of the variety VarB generated by B.

Definition. We call a variety V finitely presentable if any finite ring from V is presentable (that is, embeddable into a matrix ring over some commutative ring). A variety W is called almost finitely presentable if any proper subvariety of W is finitely presentable, while W itself is not finitely presentable.

Theorem 1. (Yu.N. Mal'tsev). The variety VarB generated by Bergman's ring B is almost finitely presentable.

In the same paper an infinite series B_m, $B_1 = B$, of finite rings with similar properties was constructed (the ring B_2 has been constructed by M.Y. Finkelstein). It was proved that any almost finitely presentable variety V which satisfies some additional assumptions is precisely one of the varieties $\mathrm{Var}B_m$.

§2. The solution of Specht's problem

We are refering to the following well known problem raised by Specht in 1950: is it true that any (associative) algebra over a field F of zero characteristic has a finite basis of identities? A positive answer has been recently obtained by A.R. Kemer. Previously it was known that any algebra which satisfies the identity $[x_1,\dots,x_n]\dots[z_1,\dots,z_m] = 0$ has a finite basis of identities (V.N. Latyshev, G. Genov, A. Popov) and that the same is true for the matrix algebra $M_n(F)$ (A.V. Yakovlev).

An important role in Kemer's proof is played by the so called $\mathbb{Z}_2$-graded algebras $A = A_0 + A_1 = (A,A_0,A_1)$ where generally speaking A may not have a unit, and $A_0 \cap A_1 \neq (0)$. As an example of such an algebra we may consider the free algebra $\underline{F} = F\langle X\rangle = \underline{F}_0 + \underline{F}_1$ on a countable set of free generators $X = Y \dot{\cup} Z$; here Y and Z are countable sets, $\underline{F}_0$ is spanned by all monomials of even degree with respect to Z, $\underline{F}_1$ is spanned by all monomials of odd degree with respect to Z. A polynomial $f(y_1,\dots,y_k,z_1,\dots,z_m)$ is called a graded identity of a graded algebra (A, A_0, A_1) if $f(A_0,\dots,A_0,A_1,\dots,A_1) = 0$. Denote by $T_2(A,A_0,A_1) = T_2(A)$ the set of all graded identities of A (this is a so-called T_2-ideal of $\underline{F}$, i.e. an ideal which is invariant under all graded endomorphisms of $\underline{F}$).

<u>Theorem 1</u> (A.R. Kemer, 1984). The positive solution of Specht's problem follows from the ACC for those T_2-ideals which contain the graded ideal $T_2(M_n(F),M_n(F),M_n(F)) = T_2(M_n)$ for some n.

Suppose now that Specht's conjecture is false (and derive a contradiction). Then by Theorem 1 there exist an infinite chain of multilinear polynomials $f_1,f_2,\dots$ and an integer $n \geqslant 1$ such that

$$T_2(M_n) \subsetneqq T_2(M_n) + \{f_1\}^{T_2} \subsetneqq T_2(M_n) + \{f_1\}^{T_2} + \{f_2\}^{T_2} \subsetneqq \dots .$$

Denote by $\{f\}_+^{T_2}$ the set of those polynomials from $\{f\}^{T_2}$ which have degree greater than the degree of f. Then the T_2-ideal

$$\Gamma = T_2(M_n) + \{f_1\}_+^{T_2} + \{f_2\}_+^{T_2} + \dots$$

has the following properties: 1) $f_k \notin \Gamma$, 2) $\{f_k\}_+^{T_2} \subseteq \Gamma$, $k > 1$.

Definition. Say that a T_2-ideal is a degenerate counter-example (to Specht's problem) if there exists a sequence of multilinear polynomials $f_1, \dots, f_k, \dots$ with properties 1), 2).

Denote by $\underline{P}$ the set of all degenerate counter-examples to Specht's problem.

We note the following three fundamental properties of $\underline{P}$:

a) If $\Gamma \in \underline{P}$ then $\Gamma \supseteq T_2(M_n)$ for some n.

b) If (A, A_0, A_1) is a finite-dimensional (f.d.) $\mathbb{Z}_2$-graded algebra then $T_2(A, A_0, A_1) \notin \underline{P}$.

c) If $\Gamma \in \underline{P}$ and A is a f.d. graded algebra then either $\Gamma + T_2(A) \in \underline{P}$ or $\Gamma \cap T_2(A) \in \underline{P}$.

Theorem 2 (A.R. Kemer). If a set $\underline{P}_1$ of T_2-ideals satisfies a), b), c) then $\underline{P}_1 = \emptyset$.

In what follows we need a classification of simple f.d. $\mathbb{Z}_2$-graded algebras over an algebraically closed field F. It is given by the following list:

$(M_n(F), M_n(F), M_n(F)) = M_n(F)$;

$(M_n(F), M_n(F), 0)$;

$(M_n(F+cF), M_n(F), cM_n(F))$, where $c^2 = 1$; 1, c are linearly independent;

$\left(M_{n,k}(F), \begin{pmatrix} M_k & 0 \\ 0 & M_{n-k} \end{pmatrix}, \begin{pmatrix} 0 & M_{k,n-k} \\ M_{n-k,k} & 0 \end{pmatrix}\right)$, where $k < n$.

Moreover, if A is a f.d. $\mathbb{Z}_2$-graded algebra over an algebraically closed field of zero characteristic then $A = D + \mathrm{Rad}A$, where $D = \bigoplus_{i=1}^{s} D^{(i)}$, $D^{(i)}$ are simple $\mathbb{Z}_2$-graded algebras, RadA is a graded subalgebra.

Definition. Let A be a f.d. graded algebra over a field F. We call A classical if all its simple components are isomorphic to one of the algebras listed above (that is, to a matrix algebra over F or over F+cF) with the appropriate $\mathbb{Z}_2$-grading.

The following simple lemma shows that for our aims it suffices to consider classical algebras only (by algebra we mean a $\mathbb{Z}_2$-graded algebra).

Lemma 1. For any f.d. algebra A over F there exists a classical f.d. algebra B over F such that $T_2(A) = T_2(B)$.

Consider the following parameters of a classical algebra $A=D+\mathrm{Rad}A$, where $D = D_0+D_1$ is the semisimple part of A: $b_0(A) = \dim D_0$, $b_1(A) = \dim D_1$, $c(A) = \mathrm{index}(\mathrm{Rad}A)$, a(A) is the capacity of A (the maximal number of pairwise orthogonal idempotents of A_0). Set $t(A) = (a(A), b_0(A), b_1(A), c(A))$. We call t(A) the height of A and bring the lexicographic order to the set of all heights.

Now for a pair (Γ,A) of a T_2-ideal Γ and a f.d. classical algebra A such that $T(A)\subseteq\Gamma$, define

$$S_A^t(\Gamma) = \mathrm{Id}(S_{\Lambda_1} \dots S_{\Lambda_t} f,\ f\in\Gamma),$$

where $\Lambda_i\subset Y$ or $\Lambda_i\subset Z$, $|\Lambda_i| = b_0(A)+1$ for $\Lambda_i\subset Y$, $|\Lambda_i| = b_1(A)+1$ for $\Lambda_i\subset Z$ and $S_\Lambda f$ is the skew-symmetrization of the polynomial f with respect to Λ.

Lemma 2. $S_A^{C(A)}(\Gamma) \subseteq T_2(A)$.

From this lemma it follows that $d(A,\Gamma) = \min\{t,\ S_A^t(\Gamma) \subseteq T_2(A)\}$ is finite and satisfies the inequality $0 \leqslant d(A,\Gamma) \leqslant c(A)$ (we assume that $S_A^0(\Gamma) = \Gamma$).

The idea of obtaining a contradiction with the assumption $\underline{P}_1 \neq \emptyset$ is the following. For a pair (Γ,A), $\Gamma\in\underline{P}$, $A\in\Gamma$ we construct a new pair (Γ_1,A_1) with the same properties such that $t(A_1) < t(A)$. Then it remains to consider the pair (Γ,A) with minimal height $t(A)$. An important role in this construction is played by

Proposition 1. Let Γ be a T_2-ideal, let A be a f.d. classical algebra, such that $T(A) \subseteq \Gamma$. If $S(A) \geqslant 2$ and $d(A,\Gamma) = c(A)$ then there exist such f.d. classical algebras $B, B^{(1)},\dots,B^{(k)}$ that $\Gamma \supseteq T_2(B) \supseteq T_2(A)$, $a(B^{(i)}) < a(A)$, $i = 1,\dots,k$, $S_A^{d(A,\Gamma)-1}(\Gamma\cap\bigcap_i T_2(B^{(i)})) \subseteq T_2(B)$.

Similar assertions are valid also for $S(A) = 1$ and for $S(A) \geqslant 2$, $d(A,\Gamma) < c(A)$.

In the proof of Proposition 1 and similar assertions A.R. Kemer used in an essential way the so called trace identities. Let us again consider the free $\mathbb{Z}_2$-graded algebra $\underline{F} = F\langle X\rangle = \underline{F}_o+\underline{F}_1$, $X = Y\dot{\cup}Z$. Denote by $S^{\underline{F}_o}$ the commutative algebra of formal polynomials in the variables Spu, where u is a monomial from $\underline{F}_o$. The algebra $\tilde{F}\langle X\rangle = F\langle X\rangle \otimes S^{\underline{F}_o} = (\underline{F}_o\otimes S^{\underline{F}_o}) + (\underline{F}_1\otimes S^{\underline{F}_o})$ with the mapping $Sp:\underline{F}_o\otimes S^{\underline{F}_o} \to S^{\underline{F}_o}$ (here $Sp(u\otimes Spv) = Spu.Spv$) is called the free algebra with a trace function. Generally speaking, by an R-trace of a $\mathbb{Z}_2$-graded algebra (A,A_o,A_1), where A_o, A_1 are R-modules over a commutative unitary ring of scalars R, we mean an arbitrary R-linear mapping $Sp:A_o\to R$. By a polynomial of Hamilton-Cayley type we mean a polynomial $f = y^n+a_1(y)y^{n-1}+\ldots+a_{n-1}(y)y$, where $y\in Y$, $a_i(y)\in F[Sp(y^k)]$. The following proposition is essential for the proof of Theorem 2.

Proposition 2. For a T_2-ideal Γ, $\Gamma\supset T_2(M_n)$, the following assertions are equivalent:

1) $\Gamma = T_2(A)$, where A is a f.d. classical $\mathbb{Z}_2$-graded algebra.

2) $F\langle X\rangle/\Gamma\subset B$, where B is a $\mathbb{Z}_2$-graded algebra with a trace function which satisfies an identity of Hamilton-Cayley type.

3) There exists an R-algebra A over a commutative unitary ring of scalars R, such that $F\langle X\rangle/\Gamma\subset A$ and every element $a\in \underline{F}_o+\Gamma/\Gamma$ is algebraic (with R-coefficients) in A.

4) The algebra $F\langle X\rangle/\Gamma$ is locally presentable (that is, any graded finitely generated subalgebra of $F\langle X\rangle/\Gamma$ is embeddable into a matrix algebra over a suitable field).

§3. The group of units of a coproduct of rings.

Consider a skew-field Λ and a Λ-ring A (that is $A\supset\Lambda$, $1_A = 1_\Lambda$). The set $\underline{A} = \{A_i;\ A_i$ is a Λ-ring$\}$ is called an amalgam if $A_i\cap A_j = \Lambda$, $i \neq j$. Let $C = *_\Lambda\underline{A}$ be a coproduct in the category of Λ-rings (its existence and good properties were established by P.M. Cohn in 1961). Denote by U(C) the group of units of the ring C. Then $U(C)\supset U(A)$, $A\in\underline{A}$ and $U(C)\supset E(A,C) = \{1+ycx;\ x,y\in A,\ c\in C\}$ (here $1 + ycx\in A$ is a transvection). Consider the subgroup $G(A,C) = U(A)E(A,C)$ of U(C). G.Bergman (1972) proved that if all A_i are weakly finite (i.e. $xy = 1$ implies $yx = 1$), then the group U(C) is generated by the subgroups G(A,C).

Theorem 1 (V.N. Gerasimov). Under these assumptions the group U(C) is the coproduct of the groups G(A,C), $A\in\underline{A}$ with amalgamated subgroup

$U(\Lambda) : U(C) = *_{U(\Lambda)}\{G(A,C);A\in \underline{A}\}$.

Now consider a group U and a system $\underline{H}$ of its subgroups, $H\subset\cap\underline{H}$. Call a function $\nu:U\to\underline{H}\cup\{H\}$ an origin function if

a) $x\in H \to \nu(x)\in H$

b) $\nu(x) \neq G\in\underline{H},\ y\in G\setminus H \to \nu(yx) = G$.

Lemma 1. Suppose that U is generated by $\underline{H}$. If there exists an origin function $\nu:U\to\underline{H}\cup\{H\}$ then the set $\underline{H}$ is an amalgam of groups over H and U is the coproduct of $\underline{H}$ with the amalgamated subgroup H.

Thus for the proof of Theorem 1 it is sufficient to construct an origin function $\nu:U\to\{G(A,C)\}\cup\{U(\Lambda)\}$.

Let $M = Ce$ be the free C-module of rank 1. Consider an infinite sequence

$\ldots,F_3,F_2,F_1,F_0 = \Lambda$

of rings $F_i\in\underline{A}$ such that $F_{s+1} \neq F_s$ and for any $A\in\underline{A}$ the set $J_A = \{s;F_s=A\}$ is infinite. It defines a filtration on M:

$M_{-1}=(0)\subset M_0=F_0e\subseteq\ldots\subseteq M_s=F_sM_{s-1}\subseteq\ldots,\ \cup M_s = M.$

Set $d(m) = \min\{s;m\in M_s\}$, where $m\in M$.

Proposition 1. Let $m\in M$, $\mathrm{Ann}_C(m) = 0$ (for example $m = ge$, $g\in U(C)$). If $g\in G(A,C)\setminus U(\Lambda)$ and $d(m)\notin J_A$ then $d(gm)\in J_A$.

This proposition immediately gives an origin function of U(C) via $\nu(x) = G(F_{d(xe)},C)$, where $x\in U(C)$.

Basic definition. Let $A\in\underline{A}$. The subgroup $B = \sum_{i\in J_A} S_iM_{i-1}$ of M, where S_i are right Λ-subspaces of A and almost all $S_i = 0$, is called a block. If $s \geq 0$ then denote by s_+ the smallest integer such that $s_+>s$ and $s_+\in J_A$. The expression above for a block B is called its normal form if $S_i\supseteq S_{i_+}A$ for all i. Any block has some normal form (it is sufficient to replace S_i by $S_i+S_{i_+}A$).

Lemma 2. Any block has only one normal form. In other words, for any block B there are uniquely determined right Λ-subspaces $S_i(B)\subset A$ such that $B = \Sigma S_i(B)M_{i-1}$ ($i\in J_A$ for almost all i).

For any block B we shall construct a subgroup $N(B)\subseteq E(A,C)$. Let

$L_i = L_i(B) = \{x\in A;\ xS_{i_+}(B) = 0\}$, $P_{s,t} = F_sF_{s-1}\cdots F_t$ if $s > t$ and $P_{s,t} = 0$ if $s < t$. Denote $H_s(B) = \sum_{s>i} P_{s,i}L_i$, $R(B) = \sum_{i\in J_A} S_iH_{i-1}(B)$. Then $R(B)$ is a nilpotent ring. Let $N(B) = 1+R(B)\subseteq E(A,C)$. Then $N(B)B\subseteq B$. The orbits $K = N(B)m$, $m\in B$, will be called the cells of the block B.

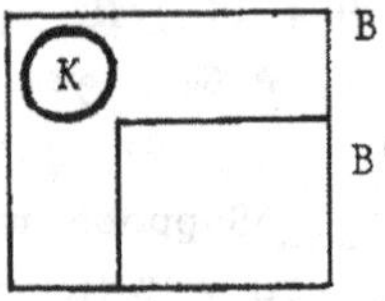

Definition. A cell K of a block B is called minimal if for any block $B'\subsetneq B$ the intersection $K\cap B'$ is empty:

Proposition 2. Let K_1, K_2 be minimal cells of a block. If $K_1 = gK_2$, $g\in G(A,C)$ then $K_1 = aK_2$, $a\in U(A)$ (in this case we write $K_1 \sim K_2$).

Proposition 1 follows from Proposition 2. The proof of Proposition 2 consists of several lemmas, among them:

Intersection lemma. Let $\{B^\alpha\}$ be a set of blocks. Then $\bigcap_\alpha B^\alpha = B$ is a block and $S_i(B) = \bigcap_\alpha S_i(B^\alpha)$.

Minimal block lemma. Let K be a cell of a block B. Then the set of blocks $\{B_1;\ K\cap B_1 \neq \emptyset,\ B_1 \supseteq \sum_{i\in J_A} S_{i_+}AM_{i-1}\}$ has a minimal element $B'(B,K)$.

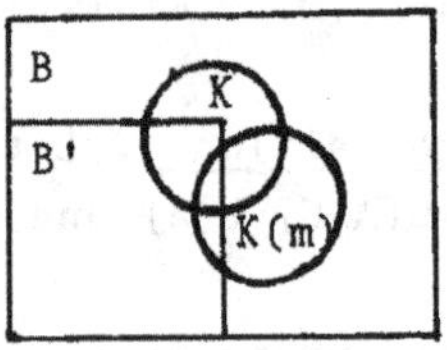

Triangle lemma. Let K be a cell of a block B, $m\in K$. Consider the smallest block $B(m)$ which contains the element $m\in K$ and consider the cell $K(m) = N(B(m))m$, $m\in K$. Then $K(m)\cap B'(B,K)\cap K \neq \emptyset$.

Connectedness lemma. Let $v = 1+ycx$, $xy = 0$, be an essential A-transvection (that is $c\in A$). Then $K(m)\cap K(vm) \neq \emptyset$ for any $m\in M$.

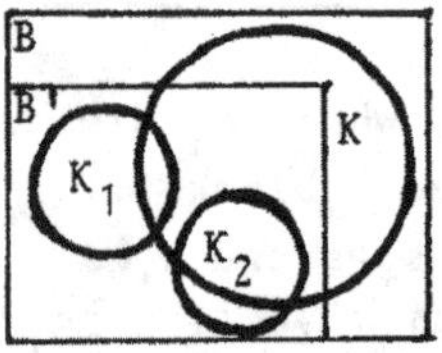

Proposition 3. (The basic one). Let K be a cell of a block B and let K_1, K_2 be cells of the block $B'(B,K)$, such that $K_i\cap K \neq \emptyset$, $i = 1,2$. Then $K_1 \sim K_2$.

Proposition 3 implies Proposition 2.

Let us now formulate some corollaries of Gerasimov's theorem (they were announced by V.N. Gerasimov in 1982). For a field F consider the free linear group $GL_n(X,F)$ over a set X (it means that any mapping $X \to GL_n(R)$, where R is an F-algebra, may be extended to a homomorphism $GL_n(X,F) \to GL_n(R)$). Then $GL_n(X,F) = GL_n(A)$ where A is the algebra of generic invertible $n\times n$ matrices $X = \{X_i\}$, $X_iY_i = E$. It is easy to verify that $GL_n(X,F) = U(F(Gr\langle X\rangle) *_F M_n(F))$, where $Gr\langle X\rangle$ is the free group on the set of generators X. From Gerasimov's theorem it follows that $GL_n(A) \cong U(F) Gr\langle X\rangle *_{U(F)} E$, where $E = E(GL_n(A))$ is the subgroup generated by the elementary matrices (transvection and scalar matrices). Hence E is not a normal subgroup of $GL_n(A)$ and any permutation of X induces a "nonstandard" automorphism of $GL_n(A)$. Thus, we have

Corollary 1. For a noncommutative algebra R the elementary linear group $E(GL_n(R))$ is not necessarily a normal subgroup of $GL_n(R)$.

Recall that for a commutative algebra R it is always a normal subgroup (A.A. Suslin).

Corollary 2. There exists a noncommutative algebra R whose general linear group $GL_n(R)$ has nonstandard automorphisms.

For contrast recall that any automorphism of $GL_n(R)$ is standard on $E(GL_n(R))$ (I.Z. Golubchik, A.V. Mikhalev [8], E.I. Zel'manov [9]).

§4. Embeddings into matrix algebras

Let F be a field. A variety $\underline{M}$ of F-algebras is called (triangularly) presentable if any algebra from $\underline{M}$ is embeddable into the algebra of (upper triangular) matrices over some commutative F-algebra.

Theorem 1 (A.Z. Anan'in). Let F be a field of characteristic 0. A variety $\underline{M}$ is (triangularly) presentable if and only if it satisfies the identities $[x_1,y_1]\dots[x_n,y_n] = 0$, $[x_1,\dots,x_n]y_1\dots y_n[z_1,\dots,z_n] = 0$ for some $n \geq 1$.

Now let R be an algebra over a field F, and let X be an infinite set. We call an element $f \in R\langle X\rangle$ n-polynomial if there exists such a disjunction $X = X_1 \dot\cup X_2 \dot\cup \dots \dot\cup X_n$ that $f = \Sigma a_1 x_1 a_2 x_2 \dots a_n x_n a_{n+1}$, $x_i \in X_i$, $a_j \in R$.

The following theorem proves a conjecture by A.Z. Anan'in.

Theorem 2 (N.G. Nesterenko, 1986). An algebra R is n-triangularly presentable if and only if for any n-polynomial f the following quasi-identity holds in R:

$$\bigwedge_{k\geq 1} \frac{\partial^k f}{\partial x_{i_1}\dots\partial x_{i_k}} = 0 \rightarrow f = 0.$$

Theorem 3. (A.Z. Anan'in). Any finitely generated right Noetherian PI-algebra over a field is presentable.

This theorem answers a question of I.V. L'vov and V.T. Markov.

§5. Some examples of groups with torsion, and without it.

Example 1. (L.A. Bokut'). The group $G = \langle a,b_1,b_2;\ (b_1b_2^{-1})^2ab_2^{-1}b_1b_2^{-1}a = ab_2^{-1}b_1b_2^{-1}a(b_2^{-1}b_1)^2,\ (b_1b_2^{-1})^2a(b_2^{-1}b_1)^2a^{-1} = a(b_2^{-1}b_1)^2a^{-1}(b_1b_2^{-1})^2\rangle$ is free of torsion. To prove this we have constructed a normal form for the words in G. It is not known whether the group algebra of G (over a field) has zero divisors.

Example 2. (L.A. Bokut'). The group $G = \langle a,b; aba^{-1} = ba^{-1}b,\ b^{s_k}a^{t_k}\ldots b^{s_1}a^{t_1} = (ba^{-1})^n,\ s_i, t_i > 0,\ k \geq 1,\ n \in \mathbb{Z}\rangle$ has torsion. To prove this we used the equality $a\alpha = \alpha^2 a$, where $\alpha = ba^{-1}$. Let H be the group generated by elements $a,b,b_1,\ldots,b_{t_1},t_{1+t_1},\ldots,b_{s_1+t_1},\ldots,b_{s_k+t_k+\ldots+s_1+t_1}$, where $s_i, t_i > 0$, $k \geq 1$ and which has the following defining relations (given in the form of a Cayley graph):

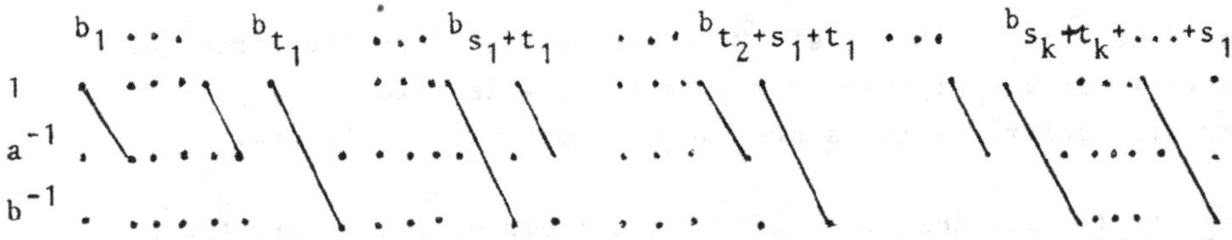

If we require in addition that the following two identities

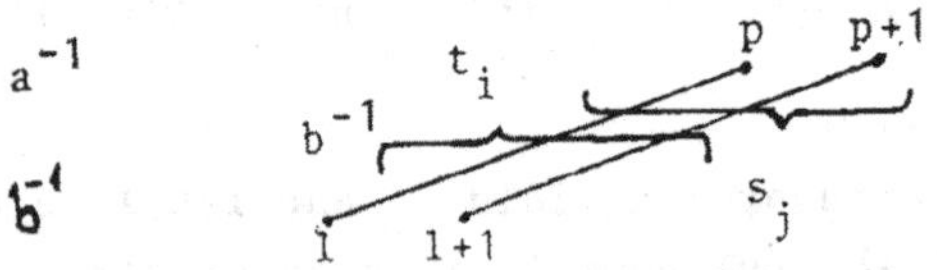

hold, then H has torsion (or is equal to ⟨1⟩).

B.V. Tarasov used the Gauss elimination algorithm for the group equalities to make several programs for the search of "pathological" examples of groups without torsion. One of them is based on the idea of the work [12].

References

1. G. Bergman. Some examples in PI-ring theory. Isr. J. Math., 18 (1974), 257-277.

2. Yu.N. Mal'tsev. On representations of finite rings by matrices over a commutative ring (in Russian). Mat. Sb., 128(1985), 383-402. (English translation: Math. USSR, Sb., 56 (1987), 379-402).

3. A.R. Kemer. Varieties and $\mathbb{Z}_2$-graded algebras (in Russian). Izv. Akad. Nauk SSSR, Ser. Mat., 48 (1984), 1042-1059. (English translation: Math. USSR, Izv., 25 (1985), 359-374).

4. G. Bergman. Coproducts and some universal ring constructions. Trans. Am. Math. Soc., 200 (1974), 33-88.

5. V.N. Gerasimov. On free linear groups (in Russian). The 17-th All-Union Algebraic Conference. Minsk, 1983, part 1, 52-53.

6. V.N. Gerasimov. Groups of units of free products of rings (in Russian). Mat. Sb., 134, (1987), 42-65. (English translation: Math. USSR, Sb. (to appear)).

7. A.A. Suslin. On the structure of the special linear group over polynomial rings (in Russian). Izv. Akad. Nauk SSSR, Ser. Mat., 41 (1977), 235-252. (English translation: Math. USSR, Izv., 11 (1977), 221-238).

8. I.Z. Golubchik, A.V. Mikhalev. Elementary subgroup of a unitary group over a PI-ring (in Russian). Vest. Mosk. Univ., Ser. I, No. 1 (1985), 30-36. (English translation: Mosc. Univ. Math. Bull. 40, No. 1, 44-54 (1985)).

9. E.I. Zel'manov. Isomorphisms of linear groups over associative rings (in Russian). Sib. Mat. Zh., 26 (1985), No. 4, 49-67. (English translation: Sib. Math. J., 26 (1986), 515-530).

0. A.Z. Anan'in. Presentable varieties of algebras (in Russian). Novosibirsk, 1986, 21 pp. Complete text: Depon. VINITI 13.05.1986, No. 3471-B (Reviewed in R.Zh. Mat., #8A210 (1986) (in Russian)).

1. N.G. Nesterenko. Representations of algebras by triangular matrices (in Russian). Algebra Logika, 24 (1985), 65-86. (English translation: Algebra Logic, 24 (1985), 41-55).

2. E. Rips, Y. Segev. Torsion-free group without unique product property. J. Algebra, 108 (1987), 116-126.

"This paper is in final form and no version of it will be submitted for publication elsewhere"

ON REPRESENTATIONS OF INFINITE GROUPS*

K. Buzási
Mathematical Institute of L. Kossuth University
Debrecen, H:4010 Pf. 12
Hungary

Introduction

First of all I should like to present the Department of Algebra and Number Theory of L.Kossuth University and to report shortly on the algebraic investigations in this department. Debrecen is a small town in Hungary. It is an old cultural center. The Kossuth Lajos University is one of the high schools of the town. The Department of Algebra and Number Theory belongs to the five departments of the Mathematical Institute of the University. Algebraic investigations have an old tradition in the University. In this department worked Tibor Szele (1918-1955) and Andor Kertész (1929-1974), both of them well known mathematicians.

In the present survey I shall deal with algebraic investigations of the last 10 years. This research belongs to the areas of group theory, representation theory of groups, algebraic coding theory and its applications, dimension theory of topological groups and uniform spaces. Among these topics, group representation theory can be considered to be the central one. It has influence on the investigations in group theory and in algebraic coding theory. Since many of my colleagues from the department give at this School lectures on their own results, I shall provide more details on the results in representation theory.

A part of the group theoretical investigations refers to the characterization of pairs of abelian groups by invariants. A group G together with its subgroup H is called a pair (G,H) of groups. Two pairs (G,H) and (G_1,H_1) are said to be isomorphic if there exists an isomorphism $\varphi:G\to G_1$ for which $\varphi(H) = H_1$. K.Buzási showed (see [1], [2]) that if G is a finite or infinite abelian p-group, $H\subseteq G$ is a cyclic subgroup, then the pair (G,H) can be characterized (up to isomorphism) by inva-

* Research (partially) supported by the Hungarian National Foundation for Scientific Research, Grant no. 1813.

riants of G and G/H. E.Szabó ([3]) proved that if G is a finite abelian p-group and H is a non-cyclic subgroup, then the pair (G,H) cannot be characterized by invariants of G and G/H. She examined the restrictions imposed on the subgroup H when such a characterization holds.

Another part of the group theoretical investigations applies methods of group representation theory. Many years ago K.Buzási described all the normal subgroups of the wreath product of n cyclic groups of prime order (see [4], [5], [6]). P.Lakatos examined the upper and lower central series of wreath products of two or more cyclic groupsof order p^n (p is a prime). She described all the characteristic subgroups of such wreath products (see [7], [8]).

S.Buzási has several results in the dimension theory of topological groups and uniform spaces. These investigations are too far from the direction of my lecture, so I do not give details.

Algebraic coding theory is another important area of applications of representation theory. We studied group codes, that is ideals of group algebras over fields. K.Buzási, A.Pethö and P.Lakatos investigated the Hamming-distances in group codes of (2,2,...,2)-groups. They confuted and corrected a classical conjecture on these codes (see [9]). B.Brindza confuted a remarkable conjecture on the Hamming-distance of (p,p)-group codes by using ingenious representation theoretic and number theoretic arguments (see [10]).

Investigations of G.Fazekas give an entirely practical application of algebraic coding theory. He translates the tone and picture information by permutation coding methods (see [19],[20]).

Now I come to the investigations in the theory of group representations. It is a very interesting fact that although the results in the representation theory of finite and topological groups could fill a rich library, there is a rather modest literature on the representation theory of infinite discrete groups. It has been well known for a long time that the representation theory of an infinite cyclic group (a) over an arbitrary field K is equivalent to the theory of modules over the ring K[x] of polynomials. This makes it possible to describe the finite dimensional K-representations of the infinite cyclic group.

The next result in this direction was obtained in 1981, when S.D. Berman and K.Buzási (see [11]) described the structure of modules over a semisimple group algebra of the infinite dihedral group. Their results can be summarized as follows.

Let D be the infinite dihedral group:

$$D=(a).(b);\ b^2=1;\ b^{-1}ab=a^{-1},$$

and let K be an arbitrary field with Char K $\neq$ 2. The following theorems then hold:

THEOREM 1. $KD = I_1 \oplus I_2 = I_3 \oplus I_4$, where
$I_1 = KD(1+b)$; $I_2 = KD(1-b)$; $I_3 = KD(1+ab)$; $I_4 = KD(1-ab)$.

THEOREM 2. Every finitely generated K(a)-torsion-free KD-module is a direct sum of indecomposable cyclic submodules which are isomorphic to one of the modules $I_1, \ldots, I_4$ defined above.

THEOREM 3. Let M_1 and M_2 be finitely generated K(a)-torsion-free KD-modules, let $\overline{M}_i$ (i=1,2) be the submodule of M_i generated by the elements $(1+b)x$, $x \in M_i$. Then M_1 and M_2 are KD-isomorphic if and only if they are K(a)-isomorphic and the factor modules $M_1/\overline{M}_1$ and $M_2/\overline{M}_2$ are K(a)-isomorphic.

THEOREM 4. The Krull-Schmidt theorem does not hold for the finitely generated K(a)-torsion-free KD-modules, but each such module M can be defined uniquely up to isomorphism by the vector $(M) = (r, n_1, n_2)$, where r is the K(a)-rank of M, n_1 is the number of submodules isomorphic to $K(a)/(a+1)$ in the direct decomposition of the K(a)-module $M/\overline{M}$, and n_2 is the number of submodules isomorphic to $K(a)/(a-1)$ in the direct decomposition of $M/\overline{M}$.

THEOREM 5. Every indecomposable finite K-dimensional KD-module M corresponds to a polynomial $\varphi = [f(x)]^n$, where $f(x) = x^m + a_{m-1}x^{m-1} + \ldots + a_0 \in K[x]$. If f(x) is an irreducible polynomial, and $f(x) \neq x+1, x-1$, then (up to isomorphism) exactly one indecomposable KD-module M corresponds to φ.

Let $I \subseteq K(a)$ be the ideal generated by $\varphi(a) = [f(a)]^n$. If the polynomial f(x) is not symmetric (a polynomial $g(x) \in K[x]$ is called symmetric if $g(x^{-1}) = x^{-m}g(x)$ for some integer m), then $M_\varphi = (K(a)/I)^D$ is an induced KD-module. If f(x) is symmetric, then M_φ is isomorphic to the ideals $A(e_1+I) \cong A(e_2+I)$, where $A = KD/I$, $e_1 = (1+b)/2$; $e_2 = (1-b)/2$.

If $f(x) = x+1$ (or $x-1$) and $\varphi = (x+1)^n$ (or $(x-1)^n$) then M_φ is isomorphic to one of ideals $A(e_1+I)$, $A(e_2+I)$, which are non-isomorphic.

If $\varphi_1 \neq \varphi_2$ then $M_{\varphi_1} \not\cong M_{\varphi_2}$.

S.D.Berman and K.Buzási began the study of the representations of a class of groups which contain infinite cyclic subgroups of finite index. The infinite cyclic group and the infinite dihedral group belong to this class. They published in 1982 (see [12]) the following results:

THEOREM 6. Let G be a group containing an infinite cyclic subgroup of finite index. Then G has a normal subgroup H such that $(G:H)=2^k$, where $k=0,1$, and $H=F.(a)$, F is a finite normal subgroup of H and (a) is the infinite cyclic group.

Let K be an algebraically closed field with Char $K \not\equiv 0 (\mathrm{mod}\ |F|)$, and Char $K \neq 2$. Let t be the number of conjugacy classes of H lying in F and m be the number of those among them which are, at the same time, classes in G. Then the group algebra KG is decomposed into the direct sum of $s = m+(t-m)/2$ two-sided ideals

$$KG=I_1 \oplus \dots \oplus I_s,$$

where I_j is isomorphic to a full matrix ring over some group algebra KG_j, and G_j is either the infinite cyclic group or the infinite dihedral group. If the group G acts on H/F trivially, then every G_j is the infinite cyclic group. If the action of G on H/F is non-trivial, then there are exactly m dihedral groups among the G_j-s, and $(t-m)/2$ infinite cyclic groups.

The next lemma makes the investigation of the structure of modules over group algebras easier.

LEMMA 1. Let B be the tensor product of a full matrix ring L_n over a field L and an algebra A with identity over the field L. Then every B-module M is isomorphic to a tensor product $M=M_1 \otimes_L M_2$, where M_1 is an irreducible L_n-module, and M_2 is an A-module. Two B-modules M and N are isomorphic if and only if the corresponding A-modules M_2 and N_2 are isomorphic. In particular, a B-module M is decomposable if and only if the A-module M_2 is decomposable.

THEOREM 7. Every finitely generated KG-module is decomposed into the direct sum of indecomposable cyclic submodules. A KG-module is indecomposable if and only if it is an indecomposable I_j-module for some j (see Theorem 6). Indecomposable I_j-modules correspond bijectively to indecomposable KG_j-modules, and these can be obtained by means of Lemma 1.

The number of indecomposable infinite K-dimensional KG-modules equals s if G acts on H/F trivially, and equals $4m+(t-m)/2$ if the action of G on H/F is non-trivial.

By these results the study of the structure of finitely generated

KG-modules has been reduced to the study of finitely generated modules over group algebras of the infinite cyclic group and the infinite dihedral group.

The next step in the study of the representations of this class of groups was taken by S.D.Berman and K.Buzási in [13] and appeared in 1981. In that paper K is an arbitrary field with Char K $\not\equiv 0 \pmod{|F|}$. The more important results of that work are as follows:

DEFINITION. Let L be a field, let T be a (skew) field containing L in its center, and assume $(T:L) < \infty$. The twisted group algebras

$$\mathcal{E} = (T,a);\ a\lambda = \lambda^{\varphi}a$$

and

$$\Phi = (T,a,b);\ a\lambda = \lambda^{\varphi}a;\ b\lambda = \lambda^{\psi}b;\ b^{-1}ab = \gamma a^{-1};\ b^2 = \mu$$

of the infinite cyclic group and the infinite dihedral group, where $\lambda \in T$, γ and μ are fixed elements in T, φ and ψ are fixed L-automorphisms of T, are called algebras of type E over the field L (with respect to T).

THEOREM 8. Let G be a group containing an infinite cyclic subgroup of finite index and let K be an arbitrary field with Char K $\not\equiv 0 \pmod{|F|}$. Then the group algebra KG is decomposed into a direct sum

$$KG = I_1 \oplus \dots \oplus I_s,$$

where for $j = 1,\dots,s$ the minimal ideal I_j is isomorphic to a full matrix ring over some algebra of type E over K.

By means of Lemma 1, the study of KG-modules can be reduced to the study of modules over the algebras of type E over K. The algebras $\mathcal{E}$ (see the definition above) are principal ideal rings. Since the modules over such rings are known, we have to study only the modules over algebras Φ.

THEOREM 9. Let $A = (T,a,b)$, $b^2 = 1$, be an algebra with zero divisors of type E over K, and let M be a finitely generated torsion-free A-module. If Char K $\neq 2$, then M is decomposed into a direct sum of A-modules, which are free cyclic T(a)-modules, and each one of them is isomorphic to some of the modules

$$J_1 = A(1+b);\ J_2 = A(1-b);\ J_3 = A(1+\delta ab);\ J_4 = A(1-\delta ab),$$

where $\delta \in T$; $(\delta ab)^2 = 1$. Furthermore, J_1 and J_2 (resp. J_3 and J_4) are nonisomorphic if and only if the element b (resp. δab) commutes with every element of T. The ideals J_1 and J_2 are not isomorphic to J_3 and J_4.

If Char K = 2, then M is decomposed into a direct sum of cyclic submodules, which are isomorphic to A, J_1 or J_3.

By the results of Lemma 1 and of [13], in order to describe the structure of finitely generated KG-modules in the case of different special fields K, it is necessary to solve the following problems:

Problem 1. Describe all algebras of type E over different concrete fields K.

Problem 2. Investigate the isomorphism classes of algebras of type E over K.

Problem 3. Study the question: If A is an algebra of type E over a field K, is A isomorphic to an ideal of KG for some group G which contains an infinite cyclic subgroup of finite index?

Problem 4. Find out whether any algebra without zero divisors of type E over K is a principal left ideal ring.

In case of positive solution of Problem 4 the algebra A of type E over K would be a principal left ideal ring, consequently, the structure of modules over A could be considered known.

If the solution of Problem 4 is negative, then:

Problem 5. Describe all the left ideals of those algebras of type E over K which are not principal left ideal rings.

Problem 6. Describe the isomorphism classes of ideals in such algebras.

Problem 7. Investigate the structure of the finitely generated modules over such algebras.

S.D.Berman and K.Buzási in 1984 (see [14]) described all the algebras of type E over the field $\mathbb{R}$ of real numbers. They can be defined by the following relations, where $\mathbb{C}$ denotes the field of complex numbers, and Q the skew field of quaternions:

$$A_1 = \mathbb{R}(a);\ A_2 = \mathbb{C}(a);\ A_3 = Q(a),$$

are group algebras of the infinite cyclic group;

$$A_4 = (\mathbb{C}, a);\ \lambda a = a\bar{\lambda},$$

($\lambda \in \mathbb{C}$, $\bar{\lambda}$ is the complex conjugate of λ) is the twisted group algebra of the infinite cyclic group;

$A_5=\mathbb{R}D$; $A_6=\mathbb{C}D$, $A_7=QD$,

are group algebras of the infinite dihedral group D,

$A_8=(\mathbb{C},a,b)$; $\lambda a=a\bar{\lambda}$; $\lambda b=b\lambda$; $b^{-1}ab=a^{-1}$; $b^2=1$ $(\lambda\in\mathbb{C})$;

$A_9=(\mathbb{R},a,b)$; $\lambda a=a\lambda$; $\lambda b=b\lambda$; $b^{-1}ab=a^{-1}$; $b^2=-1$ $(\lambda\in\mathbb{R})$;

$A_{10}=(\mathbb{R},a,b)$; $\lambda a=a\lambda$; $\lambda b=b\lambda$; $b^{-1}ab=-a^{-1}$; $b^2=1$ $(\lambda\in\mathbb{R})$;

$A_{11}=(\mathbb{C},a,b)$; $\lambda a=a\bar{\lambda}$; $\lambda b=b\lambda$; $b^{-1}ab=-a^{-1}$; $b^2=1$ $(\lambda\in\mathbb{C})$;

$A_{12}=(\mathbb{C},a,b)$; $\lambda a=a\lambda$; $\lambda b=b\bar{\lambda}$; $b^{-1}ab=a^{-1}$; $b^2=1$ $(\lambda\in\mathbb{C})$;

$A_{13}=(\mathbb{C},a,b)$; $\lambda a=a\lambda$; $\lambda b=b\bar{\lambda}$; $b^{-1}ab=a^{-1}$; $b^2=-1$ $(\lambda\in\mathbb{C})$;

$A_{14}=(\mathbb{C},a,b)$; $\lambda a=a\lambda$; $\lambda b=b\bar{\lambda}$; $b^{-1}ab=-a^{-1}$; $b^2=1$ $(\lambda\in\mathbb{C})$;

$A_{15}=(Q,a,b)$; $\lambda a=a\lambda$; $\lambda b=b\lambda$; $b^{-1}ab=a^{-1}$; $b^2=-1$ $(\lambda\in Q)$.

The algebras $A_8,\dots,A_{15}$ are twisted group algebras of the infinite dihedral group.

The authors described all the finite $\mathbb{R}$-dimensional A_i-modules for all $i=1,\dots,15$.

The algebras $A_1,\dots,A_4$ are principal ideal rings, while the algebras A_5, A_6, A_7 are group algebras of the infinite dihedral group D. Therefore the finitely generated torsion-free modules over $A_1,\dots,A_7$ are described by classical results and [13].

The algebras A_i $(i=8,\dots,15;\ i\neq 9,13)$ contain zero divisors, therefore the finitely generated torsion-free A_i-modules are described by [13].

It has been proved that the algebras A_9, A_{13} contain no zero divisors, so the finitely generated torsion-free modules over A_9 and A_{13} are not described by [13].

K.Buzási (see [15]) in 1985 gave a positive solution to Problem 3 in the case of $\mathbb{R}$. He constructed for every algebra A of type E over $\mathbb{R}$ a group containing an infinite cyclic subgroup of finite index, such that the group algebra $\mathbb{R}G$ contains an ideal I isomorphic to A.

K. Buzási, T.Krausz and Z.M.Moneim (see [16], [17]) in 1986 described all the algebras of type E over finite fields (with respect to a finite extension field). Let K be a finite field of order p^m and let L be an extension field with $(L:K)=n$. If n is an even number, then the algebras of type E over K with respect to L are as follows:

$E_i=(L,a)$; $\lambda a=a\lambda^{p^{im}}$; $(i=1,\dots,n/2)$;

$A_1=(L,a,b)$; $\lambda a=a\lambda$; $\lambda b=b\lambda$; $b^{-1}ab=a^{-1}$; $b^2=1$;

$A_2=(L,a,b);\ \lambda a=a\lambda;\ \lambda b=b\lambda;\ b^{-1}ab=\xi a^{-1};\ b^2=1;$

$A_3=(L,a,b);\ \lambda a=a\lambda;\ \lambda b=b\lambda;\ b^{-1}ab=a^{-1};\ b^2=\xi;$

$A_4=(L,a,b);\ \lambda a=a\lambda;\ \lambda b=b\bar{\lambda};\ b^{-1}ab=a^{-1};\ b^2=1;$

$A_5=(L,a,b);\ \lambda a=a\bar{\lambda};\ \lambda b=b\lambda;\ b^{-1}ab=a^{-1};\ b^2=1;$

$A_6=(L,a,b);\ \lambda a=a\bar{\lambda};\ \lambda b=b\bar{\lambda};\ b^{-1}ab=\xi a^{-1};\ b^2=1,$

where $\lambda\in L$, ξ is a fixed non-square element in L, and $\bar{\lambda}=\lambda^{p^{nm/2}}$.

If n is an odd number, then all the algebras of type E over K with respect to L are E_i, A_1, A_2 and A_3 $(i=1,\dots,[n/2]+1)$.

They showed (see [17]) that the algebras E_i, $A_1,\dots,A_6$ are pairwise non-isomorphic, and gave a positive answer to Problem 3 in case of finite fields (see [18]).

At the end of my lecture, I stated that the Problems 5, 6, and 7 were open in the case of the real twisted group algebras A_9 and A_{13}. Recently, I solved these problems for the algebra A_9. The remaining part of this work contains an investigation of the structure of this algebra and shows that A_9 is not a principal left ideal ring.

1. The structure of left ideals

For the rest of this paper, A denotes the algebra A_9. It was shown in [14] that the group algebra $\mathbb{R}(a)\subset A$ of the infinite cyclic group (a) is an Euclidean ring with respect to the following norm: for $f(a)\in\mathbb{R}(a)$

$$|f(a)| = |\lambda_n a^n+\dots+\lambda_m a^m| = n-m\ (\lambda_i\in\mathbb{R};\ n\geq m;\ n,m\in\mathbb{Z}).$$

It is easy to see that for $f(a)$, $g(a)\in\mathbb{R}(a), g(a)\neq 0$ there exist $h(a)$, $r(a)\in\mathbb{R}(a)$ such that

$$f(a) = g(a).h(a)+r(a),$$

where $r(a) = 0$ or $|g(a)| > |r(a)|$, but the elements $h(a)$ and $r(a)$ are not uniquely determined.

We set $\overline{f(a)} = f(a^{-1})$. The element $f(a)\in\mathbb{R}(a)$ is called symmetric if $\overline{f(a)} = \mu a^{-m}f(a)$ for some $m\in\mathbb{Z}$ and $\mu\in\mathbb{R}$. The units of $\mathbb{R}(a)$ are exactly the elements μa^m $(\mu\in\mathbb{R};\ m\in\mathbb{Z})$.

LEMMA 1.1. Let $f(a)$, $g(a)\in\mathbb{R}(a)$; $f(a)\neq 0$ and $|f(a)|\geq|g(a)|$. Then there exist elements $h(a)$, $r(a)\in\mathbb{R}(a)$ such that

$$f(a) = g(a).h(a)+r(a),$$

where $r(a) = 0$ or $|r(a)| < |g(a)|$ and $|f(a)| = |g(a).h(a)|$.

Proof. At first let $f(a) = \alpha_n a^n+\ldots+\alpha_0$, $g(a) = \beta_m a^m+\ldots+\beta_0$, where α_i, $\beta_j \in \mathbb{R}$, $\alpha_0 \neq 0$, $\beta_0 \neq 0$. Then there exist elements $h_1(a) = \gamma_k a^k+\ldots+\gamma_0$ and $r_1(a) = \lambda_s a^s+\ldots+\lambda_0$ in $\mathbb{R}(a)$ such that

$$f(a) = g(a).h_1(a)+r_1(a), \tag{1.1}$$

where $r_1(a) = 0$ or $\deg r_1(a) < \deg g(a)$. Formula (1.1) implies

$$\alpha_0 = \beta_0.\gamma_0+\lambda_0. \tag{1.2}$$

If $\gamma_0 \neq 0$, then $|h_1(a)| = \deg h_1(a)$, and since $n > m$,

$$|f(a)| = \deg f(a) = \deg(g(a).h_1(a)) = |g(a).h_1(a)|.$$

If $\gamma_0 = 0$, then (1.2) implies $\lambda_0 \neq 0$. Consider the element

$$h(a) = h_1(a) - \frac{\lambda_0}{\beta_0}.$$

Then the equality

$$f(a) = g(a).h(a) + [r_1(a) - \frac{\lambda_0}{\beta_0} g(a)]$$

holds. It is clear that the element

$$r(a) = r_1(a) - \frac{\lambda_0}{\beta_0} g(a)$$

has no constant term, so $|r(a)| < \deg r(a)$. Consequently, for $\deg r_1(a) < \deg g(a)$ one has $\deg r(a) = \deg r_1(a)$ and

$$|r(a)| < \deg r(a) = \deg g(a) = |g(a)|.$$

So we obtain the equation

$$f(a) = g(a).h(a)+r(a), \tag{1.3}$$

where $|r(a)| < |g(a)|$ and

$$|f(a)| = \deg f(a) = \deg(g(a).h(a)) = |g(a).h(a)|.$$

Now let $f_1(a) = \alpha'_{n_1} a^{n_1}+\ldots+\alpha'_{n_2} a^{n_2}$; $g_1(a) = \beta'_{m_1} a^{m_1}+\ldots+\beta'_{m_2} a^{m_2}$;

α'_i, $\beta'_j \in \mathbb{R}$, $n_i, m_j \in \mathbb{Z}$, $n_1 > n_2$, $m_1 > m_2$. Then

$$f_1(a) = a^{n_2}.f(a);\ g_1(a) = a^{m_2}.g(a),$$

where

$$f(a) = \alpha_n a^n+\ldots+\alpha_0;\ g(a) = \beta_m a^m+\ldots+\beta_0,$$

$\alpha'_{n_1} = \alpha_n, \ldots, \alpha'_{n_2} = \alpha_0$; $n = n_1 - n_2$; $\beta'_{m_1} = \beta_m, \ldots, \beta'_{m_2} = \beta_0$, $n = m_1 - m_2$. As was shown above the equality (1.3) holds for the elements f(a) and g(a). Then, applying (1.3),

$$f_1(a) = a^{n_2}.f(a) = a^{n_2}[g(a).h(a)+r(a)] =$$
$$= a^{n_2}.g(a).a^{n_2-m_2}.h(a)+a^{n_2}.r(a) = g_1(a).a^{n_2-m_2}.h(a)+a^{n_2}.r(a).$$

Since $|f_1(a)| = |f(a)|$; $|g_1(a)| = |g(a)|$; $|r(a)| = |a^{n_2}.r(a)|$, the inequality $|g_1(a)| > |a^{n_2}.r(a)|$ follows from $|g(a)| > |r(a)|$, consequently we obtain

$$|f_1(a)| = |f(a)| = |g(a).h(a)| = |g_1(a).a^{n_2-m_2}.h(a)|.$$

<u>LEMMA 1.2</u>. Let $I \subseteq A$ be a left ideal generated by elements p and 1+qb, where $p,q \in \mathbb{R}(a)$ and p generates the ideal $I \cap \mathbb{R}(a)$. Then p is symmetric.

<u>Proof</u>. Since $p.(1+qb) = p+pqb \in I$, we have $pqb \in I$. This implies $b\overline{pq} \in I$ and $\overline{p}.\overline{q} \in I \cap \mathbb{R}(a)$. The element p generates the ideal $I \cap \mathbb{R}(a)$, so one has

$$\overline{p}.\overline{q} \equiv 0 \pmod p \tag{1.4}$$

But $(1-qb)(1+qb) = 1+q.\overline{q} \in I \cap \mathbb{R}(a)$, hence

$$1 + q.\overline{q} \equiv 0 \pmod p. \tag{1.5}$$

So the congruence (1.4) implies $\overline{p} \equiv 0 \pmod p$. Then $\overline{p} = s.p$ for an element $s \in \mathbb{R}(a)$. By the action of the automorphism $f \to \overline{f}$ this equation implies $p = \overline{s}.\overline{p} = \overline{s}.s.p$. As the ring $\mathbb{R}(a)$ has no zero divisors, it follows from this equation that s is a unit in $\mathbb{R}(a)$, and p is a symmetric element.

<u>LEMMA 1.3</u>. Every left ideal I of the algebra A can be generated by elements p, s_0+s_1b, with $p, s_0, s_1 \in \mathbb{R}(a)$, where p is a symmetric element and generates the ideal $I \cap \mathbb{R}(a)$.

<u>Proof.</u> Every element of A can be expressed in the form f+gb; $f,g \in \mathbb{R}(a)$. Consider all the elements $x = t_0+t_1b$ of the left ideal I. Then x ranges over I, the corresponding element t_1 ranges over an ideal L_1 in $\mathbb{R}(a)$. Since $\mathbb{R}(a)$ is a principal ideal ring, $L_1 = (s_1)$ for some element $s_1 \in \mathbb{R}(a)$. Consider a fixed element $x_0 = s_0+s_1b \in I$ and let $x = \lambda_0+\lambda_1 b$ be an arbitrary element of I. Here $\lambda_1 \in L_1$, so $\lambda_1 = ts_1$ for some $t \in \mathbb{R}(a)$. Consider the element

$$x-tx_0 = (\lambda_0+ts_1b)-t(s_0+s_1b) = \lambda_0-ts_0 = p_0 \in I.$$

If the element x ranges over I, then the element p_0 ranges over an ideal L_0 of $\mathbb{R}(a)$. Let $L_0 = (p_1)$. Then $p_0 = t_0 p_1$ for some $t_0 \in \mathbb{R}(a)$ and the element x is expressed in the form $x = t_0 p_1 + t x_0$, that is the elements p_1, $s_0 + s_1 b$ generate the left ideal I. If $I \cap \mathbb{R}(a) = (p)$, then $p_1 = t_1 p$ for some $t_1 \in \mathbb{R}(a)$, consequently I can be generated by the elements p, $s_0 + s_1 b$. By Lemma 1.2 the element p is symmetric.

LEMMA 1.4. Let the left ideal $I \subseteq A$ be generated by elements p, $s_0 + s_1 b$, where $p, s_0, s_1 \in \mathbb{R}(a)$, $(p) = I \cap \mathbb{R}(a)$. If either $(p, s_0) = 1$, or $(p, \bar{s}_1) = 1$, then there exists an element $q \in \mathbb{R}(a)$ such that the left ideal I can be generated by the elements $p, 1+qb$.

Proof. Let $(s_0, p) = 1$. One can assume that $|p| > |s_0|$. Indeed, if $|p| < |s_0|$ then $s_0 = ph + r$ for some $h, r \in \mathbb{R}(a)$, where $r = 0$ or $|r| < |p|$. We have

$$(s_0 + s_1 b) - hp = ph + r + s_1 b - hp = r + s_1 b,$$

and the elements $p, r + s_1 b$ generate the left ideal I, where $|p| > |r|$. Applying the Euclidean algorithm to the elements p and s_0 we obtain:

$$\begin{aligned}
p &= s_0 h_0 + r_0, & |r_0| &< |s_0|;\\
s_0 &= r_0 h_1 + r_1, & |r_1| &< |r_0|;\\
r_0 &= r_1 h_2 + r_2, & |r_2| &< |r_1|;\\
&\vdots & &\vdots\\
r_{k-2} &= r_{k-1} h_k + r_k, & |r_k| &< |r_{k-1}|;\\
r_{k-1} &= r_k h_{k+1},
\end{aligned}$$

hence $r_k = (p, s_0) = 1$. We use this algorithm in the following way: At the first step we form

$$p - h_0(s_0 + s_1 b) = r_0 - h_0 s_1 b = r_0 + m_0 b \qquad (m_0 \in \mathbb{R}(a))$$

and change the generating elements of I with the generators

$$r_0 + m_0 b, \ s_0 + s_1 b.$$

At the second step we form

$$(s_0 + s_1 b) - h_1(r_0 + s_1 b) = r_1 + m_1 b \qquad (m_1 \in \mathbb{R}(a))$$

and change the generating elements with

$$r_0 + m_0 b, \ r_1 + m_1 b.$$

And so on. At the last step we get the generating elements

$$m_k b, \ 1 + m_{k+1} b \qquad (m_k, m_{k+1} \in \mathbb{R}(a)).$$

Since b is invertible and $m_k \in I \cap \mathbb{R}(a) = (p)$, we obtain the generators $p, 1+qb$, where $m_{k+1} = q$.

If $(p, \bar{s}_1) = 1$, then the element $b(s_0+s_1b) = \bar{s}_1 + \bar{s}_0 b$, also is a generator and we are back in the case considered above.

<u>THEOREM 1.1.</u> Every left ideal $I \subseteq A$ can be expressed in the form $I = I_1 d$, where I_1 is a left ideal generated by elements $p, 1+qb; p, q \in \mathbb{R}(a)$; $(p) = I_1 \cap \mathbb{R}(a)$ and $d \in \mathbb{R}(a)$.

<u>Proof</u>. By Lemma 1.3 $I = (p_1, s_0+s_1b)$, where $p_1, s_0, s_1 \in \mathbb{R}(a)$ and $(p_1) = I \cap \mathbb{R}(a)$. If $(p_1, s_0) = 1$ or $(p_1, \bar{s}_1) = 1$, then by Lemma 1.4 we obtain the theorem with $d = 1$.

Let us consider the set of all elements $x = \mu_0 + \mu_1 b$ $(\mu_0, \mu_1 \in \mathbb{R}(a))$ of I. When the element x ranges over I, the corresponding elements μ_i form an ideal L_i in $\mathbb{R}(a)$ $(i = 0,1)$. Let $L_0 = (d)$. Then $L_1 = (\bar{d})$. Indeed, $bx = -\bar{\mu}_1 + \bar{\mu}_0 b \in I$, consequently if $\mu_0 \in L_0$ then $\bar{\mu}_0 \in L_1$. Since $p \in L_0$, then $p_1 = pd$ for some $p \in \mathbb{R}(a)$. Since $L_0 = (d)$, applying the Euclidean algorithm, as in the proof of Lemma 1.4, we can replace the generator $s_0 + s_1 b$ by an element $d + s_1'b$. For $s_1' \in L_1$, one has $s_1' = q.\bar{d}$ with some $q \in \mathbb{R}(a)$. Consequently, every $y \in I$ can be expressed in the form

$$y = (\lambda_0 + \lambda_1 b)p_1 + (\lambda_0' + \lambda_1' b)(d + s_1'b) = (\lambda_0 + \lambda_1 b)pd + (\lambda_0' + \lambda_1' b)(d + q.\bar{d}b) =$$

$$= [(\lambda_0 + \lambda_1 b)p + (\lambda_0' + \lambda_1' b)(1+qb)]d,$$

where $\lambda_0 + \lambda_1 b$, $\lambda_0' + \lambda_1' b \in A$. The elements $(\lambda_0 + \lambda_1 b)p + (\lambda_0' + \lambda_1' b)(1+qb)$ form a left ideal I_1 generated by p and $1+qb$.

<u>LEMMA 1.5.</u> Let the left ideal $I \subseteq A$ be generated by two pairs of elements $p, 1+qb$ and $p, 1+q_1 b$, where $(p) = I \cap \mathbb{R}(a)$; $q, q_1 \in \mathbb{R}(a)$. Then $q \equiv q_1 \pmod{p}$.

<u>Proof.</u> Clearly we have $(1+qb) - (1+q_1 b) = (q - q_1)b \in I$, which implies $\bar{q} - \bar{q}_1 \in I$. But $\bar{q} - \bar{q}_1 \in \mathbb{R}(a)$, hence $\bar{q} \equiv \bar{q}_1 \pmod{p}$. Since p is a symmetric element, the lemma follows from this congruence.

2. <u>Construction of a left ideal which is not a principal left ideal</u>

We define for the element $x = f + gb$; $f, g \in \mathbb{R}(a)$ of A a norm N(x) by the formula

$$N(x) = (\bar{f} - gb)(f + gb) = f\bar{f} + g\bar{g}.$$

It is easy to see that $N(x) \in I \cap \mathbb{R}(a)$ and $N(x.y) = N(x).N(y)$ for all $x,y \in A$.

LEMMA 2.1. Let $I \subseteq A$ be a principal left ideal generated by the element s_0+s_1b; $s_0,s_1 \in \mathbb{R}(a)$. If $(p) = I \cap \mathbb{R}(a)$, then the elements $d.p$ and $N(s_0+s_1b)$ are associated, where $d = (\bar{s}_0,s_1)$.

Proof. First let $(\bar{s}_0,s_1) = 1$. One has

$$p = (\lambda_0 + \lambda_1 b)(s_0 + s_1 b) \tag{2.1}$$

for some $\lambda_0 + \lambda_1 b \in A$. This implies

$$p = \lambda_0 s_0 - \lambda_1 \bar{s}_1; \quad 0 = \lambda_0 s_1 + \lambda_1 \bar{s}_0. \tag{2.2}$$

Because of $(\bar{s}_0,s_1) = 1$, the second equality of the system (2.2) yields $\lambda_0 = t\bar{s}_0$, $\lambda_1 = -ts_1$ $(t \in \mathbb{R}(a))$. Then (2.2) implies $p = t(s_0\bar{s}_0 + s_1\bar{s}_1)$, that is $p \equiv 0 \pmod{N(s_0 + s_1 b)}$. On the other hand, $N(s_0 + s_1 b) \in I \cap \mathbb{R}(a)$, so $N(s_0 + s_1 b) \equiv 0 \pmod p$, that is p and $N(s_0 + s_1 b)$ are associated. Now let

$$(\bar{s}_0,s_1) = d \neq 1; \; \bar{s}_0 = \bar{h}_0 d; \; s_1 = h_1 d \; (h_0,h_1 \in \mathbb{R}(a); \; (h_0,h_1) = 1). \tag{2.3}$$

In this case $s_0 + s_1 b = (h_0 + h_1 b)\bar{d}$, that is $I = I_1\bar{d}$, where I_1 is a principal left ideal generated by $h_0 + h_1 b$. Here $(h_0,h_1) = 1$, and if $(p_1) = I_1 \cap \mathbb{R}(a)$, then p_1 and $N(h_0 + h_1 b)$ are associated. At the same time $p = p_1\bar{d}$, so $pd = p_1 d\bar{d}$ and

$$N(s_0 + s_1 b) = s_0\bar{s}_0 + s_1\bar{s}_1 = (h_0\bar{h}_0 + h_1\bar{h}_1)d\bar{d} = N(h_0+h_1 b)d\bar{d}$$

are associated, too.

LEMMA 2.2. Let $I \subseteq A$ be a left ideal generated by elements $p, 1 + qb$, where $(p) = I \cap \mathbb{R}(a)$, $q \in \mathbb{R}(a)$. Then every element of I can be expressed in the form $xbp+y(1+qb)$, where $x,y \in \mathbb{R}(a)$.

Proof. Let $s_0+s_1 b \in I$ be an arbitrary element of I. Then

$$s_0 + s_1 b - s_0(1+qb) = (s_1 - s_0 q)b \in I \tag{2.4}$$

that is

$$b(s_1 - s_0 q)b = -\bar{s}_1 + \bar{s}_0\bar{q} \in I \cap \mathbb{R}(a).$$

Since $(p) = I \cap \mathbb{R}(a)$, then $\bar{s}_1 - \bar{s}_0\bar{q} = p\bar{x}$ for some $\bar{x} \in \mathbb{R}(a)$. This implies $s_1 - s_0 q = px$, because the element p is symmetric by Lemma 1.2. By (2.4) we have $xbp + s_0(1+qb) = s_0 + s_1 b$, which proves the lemma.

<u>THEOREM 2.1</u>. The algebra A is not a principal left ideal ring.

<u>Proof</u>. We shall construct a left ideal I, generated by some elements p and 1+qb, which is not a principal left ideal.

Let $q = a^3+1$. Since $(p) = I\cap\mathbb{R}(a)$, the element p divides $N(1+qb) = 1+q\bar{q}$, which is expressed as a product of prime elements

$$1+q\bar{q} = (a-\alpha)(a^{-1}-\alpha)(a^2-\alpha a+\alpha^2)(a^{-2}-\alpha a^{-1}+\alpha^2)$$

where α is a real value of $\sqrt[3]{\frac{-3+\sqrt{5}}{2}}$. Set $p = (a^2-\alpha a+\alpha^2)(a^{-2}-\alpha a^{-1}+\alpha^2)$. For p and q there exist elements h and r such that

$$p = qh+r. \tag{2.5}$$

Here

$$h = \alpha^2 a^3-(\alpha+\alpha^3)a^2;$$

$$r = (1+\alpha^2+\alpha^4)a^4-(\alpha+\alpha^2+\alpha^3)a^3+(\alpha+\alpha^2+\alpha^3)a^2. \tag{2.6}$$

It is true in (2.5) that $|r| = 2 < |q|$; $|h| = 1$. For the element

$$u = bp-h(1+qb) = (qh+r)b-h(1+qb) = -h+rb\in I \tag{2.7}$$

one has $N(u) = h\bar{h}+r\bar{r} = 4 = |p|$.

We show that the element u does not generate the left ideal I. Indeed, assume that I = (u). Then (2.7) implies

$$u-bp = -h(1+qb). \tag{2.8}$$

Since $N(u)\in I\cap\mathbb{R}(a)$, it follows that $N(u)\equiv 0 \pmod{p}$. However, $|N(u)| = |p|$, so the elements N(u) and p are associated. One can assume that $p = N(u)$. This means $p = (-\bar{h}-rb)(-h+rb)$, hence (2.8) implies

$$u-b(-\bar{h}-rb)u = -h(1+qb). \tag{2.9}$$

Since I = (u), there exists an element $\mu_0+\mu_1 b\in A$ such that

$$1+qb = (\mu_0+\mu_1 b)u.$$

Then (2.9) can be expressed in the form

$$[1-b(-\bar{h}-rb)]u = -h(\mu_0+\mu_1 b)u.$$

But the algebra A contains no zero divisors, so we have $1-\bar{r}+hb = -h(\mu_0+\mu_1 b)$, or $1-\bar{r} = -h\mu_0$, that is

$$1-\bar{r}\equiv 0 \pmod{h}. \tag{2.10}$$

We show that the congruence (2.10) gives rise to a contradiction. Indeed, applying (2.6) we have

$$1-\bar{r} = 1-(\alpha+\alpha^2+\alpha^3)a^{-2}+(\alpha+\alpha^2+\alpha^3)a^{-3}-(1+\alpha^2+\alpha^4)a^{-4}.$$

Set

$$f(a) = a^{-4}(1-\overline{r}) = a^4-(\alpha+\alpha^2+\alpha^3)a^2 + (\alpha+\alpha^2+\alpha^3)a - (1+\alpha^2+\alpha^4),$$
$$g(a) = a^{-2}h = \alpha^2 a - (\alpha+\alpha^3). \tag{2.11}$$

It is clear that h divides the element $1-\overline{r}$ if and only if g(a) divides f(a). Since $g(a) = \alpha^2[a-\alpha^{-1}(1+\alpha^2)]$, the congruence $f(a) \equiv 0 \pmod{g(a)}$ holds if and only if $f[\alpha^{-1}(1+\alpha^2)] = 0$. It is easy to verify that this is not true for the real number α. This proves that the element $u = -h+rb$ does not generate the left ideal I.

Now let us assume that I is a principal left ideal generated by an element $z = s_0+s_1b$. Then

$$u = -h+rb = (t_0+t_1b)(s_0+s_1b) \tag{2.12}$$

for some $t_0+t_1b \in A$. Now (2.12) implies the equation

$$N(u) = N(t_0+t_1b).N(z), \tag{2.13}$$

i.e. the element N(z) divides N(u). On the other hand, $N(z) \in I \cap \mathbb{R}(a) = (p)$, that is $N(z) = mp$ for some $m \in \mathbb{R}(a)$. Then it follows from (2.13) that $N(u) = N(t_0+t_1b)mp$. Using the assumption $N(u) = p$, we see that t_0+t_1b is invertible. By (2.12) this implies that the elements u and z are associated. This is in contradiction with the fact that u does not generate the left ideal I.

References

1. K. Buzási, On invariants of pairs of finite abelian groups. (in Russian). Publ. Math. 28 (1981), 317-326.
2. K. Buzási, On the isomorphism of pairs of infinite abelian groups (in Russian). Publ. Math. 29 (1982), 139-154.
3. E. Szabó, Über isomorph Paare endlichen abelscher Grupper. Publ. Math. 33 (1986).
4. K. Buzási, On the structure of the wreath product of a finite number of cyclic groups of prime order. (in Russian). Publ. Math. 15 (1968), 107-129.
5. K. Buzási, The kernels of the irreducible representations of the Sylow p-subgroup of the symmetric group S_pn. (in Russian). Publ. Math. 16 (1969), 199-227.
6. K. Buzási, The normal subgroups of the Sylow p-subgroup of the symmetric group S_pn (in Russian). Publ. Math. 17 (1970), 333-347.
7. P. Lakatos, On the structure of the wreath product of two cyclic groups of prime power order (in Russian). Publ. Math. 22 (1975), 293-306.
8. L. Lakatos, The nilpotency class of iterated wreath products of cyclic groups of prime power order (in Russian). Publ. Math. 31 (1984), 153-156.

9. K. Buzási, A. Pethö, P. Lakatos, On the code distance of a class of group codes, Probl. Peredachi Inf. 17, No. 3 (1981), 3-12 (in Russian). (English translation: in Probl. Inf. Transm.)

10. B. Brindza. On the code distance of certain classes of (p,p)-codes, Probl. Peredachi Inf. 21, No. 2 (1985), 10-19 (in Russian). (English translation: in Probl. Inf. Transm.)

11. S.D. Berman, K. Buzási. On the representations of the infinite dihedral group. (in Russian). Publ. Math. 28 (1981), 173-187.

12. S.D. Berman, K. Buzási. On the representations of a group which contains an infinite cyclic subgroup of finite index (in Russian). Publ. Math. 29 (1983), 163-170.

13. S.D. Berman, K. Buzási. On the modules over group algebras of groups which contain an infinite cyclic subgroup of finite index (in Russian). Studia Sci. Math. Hung. 16 (1981), 455-470.

14. S.D. Berman, K. Buzási. A description of all the finite dimensional real representations of groups which contain an infinite cyclic subgroup of finite index (in Russian). Publ. Math. 31 (1984), 133-144.

15. K. Buzási. On real group algebras of groups with an infinite cyclic subgroup of finite index (in Russian). Publ. Math. 32 (1985), 267-276.

16. K. Buzási, T. Krausz, Z.M. Moneim. On twisted group algebras over finite fields (in Russian). Publ. Math. 33 (1986), 147-152.

17. K. Buzási, T. Krausz. A description of the twisted group algebras over finite fields. (in Russian). Acta Sci. Math. (to appear).

18. K. Buzási, Z.M. Moneim. On the realization of twisted algebras in group algebras over finite fields. (in Russian). Publ. Math. (to appear).

19. G. Fazekas, A. Pethö, Permutational Source Coding. The Sixth International Conference on Information Theory. Abstr. of Papers. Moscow-Tashkent, 1984.

20. G. Fazekas, Permutational Source Coding in Picture Processing. TCGT-84. Abstr. of Papers, Debrecen, 1984.

"This paper is in final form and no version of it will be submitted for publication elsewhere"

Integral Manifolds, Harmonic Mappings, and the Abelian Subspace Problem

by James A. Carlson and Domingo Toledo*
Department of Mathematics
University of Utah
Salt Lake City, Utah 84112

1. Introduction.

The purpose of this note is to draw attention to what seems to be a neglected problem of Lie theory, the *abelian subspace problem*: given a subspace $\mathfrak{s}$ of a Lie algebra $\mathfrak{g}$, determine the integer

$$a(\mathfrak{s}) = \max\{ \dim \mathfrak{a} \mid \mathfrak{a} \subset \mathfrak{s} \text{ and } \mathfrak{a} \text{ abelian } \}.$$

To compute $a(\mathfrak{s})$ is therefore to compute the dimension of spaces of commuting matrices subject to certain side conditions. If $\mathfrak{a}$ consists of semisimple elements, then these matrices can be simultaneously diagonalized, and so the result is clear. We shall therefore be concerned with spaces containing non-semisimple elements, e.g., $\mathfrak{s} = \mathfrak{so}(n, \mathbf{C})$ but not $\mathfrak{s} = \mathfrak{so}(n, \mathbf{R})$.

The interest of the problem (from our point of view) lies its connection with geometry, the most direct instance of which is the theory of distributions (differential systems). The simplest example is given by a pair of ordinary differential equations

$$\begin{aligned} \frac{dx}{dt} &= f(x,y) \\ \frac{dy}{dt} &= g(x,y), \end{aligned} \tag{1.1}$$

to which is associated a field of line elements

$$(x,y) \mapsto \text{line with slope } m(x,y) = g(x,y)/f(x,y). \tag{1.2}$$

Solutions of (1.1), viewed as parametric submanifolds of $\mathbf{R}^2$, then become solutions of (1.2). More generally, a *distribution* on a manifold is a field of subspaces $p \mapsto \mathcal{D}_p$ of the tangent bundle, and a solution, or *integral manifold* V, is a submanifold everywhere tangent to this field.

Distributions arise naturally in a wide variety of classical problems in differential geometry, e.g., the construction of conformal maps and the construction of Riemannian surfaces with prescribed metric tensor (see chapters VII and VIII of Cartan's book [11]). More recently, in algebraic geometry, Griffiths has introduced the notion of a *variation of Hodge structure*,

* Research partially supported by the National Science Foundation

defined as an integral manifold for the so-called *horizontal distribution* $\mathcal{G}$ of a period domain [12, 15–18, especially 18, pp. 21–23].

A fundamental invariant of a differential system is the dimension of the largest integral manifold,

$$m(\mathcal{D}) = \max\{ \dim V \mid V \text{ is an integral submanifold of } \mathcal{D} \}.$$

For integrable systems this is given by Frobenius' theorem:

$$m(\mathcal{D}) = \dim(\mathcal{D}),$$

where by $\dim(\mathcal{D})$ we mean $\dim(\mathcal{D}_p)$, which we assume here to be constant, so that $\mathcal{D}$ is defined by a subbundle of the tangent bundle. By *integrable* we mean that the Lie bracket of vector fields belonging to $\mathcal{D}$ belongs to $\mathcal{D}$; a vector field X *belongs* to $\mathcal{D}$ if $X_p \in \mathcal{D}_p$ for each p in the domain of $\mathcal{D}$. The *dimension problem*, that of calculating $m(\mathcal{D})$, is therefore of interest only when $\mathcal{D}$ is nonintegrable.

Consider now a homogeneous space M and a homogeneous subbundle E of the tangent bundle. If G is the group of motions of M, then E corresponds to a subspace $\mathfrak{s}$ of the Lie algebra $\mathfrak{g}$. Since the distribution associated to E is defined Lie-theoretically, the problem of computing $m(\mathcal{D})$ should have a Lie-theoretic solution. The main result of section 3 formulates conditions under which this holds: if E is *strongly nonintegrable*, meaning in particular that $[\mathfrak{s}, \mathfrak{s}] \cap \mathfrak{s} = 0$, then tangent spaces to integral manifolds can be identified with abelian subspaces of $\mathfrak{s}$, so that

$$m(\mathcal{D}) = a(\mathfrak{s}), \tag{1.3}$$

Thus, computation of $a(\mathfrak{s})$ gives a bound on the dimension of the integral manifolds.

The problem of calculating $a(\mathfrak{s})$ arises in other situations as well, e.g., the calculation of the number of independent integrals of the motion for certain Hamiltonian systems and the determination of bounds on the dimension of the image of certain harmonic mappings (the work of Sampson [22]). We shall describe some of these applications in section 2, then discuss two cases in which the problem can be treated fully by simple techniques. These are a) the Legendre manifolds, i.e., maximal integral manifolds of the classical contact system (section 4, solvable by the calculus of differential forms and b) variations of Hodge structure of weight two (section 5, solvable by matrix theory).

As noted above, the abelian subspace problem is of interest only for spaces containing non-semisimple elements. If $\mathfrak{g}$ is a compact Lie algebra then every abelian subalgebra consists of semisimple elements and so is conjugate to a subspace of a Cartan subalgebra; consequently $a(\mathfrak{s}) \leq \operatorname{rank} \mathfrak{g}$. In general, however, $a(\mathfrak{s})$ is much larger than the dimension of a Cartan subalgebra. For example, in $\mathfrak{gl}(n)$ with $n = 2k$, the set of all matrices of the form

$$\lambda I + \begin{bmatrix} 0 & X \\ 0 & 0 \end{bmatrix},$$

where X is $k \times k$, is of dimension $1 + n^2/4$. By a result of Schur [23], this is best possible: the maximum dimension of a commutative space of $n \times n$ matrices is $1 + [n^2/4]$, where $[y]$ is the greatest integer contained in y. Schur's work is the starting point of a small literature, from which we cite [13, 19, 20, 21, 26]. Of particular note is Malcev's paper [21], which treats the abelian subspace problem for $\mathfrak{s} = \mathfrak{g}$, with $\mathfrak{g}$ complex and semisimple. The essence of the argument [21, page 293, line 6] is the following reduction: for each abelian subspace $\mathfrak{a}$ in $\mathfrak{g}$ there is an associated abelian space $\mathfrak{a}'$ of the same dimension as $\mathfrak{a}$ which has a basis of root vectors $\{ e_\alpha \mid \alpha \in \Sigma \}$. Since these root vectors commute, the sum of any two roots in Σ is not a root. Such a set of roots is called *commutative.* A combinatorial argument [21, page 293, line 6 from bottom] which bounds the size of commutative sets of roots then completes the proof. Little appears to be known about the abelian subspace problem for $\mathfrak{s} \neq \mathfrak{g}$.

We are grateful to N. Jacobson for bringing Malcev's work to our attention.

2. Applications.

The abelian subspace problem arises in Hodge theory as follows (see section 5 for more details). Consider a period domain D, i.e., a classifying space for Hodge structures, let G be its isometry group, and let $\mathfrak{g}$ be the corresponding Lie algebra. A Hodge structure H in D defines a Hodge structure on the Lie algebra $\mathfrak{g}$ with components $\mathfrak{g}^{p,-p}$, and the spaces $\mathfrak{g}^{-1,1}$ fit together to define Griffiths' horizontal tangent bundle, a subbundle of the holomorphic tangent bundle. Families of Hodge structures arising form the cohomology groups of a family of algebraic varieties define integral manifolds for the resulting distribution $\mathcal{G}$. Abstracting this notion, one defines a variation of Hodge structure to be an integral manifold of the horizontal distribution. Now let T be a tangent space to a point on such an integral manifold and let $\mathfrak{a}$ be the subspace of $\mathfrak{g}^{-1,1}$ which it defines. According to [6, 7] (or the results of the next section), the subspace $\mathfrak{a}$ is abelian, so that $m(\mathcal{G}) \leq a(\mathfrak{g}^{-1,1})$. For weight two structures one has

$$a(\mathfrak{g}^{-1,1}) \leq \frac{1}{2} h^{2,0} h^{1,1} \qquad \text{for } h^{2,0} > 1, \tag{2.1}$$

with equality when $h^{1,1}$ is even [4, Theorem 1.1; 9, Theorem 1.1]. One therefore has the following [1] :

(2.2) **Theorem.** *Let $f : V \longrightarrow \Gamma \backslash D$ be a variation of Hodge structure of weight two with $h^{2,0} > 1$. Then* $\dim f(V) \leq \frac{1}{2} h^{2,0} h^{1,1}$.

Note that D has dimension $h^{2,0} h^{1,1} + \frac{1}{2} h^{2,0}(h^{2,0} - 1)$, while the horizontal distribution has dimension $h^{2,0} h^{1,1}$.

As Sampson has shown [22], the abelian subspace problem also arises, independently of an underlying distribution, in the theory of harmonic mappings:

---

[1] Sharp bounds based on Malcev's technique have recently been obtained in all weights in joint work of A. Kasparian and the authors.

(2.3) **Theorem** (Sampson). *Let M be a compact Kähler manifold, N a locally symmetric space of non-compact type for a Lie group G, and $f : M \longrightarrow N$ a harmonic mapping. Fix a point x in M, and let $\mathfrak{k}^{\mathbf{C}} \oplus \mathfrak{p}^{\mathbf{C}}$ be a Cartan decomposition for $\mathfrak{g}^{\mathbf{C}}$, where $\mathfrak{k}$ is the isotropy algebra of $f(x)$. Finally, let $\mathfrak{a} = f_*(T^{1,0}_x M)$. Then $\mathfrak{a}$ is an abelian subspace of $\mathfrak{p}^{\mathbf{C}}$.*

Thus, any dimension constraints on abelian subspaces of $\mathfrak{p}^{\mathbf{C}}$ translate into constraints on the dimension of the images of harmonic mappings. Sampson used this result to prove that a harmonic map of a compact Kähler manifold into a manifold covered by a real hyperbolic space has image of dimension at most two. In [10] further applications of Sampson's theorem were obtained, including the following:

(2.4) **Theorem.** *Let M be a compact Kähler manifold and $f : M \longrightarrow N$ a harmonic mapping to a locally symmetric space which is not of Hermitian type. Then $\dim f(M) < \dim N$.*

If N is compact then every continuous map is homotopic to a harmonic map [14]. The preceding theorem therefore implies that there are no continuous maps $f : M \longrightarrow N$ surjective on homology. Observe that the dimension of M could be larger than the dimension of N.

The essential new ingredient in the proof of the preceding theorem is a bound $a(\mathfrak{p}^{\mathbf{C}}) < \frac{1}{2} \dim_{\mathbf{C}} \mathfrak{p}^{\mathbf{C}}$ under the stated hypotheses. In specific situations one expects much stronger dimension restrictions [2] . An example of this, which we believe to be fairly typical, is the following [10]:

(2.5) **Theorem.** *Let $f : M \longrightarrow N$ be as above. Suppose that the universal cover of N is a quaternionic hyperbolic space. Then*

$$\dim f(M) \leq \frac{1}{2} \dim N.$$

In many cases abelian subspaces of maximal dimension satisfy a rigidity condition. To make a precise statement, consider the normalizer $N(G, \mathfrak{s})$ of a subspace $\mathfrak{s}$ in G. Call a class $\mathcal{A}$ of subspaces $\mathfrak{a}$ of $\mathfrak{s}$ *rigid* if any two elements of $\mathcal{A}$ are conjugate by an element of the normalizer; let $\mathcal{A}(\mathfrak{s})$ be the class of abelian subspace of $\mathfrak{s}$ of dimension $a(\mathfrak{s})$, and ask whether $\mathcal{A}(\mathfrak{s})$ is rigid. Less precise but more important is the problem: characterize $\mathcal{A}(\mathfrak{s})$. this, in either its precise or imprecise form, is the *infinitesimal rigidity problem*. In Hodge theory it has a positive answer for variations of Hodge structure of weight two when $h^{2,0} \geq 3$ and $h^{1,1}$ is even [4, Theorem 5.3] and a negative answer when $h^{2,0} = 2$. In the theory of harmonic mappings it has a positive answer when the domain is a compact Kähler manifold and the range is a locally irreducible Hermitian symmetric space with group $G \neq SL(2, \mathbb{R})$. In this case there are only two abelian subspaces of $\mathfrak{p}^{\mathbf{C}}$ of dimension $a(\mathfrak{p}^{\mathbf{C}})$, namely $\mathfrak{p}^+$ and $\mathfrak{p}^-$, the holomorphic and antiholomorphic tangent spaces, respectively. This is Siu's rigidity theorem [24, 25].

---

[2] The authors have recently obtained sharp bounds for $\mathfrak{a} \subset \mathfrak{p}^{\mathbf{C}}$ for $\mathfrak{g} = \mathfrak{so}(p,q)$, $\mathfrak{g} = \mathfrak{sl}(n, \mathbb{R})$.

There is a related *local rigidity problem*, which is to show that germs s_1 and s_2 of integral submanifolds of dimension $m(\mathcal{D})$ are congruent under a motion of the group. By this we mean that there is a $g \in G$ such that $gs_1 = s_2$. The corresponding problem for complete integral manifolds is the *global rigidity problem.* It has a positive solution for Griffiths' distribution for weight two period domains with $h^{2,0} \geq 3$ and $h^{1,1}$ even [4, Remark 6.6], in which case $m(\mathcal{D}) = \frac{1}{2}h^{2,0}h^{1,1}$. Moreover, the complete integral manifolds of dimension $m(\mathcal{D})$ are homogeneous: they admit a transitive action of an imbedded copy of $SU(p,q)$ in $G = SO(2p, 2q)$. It appears that the integral manifolds of dimension $a(\mathcal{D})$ often (but not always) exhibit this strong rigidity behavior. The case of $h^{2,0} = 2$ is a counterexample, since all rigidity statements fail.

Finally, we would like to mention the *maximality problem.* Call an integral manifold V *maximal* if it is not contained in any other integral manifold of higher dimension. For strongly nonintegrable distributions it is enough to check maximality at the tangent space level: V is maximal if its tangent spaces are maximal as abelian subalgebras. If an abelian subalgebra of $\mathfrak{s}$ is of maximal dimension, then it is certainly a maximal abelian subalgebra. The converse, however, is not true. Indeed, the "generic" maximal abelian algebra has dimension far less than $a(\mathfrak{s})$. For a geometric example, we note that the variations of Hodge structure defined by hypersurfaces (with a small number of exceptions) are maximal but not of maximal dimension [5].

3. Split homogeneous spaces.

We shall now describe a general class of distributions for which the dimension problem reduces to the abelian subspace problem. To begin, consider a manifold M endowed with a transitive action of a Lie group G. Let B be the isotropy subgroup of a reference point, and let $\mathfrak{g}$ and $\mathfrak{b}$ be the corresponding Lie algebras, so that the tangent bundle of M is given by $TM = G \times_B \mathfrak{g}/\mathfrak{b}$. Call a G-structure on M *split* if there is a decomposition

$$\mathfrak{g} = \mathfrak{b} \oplus \mathfrak{n} \tag{3.1}$$

into subalgebras. For split actions small neighborhoods of the identity element of the group N corresponding to $\mathfrak{n}$ act transitively on small neighborhoods of M. A large class of split G-manifolds is given by flag varieties, in which case the decomposition is into a parabolic subalgebra and its opposite nil algebra, as in Figure 1. These are in fact the examples of interest to us.

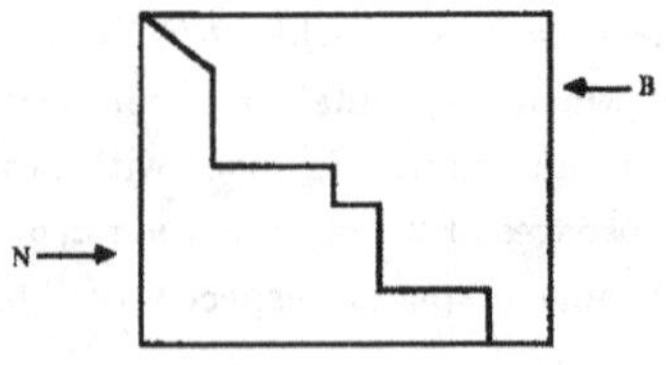

Figure 1. Split Lie algebra

For a split G-space let

$$\mathfrak{n}(\epsilon) = \{ \xi \in \mathfrak{n} \mid ||\xi|| < \epsilon \}$$

be the ϵ-ball in $\mathfrak{n}$ for some choice of norm, and let

$$M_q(\epsilon) = \{ e^{\xi} q \mid \xi \in \mathfrak{n}(\epsilon) \}.$$

Then the map $\xi \mapsto e^{\xi} q$ defines a parametrization $\mathfrak{n}(\epsilon) \longrightarrow M_q(\epsilon)$, the inverse of which gives a canonical system of coordinates. In addition, the map $n : M_q(\epsilon) \longrightarrow N$ which sends $e^{\xi} q$ to e^{ξ} defines a lifting of the identity on $M_q(\epsilon)$ with Maurer-Cartan form

$$\omega = n^{-1} dn = \mathcal{E}^1 \otimes \mathfrak{n}.$$

Here $\mathcal{E}^p$ is the space of smooth p-forms.

Now let $\mathfrak{s}$ be a subspace of $\mathfrak{n}$ such that $\mathfrak{s} + \mathfrak{b}$ is Ad_B-stable, so that the homogeneous bundle

$$E(\mathfrak{s}) = G \times_B (\mathfrak{s} + \mathfrak{b})/\mathfrak{b}$$

is defined and defines a (homogeneous) distribution on M. Fix a point q on an integral manifold $V \subset M$ of $E(\mathfrak{s})$, and let $n : M_q(\epsilon) \cap V \longrightarrow N$ be the canonical lifting. Define a lift of the tangent spaces of V into $\mathfrak{g}$ by

$$\mathfrak{a} = n_*(T_q V) \subset \mathfrak{s}.$$

Lie brackets in $\mathfrak{a}$ then remain in $\mathfrak{s}$:

(3.2) **Lemma.** *The canonical image $\mathfrak{a}$ of $T_q V$ in $\mathfrak{n}$ satisfies $[\mathfrak{a}, \mathfrak{a}] \subset \mathfrak{s}$.*

Proof: On the canonical coordinate neighborhood $\mathfrak{n}(\epsilon)$ the tangent bundle can be identified with the trivial bundle $\mathfrak{n}(\epsilon) \times \mathfrak{n}$. Under this identification the fiber of $E(\mathfrak{s})$ at ξ is identified with $Ad(e^{-\xi})(\mathfrak{s})$, and the condition that a submanifold V be tangent to $E(\mathfrak{s})$ becomes

$$i^*\omega = i^*(n^{-1} dn) \in \mathcal{E}^1 \otimes \mathfrak{s}, \tag{3.3}$$

where i is the inclusion map. Moreover,

$$\mathfrak{a} = i^*\omega(T_q M). \tag{3.4}$$

Now pull the integrability condition $d\omega - \omega \wedge \omega = 0$ back via i and evaluate on a pair of tangent vector fields to get

$$X(i^*\omega(Y)) - Y(i^*\omega(X)) - i^*\omega([X, Y]) - [i^*\omega(X), i^*\omega(Y)] = 0. \tag{3.5}$$

By (3.3) the first three terms of the preceding expression lie in $\mathcal{E}^0 \otimes \mathfrak{s}$, and so the fourth must as well, as required.

We shall say that a distribution $E(\mathfrak{s})$ on a split G-space is *strongly nonintegrable* if it satisfies $[\mathfrak{s}, \mathfrak{s}] \cap \mathfrak{s} = 0$. Opposite to this is the integrable case: $[\mathfrak{s}, \mathfrak{s}] \subset \mathfrak{s}$. From the Lemma we obtain the following:

(3.6) **Corollary.** *If $E(\mathfrak{s})$ is strongly nonintegrable, then $\mathfrak{a}$ is abelian: $[\mathfrak{a}, \mathfrak{a}] = 0$.*

This is the desired reduction.

4. Contact structures.

We shall now study a simple but instructive example of strongly nonintegrable homogeneous distributions, that of a so-called contact structure. Before doing so, however, we recall a few basic facts about the relation between distributions and differential ideals.

To begin, there is a dual description of differential systems. Let $\mathcal{D}(U)^{\perp}$ be the space of one-forms on an open set U of M which annihilate $\mathcal{D}_p$ for each $p \in U$. Call $\Delta : U \mapsto \mathcal{D}(U)^{\perp}$ the *codistribution* dual to $\mathcal{D}$. Consider the smallest subspace $I(U)$ of the algebra of forms on U containing $\Delta(U)$ which is closed under multiplication by an arbitrary form and under exterior differentiation. The resulting ideal — the *differential ideal* of $\Delta(U)$ — is graded, with $I(U)^1 = \Delta(U)$. Moreover, if $i : S \longrightarrow M$ is the inclusion map for an integral submanifold of $\mathcal{D}$, then $i^* d\omega = d i^* \omega = 0$ for all $\omega \in I(U)^1$. From this one derives the additional relations $i^* I(U)^p = 0$.

If $\Delta(U)$ is generated by closed forms $\omega_1, \ldots, \omega_k$, then the distribution is integrable, as one sees from the formula

$$d\omega(X,Y) = X\omega(Y) - Y\omega(X) - \omega([X,Y]).$$

The converse is false, since nonclosed generators are defined by the substitution $\nu_i = \sum a_{ij}\omega_j$. However, if (b_{ij}) denotes the matrix inverse to (a_{ij}), then

$$d\nu_i = \sum da_{ij} \wedge b_{jk} \wedge \nu_k = \sum \beta_{ik} \wedge \nu_k$$

for one-forms β_{ik}. In other words, the algebraic and differential ideals determined by the ν_i coincide. This leads to a dual formulation of Frobenius' theorem [2, p. 201]:

(4.1) **Theorem.** *A codistribution Δ is integrable if and only if there is a minimal set of generators $\{ \omega_i \}$ such that $d\omega_i = \sum \alpha_{ij} \wedge \omega_j$ for some matrix of one-forms (α_{ij}).*

A useful consequence of this result is the following criterion:

(4.2) **Corollary.** *Let $\{ \omega_i \mid i = 1, \ldots, k \}$ be a set of generators for a codistribution Δ. Then Δ is integrable if an only if*

$$d\omega_i \wedge \omega_1 \wedge \cdots \wedge \omega_k = 0$$

for all i.

A *contact structure* on $\mathbb{R}^{2n+1}$ is the distribution determined by the one-form

$$\omega = dr - \sum p_i dq_i, \tag{4.3}$$

where the coordinates on $\mathbb{R}^{2n+1}$ are $p_1, \ldots, p_n, q_1, \ldots, q_n, r$. Such structures arise in classical mechanics [1, appendix 4, p. 271] and in classical differential geometry. According to the above corollary, this distribution is nonintegrable, since

$$\omega \wedge (d\omega)^n = dVol$$

is a volume form.

The contact distribution arises in nature as follows. Consider a function $f : \mathbb{R}^n \longrightarrow \mathbb{R}$, and let $\tilde{\Gamma}(f) = \{ (q,p,r) \mid p_k = \partial r/\partial q_k \}$ be the graph of f in the manifold of contact elements. By this we mean the set of triples (q,p,r) where p is viewed as a possible gradient vector for a function $r = f(q)$. Clearly $\tilde{\Gamma}(f)$ is an integral manifold for the contact distribution. Conversely, one can show that a maximal integral manifold V for this distribution is, at least locally, of the form $\tilde{\Gamma}(f)$ for some function f. The first step (which is the only one we give) is to show that V has the correct dimension.

(4.4) **Proposition.** *Integral manifolds for the contact distribution on $\mathbb{R}^{2n+1}$ have dimension at most n.*

Proof: Take a basis $\{ \tau_1, \ldots, \tau_d \}$ for the tangent space of V at x, then extend it to a basis for the tangent space of $\mathbb{R}^{2n+1}$ by adjoining "normal vectors" $\nu_1, \ldots, \nu_e$. Next, evaluate $E = \omega \wedge (d\omega)^n(\tau_1, \ldots, \tau_d, \nu_1, \ldots, \nu_e)$. Because $\omega \wedge (d\omega)^n$ is a volume form, E is nonzero. But E can also be expanded as a sum of terms

$$\text{constant}\, \omega(\xi_1) d\omega(\xi_2, \xi_3) \cdots d\omega(\xi_{2n}, \xi_{2n+1}),$$

at least one of which must be nonzero. For a term to be nonzero, ξ_1 must be nontangent, and at most one element of each of the n pairs (ξ_{2i}, ξ_{2i+1}) can be tangent. Thus at most n "slots" can be filled by tangent vectors, so that $d \leq n$, as required.

Let us state what we have just done in a slightly different language. Consider the real Heisenberg group of genus n, $\mathcal{H}_{\mathbb{R}}(n)$, given by matrices

$$U(p,q,r) = \begin{pmatrix} 1 & & & & & \\ q_1 & 1 & & & & \\ \cdot & 0 & \cdot & & & \\ \cdot & \cdot & & \cdot & & \\ \cdot & \cdot & & & \cdot & \\ q_n & 0 & \cdot & \cdot & 0 & 1 \\ r & p_1 & \cdot & \cdot & \cdot & p_n & 1 \end{pmatrix}.$$

The left-invariant form on $\mathcal{H}$ which is dr at the identity matrix is given by

$$\omega = dr - \sum p_i dq_i,$$

so that the contact geometry on $\mathbf{R}^{2n+1} \cong \mathcal{H}_{\mathbf{R}}(n)$ is homogeneous.

The tangent space to the identity is just the Heisenberg Lie algebra $\mathfrak{h}$, given by matrices

$$N(p,q,r) = \begin{pmatrix} 0 & & & & & \\ q_1 & 0 & & & & \\ \cdot & 0 & \cdot & & & \\ \cdot & \cdot & \cdot & \cdot & & \\ \cdot & \cdot & & \cdot & & \\ q_n & 0 & \cdot & \cdot\ 0 & 0 & \\ r & p_1 & \cdot & \cdot & p_n & 0 \end{pmatrix}.$$

Consequently (compare with figure 1) the Heisenberg group is a split homogeneous space with $\mathfrak{b} = 0$, $\mathfrak{n} = \mathfrak{h}$, and the contact distribution is given by the space $\mathfrak{s}$ of matrices with $r = 0$. Let $\mathfrak{z}$ be the center of $\mathfrak{n}$, given by matrices with $p = q = 0$. Then $\mathfrak{n} = \mathfrak{s} \oplus \mathfrak{z}$, whereas $[\mathfrak{s}, \mathfrak{s}] = \mathfrak{z}$, so that the contact distribution is strongly nonintegrable. Corollary (3.6) now applies to show that if V is an integral manifold which passes through the identity, then its tangent space may be viewed as an abelian subspace $\mathfrak{a}$ of $\mathfrak{s}$. Our argument on the differential ideal of the contact distribution therefore gives a solution to a simple case of the abelian subspace problem:

(4.5) **Proposition.** *Let $\mathfrak{a}$ be an abelian subspace of the real Heisenberg algebra $\mathfrak{h}$ which is contained in $\mathfrak{s}$. Then* $\dim \mathfrak{a} \leq \operatorname{genus} \mathfrak{h}$.

The proposition also holds for the complex Heisenberg algebra, a situation of interest in Hodge theory: the group $\mathcal{H}_{\mathbf{C}}(n)$ acts transitively and effectively on the set of mixed Hodge structures whose graded quotients are $\mathbf{Z}$, $\mathbf{Z}(1)^n$, and $\mathbf{Z}(2)$. Therefore the complex Heisenberg group can be identified with the classifying space for these structures, and the distribution of the complex contact form can be identified with Griffiths' horizontal distribution. Consequently (see [9, Proposition 2.3]) a variation of mixed Hodge structure of this kind has dimension at most n. Integral manifolds of maximal dimension n are known classically as *Legendre manifolds* [1, appendix 4].

5. Hodge structures.

Let H be a real Hodge structure of weight n, i.e., a real vector space $H_{\mathbf{R}}$ with a direct sum decomposition of the complexification, $H_{\mathbf{C}} = \oplus_{p+q=n} H^{p,q}$, satisfying $\overline{H^{p,q}} = H^{q,p}$. Let $S = \langle \cdot, \cdot \rangle$ be a nondegenerate bilinear form, symmetric if n is even, antisymmetric if n is odd. H is *polarized* if the following hold:

a) $\langle H^{p,q}, H^{r,s} \rangle = 0$ unless (p,q) is complementary to (r,s) in the sense that $(p,q) = (s,r)$.

b) Define the *Weil operator* by $Cx = \sqrt{-1}^{\,p-q} x$ for $x \in H^{p,q}$. Then the hermitian form $h_C(x,y) = \langle Cx, \bar{y} \rangle$ is positive-definite.

If only condition (a) holds, H is called *weakly polarized.* The n-th cohomology of a projective algebraic manifold of dimension n defines a real Hodge structure of weight n, and the primitive part (the part which maps to zero in the cohomology of a hyperplane section) is a polarized structure, with bilinear form given, up to sign, by the cup product.

Let D denote the set of polarized real Hodge structures on $H_{\mathbb{R}}$ with given Hodge numbers $h^{p,q} = \dim H^{p,q}$, and let $\hat{D}$ be the corresponding set of weakly polarized structures. These are the Griffiths period domains. To define transitive G-structures on them, let $G_{\mathbb{R}} = SO(S, \mathbb{R})$ be the group of automorphisms of $H_{\mathbb{R}}$ which preserves S, and let $G_{\mathbb{C}}$ be the corresponding group of complex automorphisms. Choose a reference structure in D, and let V and B be the isotropy groups in $G_{\mathbb{R}}$ and $G_{\mathbb{C}}$, respectively. Identifying coset spaces with the orbit of the reference structure we find

$$D \cong G_{\mathbb{R}}/V$$

and

$$\hat{D} \cong G_{\mathbb{C}}/B,$$

with D open in the projective variety $\hat{D}$.

To define the horizontal distribution, recall that a Hodge structure determines (and is determined by) a natural filtration

$$F^p = \bigoplus_{a \geq p} H^{a,b}$$

which satisfies

$$H_{\mathbb{C}} = F^p \oplus \overline{F^{n-p+1}}, \tag{5.1}$$

where n is the weight. If $\{ F^p \}$ is a filtration depending holomorphically on a local parameter z, then Griffiths' infinitesimal period relation is defined by

$$\partial F^p / \partial z_i \subset F^{p-1}.$$

Velocity vectors of curves satisfying this condition span the horizontal distribution on $\hat{D}$. For a group-theoretic description, consider the Lie algebra $\mathfrak{g}_{\Lambda}$ of G_{Λ}, which we can view as a space of linear endomorphisms of H_{Λ}, where $\Lambda = \mathbb{R}$ or $\mathbb{C}$. Fix a reference structure H, define a type decomposition [16, p. 111] by

$$\mathfrak{g}^{p,-p} = \{ \phi \in \mathfrak{g}_{\mathbb{C}} \mid \phi(H^{a,b}) \subset H^{a+p,b-p} \text{ for all } (a,b) \},$$

and note that

$$[\mathfrak{g}^{p,-p}, \mathfrak{g}^{q,-q}] \subset \mathfrak{g}^{(p+q),-(p+q)}. \tag{5.2}$$

This decomposition defines a real Hodge structure on $\mathfrak{g}_{\mathbb{R}}$. Set

$$\mathfrak{b} = \sum_{p \geq 0} \mathfrak{g}^{p,-p} = F^0$$

$$\mathfrak{n} = \sum_{p \leq -1} \mathfrak{g}^{p,-p} = \overline{F^1}$$

$$\mathfrak{s} = \mathfrak{g}^{-1,1}.$$

Because of (5.2), $\mathfrak{b}$ and $\mathfrak{n}$ are subalgebras, with $\mathfrak{b} = Lie(B)$. Because of (5.1),

$$\mathfrak{g} = \mathfrak{b} \oplus \mathfrak{n} = F^0 \oplus \overline{F^1},$$

so that the $G_{\mathbf{C}}$ structure of $\hat{D}$ is split. Applying (5.2) once again, one finds that $\mathfrak{s} + \mathfrak{b}$ is Ad_B-stable and so defines a homogeneous subbundle of the tangent bundle. This is the horizontal tangent bundle. Since $[\mathfrak{g}^{-1,1}, \mathfrak{g}^{-1,1}] \subset \mathfrak{g}^{-2,2}$, one has $[\mathfrak{g}^{-1,1}, \mathfrak{g}^{-1,1}] \cap \mathfrak{g}^{-1,1} = 0$, and so the horizontal distribution is strongly nonintegrable. Therefore tangent spaces to variations of Hodge structure define abelian subspaces:

(5.3) **Proposition.** *Let V be an integral manifold for Griffiths distribution, and let T be a tangent space for V, viewed in $\mathfrak{g}^{-1,1}$. Then T is an abelian subspace of $g^{-1,1}$.*

The dimension problem for variations of Hodge structures therefore reduces to the abelian subspace problem, and so

$$\dim V \leq a(g^{-1,1}). \tag{5.4}$$

For an equivalent formulation of Proposition (5.3) in the language of infinitesimal variations of Hodge structures, see [6, p. 133 1.d.1], [7, p. 54 item (*iv*)].

Let us now translate the abelian subspace problem for weight two variations into the language of matrices. To do so, recall that a *Hodge frame* for H is a set of ordered bases $B^{p,q}$ for $H^{p,q}$ such that

i) $B^{p,q}$ is an h_C-unitary basis for $B^{p,q}$,

ii) $B^{q,p} = \overline{B^{p,q}}$.

The matrix of S has a natural block decomposition relative to a Hodge frame, with all blocks zero except for those on the anti-diagonal, which are, up to sign, identity matrices. Thus, in the weight two case,

$$S = \begin{pmatrix} 0 & 0 & I_p \\ 0 & -I_q & 0 \\ I_p & 0 & 0 \end{pmatrix},$$

where we have written p for $h^{2,0}$ and q for $h^{1,1}$. Elements of $\mathfrak{g}_{\mathbf{C}}$ also have a block decomposition (compare with figure 1), where

i) elements in $T_H^{1,0} \cong \mathfrak{n}$ are strictly block lower triangular,

ii) elements in $\mathcal{G}_H \cong \mathfrak{g}^{-1,1}$ have nonzero blocks only in positions immediately below the main diagonal.

Thus, in the weight two case

$$N = \begin{pmatrix} 0 & 0 & 0 \\ X & 0 & 0 \\ Y & Z & 0 \end{pmatrix}$$

for $N \in \mathfrak{n}$. The condition that the matrix N be in the orthogonal Lie algebra,

$${}^T\!NS + SN = 0,$$

imposes additional restrictions, namely

$$Z = {}^TX \qquad \text{and} \qquad {}^TY = -Y,$$

so that

$$N = N(X,Y) = N = \begin{pmatrix} 0 & 0 & 0 \\ X & 0 & 0 \\ Y & {}^TX & 0 \end{pmatrix},$$

with Y skew symmetric. (The formula for the dimension of a weight two period domain is now apparent). Because horizontal tangent vectors have the form $N(X,0)$, Griffiths' distribution has dimension pq. Since the Lie bracket of vectors in $\mathfrak{g}$ is given by

$$[N(A,0), N(B,))] = N(0, {}^TAB - {}^TBA),$$

the abelian subspace problem is reduced to the following:

(5.5) **Theorem.** *Let $\mathfrak{a}$ be a space of $q \times p$ matrices over a field K satisfying*

$$[X,Y] \underset{def}{=} {}^TXY - {}^TYX = 0$$

for all X, Y in $\mathfrak{z}$. Then dim $\mathfrak{a} \leq \frac{1}{2}pq$, *provided that $p > 1$.*

Proof: Let e_1 ,..., e_p be the standard basis for K^p, where we view $X \in \mathfrak{a}$ as a linear transformation $X : K^p \longrightarrow K^q$. Let

$$\mathfrak{a}_j = \{ X \in \mathfrak{a} \mid X(e_i) = 0 \ \text{ for } i \leq j \},$$

with $\mathfrak{a}_0 = \mathfrak{a}$. Let

$$S_j = \mathfrak{a}_j(e_{j+1}) = \{ X(e_{j+1}) \mid X \in \mathfrak{a}_j \} \subset K^q.$$

Then

$$\mathfrak{a} = \mathfrak{a}_0 \supset \mathfrak{a}_1 \supset \cdots \supset \mathfrak{a}_{p-1} \supset \mathfrak{a}_p = 0$$

is a decreasing filtration of $\mathfrak{a}$, with

$$\mathfrak{a} \cong \bigoplus \mathfrak{a}_j/\mathfrak{a}_{j+1} \cong \bigoplus S_j.$$

To proceed, we study the spaces S_j:

(5.6) **Lemma.** *The spaces S_j ($j = 0$,..., $p-1$) are orthogonal relative to the dot product on K^q.*

Proof of the lemma: Assume $i < j$. Take $\alpha_i \in S_i$, $\alpha_j \in S_j$, with $\alpha_i = X_i(e_{i+1})$ and $\alpha_j = X_j(e_{j+1})$. Note that $X_j(e_{i+1}) = 0$. But then

$$\begin{aligned}\alpha_i \cdot \alpha_j &= {}^T(X_j e_{j+1})(X_i e_{i+1}) \\ &= {}^T e_{j+1}{}^T X_j X_i e_{i+1} \\ &= {}^T e_{j+1}{}^T X_i X_j e_{i+1} \\ &= 0,\end{aligned}$$

as required.

To conclude, we apply the following:

(5.7) **Lemma.** *Let $\{ S_i \}_{i=0}^{p-1}$ be a set of p mutually orthogonal subspaces of K^q, with $p > 1$. Then*

$$\sum_{i=0}^{p-1} \dim S_i \leq \frac{1}{2} pq.$$

Proof: Since the bilinear form is nondegenerate, the dual space satisfies $\hat{S}_j \cong K^q/S_j^{\perp}$, so that

$$(a) \qquad q = \dim K^q = \dim S_j + \dim S_j^{\perp}.$$

The orthogonality relation $S_i \perp S_j$ then gives $S_i \subset S_j^{\perp}$, so that

$$(b) \qquad \dim S_j^{\perp} \geq \dim S_i.$$

Combine (a) and (b) to get

$$q \geq \dim S_j + \dim S_i \qquad (i \neq j).$$

Sum over all $i < j$ to get

$$(c) \qquad \sum_{i<j} (\dim S_i + \dim S_j) \leq \frac{1}{2} p(p-1)q.$$

Evaluate the sum in a second way:

$$(d) \qquad \sum_{i<j} (\dim S_i + \dim S_j) = (p-1) \sum_i \dim S_i.$$

Finally, combine (c) and (d) to obtain the required result.

(5.8) **Remark.** *If $K = \mathbb{R}$ then the orthogonality relations given by Lemma (5.7) imply that the S_i are complementary. In this case one has the much stronger result, that $\dim \mathfrak{a} \leq q$.*

(5.9) **Remark.** *If* $q = 1$, *so that the* X*'s are row vectors, the commutativity condition becomes* $[X, Y] = (X_iY_j - Y_iX_j) = 0$. *Therefore the vectors* X *and* Y *are proportional, i.e.,* $\dim \mathfrak{a} \leq 1$.

Bibliography

1. V.I. Arnold, *Mathematical Methods of Classical Mechanics*, Springer-Verlag 1978, pp. 462.
2. R. L. Bishop and S. I. Goldberg, *Tensor Analysis on Manifolds*, Macmillan 1968, pp. 280.
3. R. Bott, On a topological obstruction to integrability, *Proc. Symp. Pure Math* **26**, 127-131.
4. J. A. Carlson, Bounds on the dimension of a variation of Hodge structure, *Trans. AMS* **294**, March 1986, 45-64.
5. ——— and Ron Donagi, Hypersurface variations are maximal, *Inventiones Math.* **89** (1987), 371-374.
6. ———, M. L. Green, P. A. Griffiths, and J. Harris, Infinitesimal variations of Hodge structures I, *Compositio Math.* **50** (1983), pp. 109-205.
7. ——— and P. A. Griffiths, Infinitesimal variations of Hodge structure and the global Torelli problem, *Journées de Géometrie Algébrique d'Angers 1979*, Sijthoff & Noordhoff, Alphen aan den Rijn, the Netherlands (1980), pp. 51-76.
8. ——— and C. Simpson, Shimura varieties of weight two Hodge structures, *Hodge Theory*, Springer Verlag LMN 1246 (1987) 1-15.
9. ——— and D. Toledo, Variations of Hodge structure, Legendre submanifolds, and accessibility, to appear in *Trans. AMS*.
10. ——— and ———, Harmonic mappings of Kähler manifolds to locally symmetric spaces, preprint, University of Utah, May 1987.
11. E. Cartan, *Les Systémes Différentiels Extérieurs et Leurs Applications Géométriques*, Hermann & Cie, 1945, pp. 214.
12. M. Cornalba and P. A. Griffiths, Some transcendental aspects of algebraic geometry, *Proc. Symp. Pure Math.* **29** AMS 1974, 3-110.
13. R. C. Courter, The maximum dimension of nilpotent subspaces of K_n satisfying the identity S_d, *J. of Algebra* **91** (1984), 82-110.
14. J. Eells and J. H. Sampson, Harmonic mappings of Riemannian manifolds, *Amer. Jour. Math.* **86** (1964), 109-160.
15. P. A. Griffiths, Periods of integrals on algebraic manifolds, III, *Publ. Math. I.H.E.S.* **38** (1970), 125-180.
16. ——— and W. Schmid, Recent developments in Hodge theory: a discussion of techniques and results, *Proceedings of the Bombay Colloquium on Discrete Subgroups of Lie Groups*, Tata Institute for Fundamental Research, Bombay 1973.
17. ——— and ———, Locally homogeneous complex manifolds, *Acta Mathematical* **79** (1964), 109-326.
18. ——— and L. Tu, Variation of Hodge Structure, Topics in Transcendental Algebraic Geometry, P. A. Griffiths, ed., *Ann. of Math. Studies* **106**, Princeton University Press 1984.

19. W. H. Gustafson, On maximal commutative algebras of linear transformations, *J. of Algebra* **42** (1976), 557-563.

20. N. Jacobson, Schur's theorems on commutative matrices, *Bull. Am. Math. Soc.* **50** (1944), 431-436.

21. A. I. Malcev, Commutative subalgebras of semisimple Lie algebras, *Izvestia Ak. Nauk USSR* **9** (1945), 291-300 (Russian); AMS Translation 40 (1951) (English).

22. J. H. Sampson, Applications of harmonic maps to Kähler geometry, *Contemp. Math.* **49** (1986), pp. 125-133.

23. I. Schur, Zur Theorie vertauschbaren Matrizen, *J. Reine und Angew. Math.* **130** (1905), 66-76.

24. Y. T. Siu, Complex analyticity of harmonic maps and strong rigidity of complex Kähler manifolds, *Ann. Math.* **112** (1980), 73-11.

25. ______ , Complex analyticity of harmonic maps, vanishing and Lefschetz theorems, *J. Diff. Geom.* **17** (1982), 55-138.

26. D. A. Suprenenko and R. I. Tyshkevich, *Commutative Matrices*, Academic Press 1968 (New York), pp. 158.

VALUATIONS ON FREE FIELDS

P.M. Cohn
University College London, Gower Street
London WC1E 6BT, England

1. Introduction

A valuation on a commutative field can be used to break it up into simpler parts. Just as a group may be decomposed into a normal subgroup and its quotient group, so a (non-archimedean) valuation leads to a residue class field which in general has a simpler structure than the original field. In the non-commutative case the connexion with the residue class field still persists, but the latter may be no simpler than the original field (cf. the examples below, § 2). However, under appropriate conditions we can find valuations with simpler residue class fields and use the latter in the study of the original field. Our aim here is to show how this can be done for free fields.

After the general description of valuations on skew fields, with particular emphasis on those with commutative residue class fields (§ 2), we single out in § 3 some useful subclasses, the abelian valuations and the quasi-commutative valuations. In § 4 we recall the definition of a free field and sketch a proof of its existence, using the specialization lemma, and apply this information in § 5 to construct families of valuations on the free field.

2. Valuations: Definitions and examples

Generally we shall omit the prefix 'skew'; thus fields need not be commutative. All rings have a unit-element, the group of units of a ring R is written U(R) and the set of non-zero elements of R is R^*.

Definition. A valuation on a field K with values in an ordered group Γ (written additively) is a function v on K^* with values in Γ, together with the convention $v(0) = \infty$ (and the rule $\alpha < \infty$ for all $\alpha \in \Gamma$), such that

(i) $v(x+y) \geq \min\{v(x), v(y)\}$,

(ii) $v(xy) = v(x) + v(y)$.

This definition goes back to Krull [10], who was abstracting from the case of p-adic valuations previously considered by Hensel [8]. As Schilling [13] observed, the same definition still makes sense in skew fields, and it is in this sense that we shall understand the above definition. Now the group Γ need no longer be abelian, though we shall keep the additive notation. The valuation will be called abelian if the value group is abelian; this will be the case mainly considered below. If $v(K^*) = 0$, the valuation is said to be trivial. Two valuations v_1, v_2 on K with value groups Γ_1, Γ_2 respectively are said to be equivalent if there exists an isomorphism $\theta:\Gamma_1 \to \Gamma_2$ such that

$\theta(v_1(x)) = v_2(x)$ for all $x \in K$.

Let us define a total subring of a field K as a subring T such that for any $a \in K^*$ either $a \in T$ or $a^{-1} \in T$. If also $c^{-1}Tc \subseteq T$ for all $c \in K^*$, then T will be called a valuation ring in K; this agrees with the customary usage in the commutative case (when the second condition is vacuous). For any valuation v on a field K the set

$$V = \{x \in K \mid v(x) \geq 0\}$$

is easily seen to be a valuation ring in K. Conversely, if V is a valuation ring in K, then the set U(V) of units in V is a normal subgroup of K^*, so we can form the quotient $\Gamma = K^*/U(V)$. The image of V in Γ defines a positive cone and hence a total ordering of Γ. Now the natural homomorphism $K^* \to \Gamma$ is a valuation. In this way one obtains a correspondence between valuations on K and valuation rings in K, which becomes one-one if we identify equivalent valuations.

We note that (as in the commutative case) every valuation ring V is a local ring, with a unique maximal ideal $\underline{m}$ and the quotient $V/\underline{m}$ is just the residue class field of the corresponding valuation.

For the moment let us give a few simple examples of valuations on skew fields:

1. Let K be any skew field and x an indeterminate. The rational function field K(x) (formed in the usual way as the field of fractions of the polynomial ring K[x]) has a valuation v, obtained by writing $\varphi \in K(x)$ as $\varphi = x^v \varphi_1$, where φ_1 is finite and non-zero for $x = 0$. This is the x-adic valuation. We note that it can still be defined in the following more general situation: Let σ be an automorphism of K and write $K[x;\sigma]$ for the skew polynomial ring, consisting of all polynomials $\Sigma x^i a_i$ with the commutation rule $cx = xc^\sigma$ $(c \in K)$. This ring is a principal ideal domain (as is well known, and easily verified, using the Euclidean

algorithm), and so it has a field of fractions, denoted by $K(x;\sigma)$ (cf. [4], p. 53). We can again write $\varphi = x^v\varphi_1$ for any element φ and verify that v is a valuation on $K(x;\sigma)$, still called the x-adic valuation; its value group is $\mathbb{Z}$ and its residue class field is K.

More generally, let K be a field with a valuation v and an automorphism σ such that $v(a^\sigma) = v(a)$ for all $a \in K$, and form the skew function field $K(x;\sigma) = L$ as before. We can define a valuation w on the skew polynomial ring $K[x;\sigma]$ by selecting an element δ in the value group Γ of v (or in its division closure) and for $f = \Sigma x^i a_i$ putting

$$w(f) = \min_r\{r\delta + v(a_r)\}.$$

Extending this function to L in the usual way by putting $w(f/g) = w(f) - w(g)$, we obtain a valuation w on L which extends v on K. If the residue class field of K is K_v and $\bar{\sigma}$ is the automorphism induced on K_v by σ, then the residue class field of L is L_w, where

$$L_w = \begin{cases} K_v & \text{if } \delta \neq 0, \\ K_v(x;\bar{\sigma}) & \text{if } \delta = 0. \end{cases}$$

2. Let G be an ordered group and consider the group algebra KG over an arbitrary field K. If G is written additively, it will be convenient to introduce an auxiliary variable t and write the elements of G as exponents, in order to convert to multiplicative notation: $t^{a+b} = t^a t^b$. The group algebra KG can be embedded in the set $K((G))$ of formal power series with well-ordered support: $\Sigma\lambda_a t^a$ has support $\{a \in G \mid \lambda_a \neq 0\}$, and it can be shown that $K((G))$ is a field, the <u>Mal'cev-Neumann construction</u> (cf. [11,12], see also [4], p. 528). Moreover, the equation

$$v(f) = \min\{a \in G \mid \lambda_a \neq 0\}$$

defines a valuation on $K((G))$, as is easily checked. The residue class field is K and the value group is G. This construction shows that every ordered group can occur as the value group of a valuation.

3. Let D be a field with a non-trivial valuation v, which is trivial on a subfield k. Then $v(a) > 0$ for some $a \in D^*$; it follows that a is not algebraic over k, and hence D can be enlarged to a field E containing an element b such that $ab = ba^{-1}$ (cf. [2], p. 116). If v could be extended to E we would have $v(a^{-1}) < 0$, but $v(a^{-1}) = v(b^{-1}ab) = -v(b) + v(a) + v(b) > 0$. This contradiction shows that v cannot be extended to E. This is in sharp contrast to the commutative case, where Chevalley's lemma tells us that an extension is always possible. In § 3 below we shall meet an appropriate generalization of Chevalley's

result.

4. Let E be a free field on a free generating set x_i ($i \in \mathbb{Z}$) over a commutative ground field k (a construction for E is outlined in § 4 below, for further details see [2], [4]). In E we have a 'shift' automorphism σ, mapping x_i to x_{i+1}. We form the skew polynomial ring $E[y;\sigma]$ and its field of fractions $E(y;\sigma) = D$. Clearly D is generated by x_0 and y over k and in fact it is the free field on these generators (cf. [2], p. 131). On D we have the y-adic valuation v; its value group is $\mathbb{Z}$ and its residue class field is E, so the residue class field is (in a sense) more complicated than the original field.

Our aim below will be to construct valuations whose residue class field is simpler than the original field; of course this will require a larger value group.

3. Abelian and quasicommutative valuations

Let K be any field and v an abelian valuation on K, with value group Γ. We shall denote the derived group of K^* by K^c; thus K^c is generated by all commutators. Since v is a homomorphism to an abelian group, it must be trivial on K^c. This simple remark allows one to extend Chevalley's extension lemma for valuations on commutative fields to prove

THEOREM 3.1. Let $K \subseteq L$ be an extension of skew fields. If v is an abelian valuation on K, then v has an abelian extension to L if and only if there is no equation

$$\Sigma a_i p_i = 1, \text{ where } a_i \in K,\ v(a_i) > 0,\ p_i \in L^c. \tag{1}$$

In one direction this is easy to see: If (1) holds, then any abelian extension w of v to L satisfies $w(a_i p_i) = v(a_i) > 0$, and so $w(1) = \min_i\{w(a_i p_i)\} > 0$, which is a contradiction.

For the converse, let V be the valuation ring of v in K and $\underline{m}$ its maximal ideal. Then $\underline{m}L^c$ is a proper ideal in VL^c, essentially because there is no equation (1), and now Chevalley's lemma (suitably generalized) leads to a ring W with an ideal $\underline{n}$ such that the pair $(W,\underline{n})$ is maximal among all pairs dominating $(VL^c, \underline{m}L^c)$, and W turns out to be the desired valuation ring (satisfying $W \cap K = V$). We refer to [6] or [5] for the details (the generalization of Chevalley's lemma was obtained independently by Krasner [9]).

For a valuation on a field K to be of use, it is necessary for both the value group Γ and the residue class field k to be as simple as possible. E.g. one could ask for both Γ and k to be commutative. Actually we shall find it convenient to impose an even stronger condition. Let U be the group of units in the valuation ring; we have a homomorphism $U \to k^*$ whose kernel is U_1, the group of <u>1-units</u> (Einseinheiten), that is, elements c such that $v(1-c) > 0$. It is easily verified that for any valuation on K we have the exact sequence

$$1 \longrightarrow k^* \longrightarrow K^*/U_1 \longrightarrow \Gamma \longrightarrow 0.$$

The valuation v is said to be <u>quasicommutative</u> if the group K^*/U_1 is abelian. Clearly in this case k and Γ are both commutative, and we can easily verify the following result (cf. [5]):

THEOREM 3.2. Let K be a field with a valuation v. Then v is quasicommutative if and only if $v(1-c) > 0$ for all $c \in K^c$.

As a consequence we obtain the following analogue of Theorem 3.1.

THEOREM 3.3. Let K be a field with a quasicommutative valuation v, and let L be an extension of K. Then v extends to a quasicommutative valuation on L if and only if there is no equation in L of the form

$$\Sigma a_i p_i + \Sigma b_j (q_j - 1) = 1, \qquad (2)$$

where $a_i, b_j \in K$, $v(a_i) > 0$, $v(b_j) \geq 0$, $p_i, q_j \in L^c$.

The proof is similar to that of Theorem 3.1.

To apply this result we shall need fields infinite-dimensional over their centres, with quasicommutative valuations. They are easily constructed, as follows.

Let k be any commutative field of characteristic 0, form the rational function field k(t) and define K as the field obtained by adjoining an element x with the commutation relation

$$tx = xt + x^2. \qquad (3)$$

We can form K as subfield of the field of skew Laurent series in x over k(t), or also as the Weyl field over k on s,t with the defining relation $st - ts = 1$ and then put $x = s^{-1}$. On K we have the x-adic valuation v; every element of K^* can be written uniquely as $x^r f$, where $v(f) = 0$ and x induces the automorphism $f(x,t) \to f(x,t+x)$, by (3). By taking a Taylor expansion we seee that every commutator is a 1-unit (cf. [5]). So v is quasicommutative; moreover, the centre of K is k, and $[K:k] = \infty$, by a well known property of the Weyl field: If we put

$u = x^{-1}t$, $v = x^{-1}$, (3) takes on the form $vu = uv + v$ or $uv = v(u-1)$. Thus we have $K = E(v;\sigma)$, where $E = k(u)$ and $\sigma: u \to u-1$. Any element of K can be written as a Laurent series in v with coefficients in $E: f = \Sigma v^i a_i$, and if f lies in the centre, then $vf = fv$, i.e. $a_i(u) = a_i(u+1)$, hence $a_i \in k$, and $uf = fu$, so $ua_i = a_i(u+i)$, and it follows that $a_i = 0$ unless $i = 0$. This shows the centre of K to be k.

In the commutative case one can define the concatenation of valuations if the residue class field is again valuated. By contrast, this is possible in the general case only if a condition is satisfied.

Let K be a skew field with a valuation v, and denote the residue class field by E. We claim that the multiplicative group K^* acts by automorphisms on E. For let $a \in K^*$ and $\xi \in E$, say $\xi = \bar{x}$, for $x \in K$. Then $v(x) \geq 0$, hence $v(a^{-1}xa) = -v(a)+v(x)+v(a) \geq 0$, and so $a^{-1}xa$ maps to an element of E, which we denote by $\xi\alpha_a$. It is clear that α_a is an automorphism of E and $a \to \alpha_a$ is a homomorphism from K^* to Aut(E). We have

THEOREM 3.4. Let K be a skew field with a valuation v and with residue class field E. If w is a valuation on E, then the concatenation of v and w exists if and only if the induced automorphisms α_a $(a \in K^*)$ preserve the sign of the valuation w, i.e.

$$w(\xi) \geq 0 \Rightarrow w(\xi\alpha_a) \geq 0 \text{ for all } \xi \in E,\ a \in K^*. \tag{4}$$

Proof. Let V be the valuation ring of v on K, W the valuation ring of w on E and put W_o for its inverse image in K. We claim that W_o is a valuation ring precisely when (4) holds. Let $x \in K$; if $v(x) > 0$, then $\bar{x} = 0$ and so $\bar{x} \in W$, therefore $x \in W_o$. Similarly, if $v(x) < 0$, then $x^{-1} \in W_o$. If $v(x) = 0$, then $\bar{x}$ is defined and either $\bar{x}$ or $\overline{x^{-1}}$ lies in W; accordingly x or x^{-1} lies in W_o; thus W_o is a total subring. Now for any $a \in K^*$, $x \in W_o$, we have $\bar{x} \in W$, hence $\bar{x}\alpha_a \in W$ by (4) and so $\overline{a^{-1}xa} \in W$, i.e. $a^{-1}xa \in W_o$. This shows W_o to be invariant and hence a valuation ring. Now the concatenation of v and w is the valuation corresponding to W_o. Conversely, if the concatenation of v and w is defined, then W_o as defined above must be invariant, i.e. $x \in W_o \Rightarrow a^{-1}xa \in W_o$. In terms of W this states that $\xi \in W \Rightarrow \xi\alpha_a \in W$, i.e. (4), and the proof is complete.

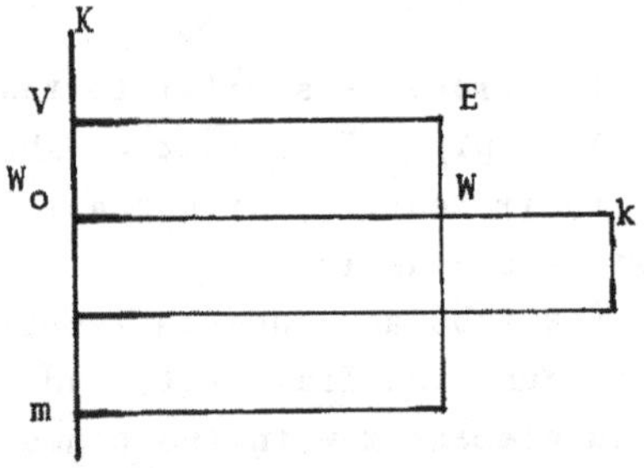

4. Free rings, completions and free fields

Throughout this section k will be a commutative field and all rings are assumed to be k-algebras; this means that every non-zero ring contains k in its centre. If A is a ring, then by an A-ring we understand a ring R with a homomorphism $A \to R$. Sometimes the role of k is emphasized by calling this an A_k-ring.

The free algebra on a set X, written $k\langle X\rangle$, is the algebra generated by X over k with defining relations

$$\alpha x = x\alpha \quad \text{for all } x \in X,\ \alpha \in k. \tag{1}$$

We shall need a more general type of ring, in which the free generators do not commute with all the coefficients. Let E be any ring; to avoid trivialities we shall assume that $E \neq \{0\}$ (later on E will usually be a field). The tensor E_k-ring on a set X, denoted by $E_k\langle X\rangle$, is defined as the E-ring generated by X with the defining relations (1). Thus X centralizes k (as is necessary to ensure a k-algebra) but nothing else. There is an obvious homomorphism

$$\alpha : E_k\langle X\rangle \to E, \tag{2}$$

obtained by mapping X to 0. This is the augmentation map, its kernel $\underline{a}$ is the augmentation ideal. Since α reduces to the identity on E, we have

$$E_k\langle X\rangle = E \oplus \underline{a}. \tag{3}$$

When E is a skew field, then by Cor. 2.9.13, p. 132 of [4], $E_k\langle X\rangle$ satisfies the inverse weak algorithm relative to the $\underline{a}$-adic filtration. The latter is just the usual filtration by order (= least degree in X of the terms occurring) of a polynomial, and it gives rise to the $\underline{a}$-adic topology on $E_k\langle X\rangle$. Here the value 1 has been assigned to each element of X. When X is finite, the precise values assigned to the elements of X are immaterial, but for infinite X the topology (and hence the completion) depend on whether or not the degrees of the elements of X are bounded. To avoid all ambiguity we shall usually assume X to be finite.

The completion of $E_k\langle X\rangle$ in the $\underline{a}$-adic topology is called the free power series ring, more precisely the free power series E-ring in X, or also the completion of $E_k\langle X\rangle$ at X = 0, and is written $E_k\langle\langle X\rangle\rangle$. Each element of $E_k\langle\langle X\rangle\rangle$ is a power series in X, which is a unit precisely when its image under the augmentation map α, i.e. its constant term, is a unit in E; the same statement holds for square matrices over $E_k\langle\langle X\rangle\rangle$.

In a similar way we can define the completion of $E_k\langle X\rangle$ at $x_\lambda = a_\lambda \in E$ (where $X = \{x_\lambda\}$) as the power series completion with respect to the kernel of the mapping $E_k\langle X\rangle \to E$ given by $x_\lambda \to a_\lambda$.

To obtain a structure which can lay claim to be called a 'free field' we need to embed $E_k\langle X\rangle$ in a field. As is well known, a commutative ring R can be embedded in a field if and only if R is an integral domain (i.e. it has no zero-divisors apart from 0, and $1 \neq 0$); further, when this condition is satisfied, then the least field containing R is determined up to isomorphism. In the non-commutative case the necessary and sufficient conditions for an embedding in a field are more complicated (cf. [4], Ch. 7). The precise conditions need not concern us here, but we note that when such an embedding exists, it will not generally be unique up to isomorphism. In order to achieve uniqueness we ask for an embedding which is universal with respect to 'specialization'. Without going into the details of defining non-commutative specializations we can give a sufficient condition for the existence of a universal field of a ring as follows.

A matrix A over a ring R is said to be <u>full</u> if it is square, say $n\times n$, and it cannot be written as a product $A = PQ$, P $n\times r$, Q $r\times n$, where $r < n$. It is clear that if A is non-full, then so is its image under any ring-homomorphism. Further, over a field a non-full matrix cannot have an inverse. Generally, under a homomorphism of a ring into a field, certain matrices will be inverted (i.e. have an inverse over the field), and from what has just been said, the most we can ever hope to invert are the full matrices. A ring with a homomorphism into a field inverting all full matrices is called a <u>Sylvester domain</u>. These rings were introduced by Dicks and Sontag [7] (see also [4], Ch. 5; the name arises because these rings satisfy a form of Sylvester's law of nullity). We observe that such rings have an embedding into a field, because every non-zero element, as a 1×1 matrix is full and so does not map to 0 in the field; this also shows the ring to be an integral domain. Of course the class of rings embeddable in fields is much larger than the class of Sylvester domains; this is already clear in the commutative case, where the polynomial ring $\mathbb{Z}[x,y]$ fails to be a Sylvester domain (cf. [4]). But for our present purposes the class is wide enough, in fact we shall only be concerned with a subclass. Let us define a <u>semifir</u> as a ring R in which every finitely generated right ideal is free as right R-module, of unique rank. The same property then holds for finitely generated left ideals. Every semifir is a Sylvester domain and hence has a universal field inverting all full matrices.

A ring homomorphism is said to be <u>honest</u> if it keeps all full matrices full. Using this definition we can describe the Sylvester domains as the rings with an honest homomorphism into a field. The power series ring $G = E_k\langle\langle X\rangle\rangle$, like $F = E_k\langle X\rangle$ is a semifir ([4], Th. 2.4.1, p. 105, Th. 2.9.4, p. 128); so both these rings have universal fields. Below we shall give an independent proof that F has a universal field, using only the specialization lemma (and the ultraproduct theorem). Assuming this for the moment, we can prove very simply that the natural embedding

$$F = E_k\langle X\rangle \to G = E_k\langle\langle X\rangle\rangle \tag{4}$$

is honest. Let us denote by U the universal field of F; as stated earlier (and as will be shown soon) every full matrix over F is invertible over U.

Let us form the power series ring $F[[t]]$ in a central indeterminate t. We have E-linear homomorphisms

$$\begin{array}{lll} F \to G \to F[[t]] & & (5) \\ x \to x \to xt & (x \in X). & \end{array}$$

Thus by mapping $x \in X$ to xt we can embed G in $F[[t]]$. Let $A = A(x)$ be a matrix over F; under the mapping (5) this becomes $A(xt)$. If A is full over F, it is invertible over U, hence $A(xt)$ is invertible over $U(t)$, and so is also invertible over $U((t))$, the field of formal Laurent series in t. Since $F[[t]] \subseteq U((t))$, it follows that $A(xt)$ is full over $F[[t]]$, and going back along the map (5), we find that $A(x)$ is full over G. This proves the homomorphism $F \to G$ to be honest.
We remark that this proof still applies when X is infinite, since A can only involve a finite part of X.

In order to establish the existence of U we shall need the specialization lemma. This generalizes the elementary fact that a non-zero polynomial over an infinite field k assumes non-zero values f(a) for some $a \in k$.

<u>Specialization lemma</u> ([4], Lemma 5.9.5, p. 285). Let E be a field with infinite centre C, infinite-dimensional over C. Then any full matrix over $E_C\langle X\rangle$ is non-singular for some set of values of X in E.

The proof relies on Amitsur's theorem on generalized polynomial identities (cf. [1]). The assumption that $[E:C] = \infty$ cannot be omitted, because a finite-dimensional algebra satisfies a polynomial identity. Whether the assumption that C be infinite can be omitted is not known.

THEOREM 4.1. Let E be a field with infinite centre C, infinite-dimensional over C, and let X be any set. Then there is an honest homomorphism of $E_C\langle X\rangle$ into a field.

We outline the proof (cf. [4], p. 286). There is a natural homomorphism

$$E_C\langle X\rangle \to E^{(E^X)}, \tag{6}$$

in which $f\in E_C\langle X\rangle$ is mapped to $f_\alpha = f(X\alpha)$ for any $\alpha\in E^X$. Thus we have a homomorphism of $E_C\langle X\rangle$ into a direct power of E. To reach our conclusion we replace the latter by a suitable ultrapower; for this we need to define an ultrafilter on the index set E^X. Let A be a square matrix over $E_C\langle X\rangle$ and define D(A), the <u>singularity support</u> of A, as the subset of E^X consisting of all $\varphi\in E^X$ such that $A(X\varphi)$ is non-singular. By the specialization lemma $D(A) \neq \emptyset$ whenever A is full, and of course otherwise $D(A) = \emptyset$. Let us define the diagonal sum of matrices as

$$A \oplus B = \begin{pmatrix} A & 0 \\ 0 & B \end{pmatrix}.$$

It is clear that over a field the diagonal sum of A and B is non-singular precisely when both A and B are, hence we have

$$D(A)\cap D(B) = D(A \oplus B).$$

Moreover, the diagonal sum of any full matrices over $E_C\langle X\rangle$ is full; this is not true in all rings, but it does hold for semifirs (see [4], Lemma 5.5.3, p. 253); therefore the family (D(A)) of singularity supports of full matrices is closed under finite intersections, and it does not include the empty set. It follows that there is an ultrafilter F on E^X containing all the D(A) (A full), and this provides a homomorphism to an ultrapower

$$E_C\langle X\rangle \to E^{(E^X)}/F. \tag{7}$$

Since E is a field, the ultrapower is a field (cf. e.g. [3], p. 210), and by definition every full matrix A over $E_C\langle X\rangle$ is non-singular on the set $D(A)\in F$, hence A is non-singular in the ultrapower. Thus (7) is an honest homomorphism to a field.

The least field inverting all full matrices over $E_C\langle X\rangle$ is denoted by $E_C\langle\!\langle X\rangle\!\rangle$ or U and is called the <u>free field</u> on X over E_C. With the help of the general embedding theorems of Ch. 7 of [4] one can prove that the rings $F = E_k\langle X\rangle$ and $G = E_k\langle\!\langle X\rangle\!\rangle$ have honest homomorphisms into fields, for any field E, based on the fact that these rings are semifirs. This will not be needed here; we shall merely indicate how any finite family of elements of U can be defined in a suitable power series completion of F. In order to do so we need to recall (from Ch. 7 of [4]) some details of the explicit construction of U.

Each element p of U is obtained as a component of the solution of a system of linear equations over F; the solution can be expressed in terms of elements of F and the entries of the inverse of a certain square matrix over F, called the <u>denominator</u> of p in the given system. Thus any homomorphism from F to another ring which inverts the denominator of p (i.e. maps it to an invertible matrix) can be defined on the element p of U. We shall apply this fact as follows. Let E be a field with centre C and assume that $|C| = \infty$, $[E:C] = \infty$. Given any finite set $p_1,\dots,p_r$ of elements in the free field $U = E_C\langle\!\langle X \rangle\!\rangle$, denote the denominator of p_i by A_i and form the diagonal sum $A = A_1 \oplus \dots \oplus A_r$. This matrix A is again full (as diagonal sum of full matrices over F, see above), hence by the specialization lemma, A specializes to a nonsingular matrix for suitable values of X in E, say for $x_\lambda \mapsto a_\lambda$. Put $x'_\lambda = x_\lambda - a_\lambda$ $(x_\lambda \in X)$ and write $X' = \{x'_\lambda\}$. Then clearly, $E_C\langle X' \rangle = E_C\langle X \rangle$ and in terms of X', A reduces to an invertible matrix for $X' = 0$. Hence A is invertible over the power series ring relative to X': $E_C\langle\langle X' \rangle\rangle$, and so we have found a power series ring in which $p_1,\dots,p_r$ are defined. We can state the result as

THEOREM 4.2. Let E be a field with infinite centre C such that $[E:C] = \infty$, and let $\{p_1,\dots,p_r\}$ be any finite family of elements of $E_C\langle\!\langle X \rangle\!\rangle$. Then the p_i all belong to the power series completion of $E_C\langle X \rangle$ at a certain point $x_\lambda = a_\lambda \in E$.

5. Valuations on a free field

Let E be any skew field and consider the free field $U = E_k\langle\!\langle X \rangle\!\rangle$. We can define a valuation of U as follows. Let t be a central indeterminate and consider the homomorphism φ of U into U(t) which is E-linear and maps $x \in X$ to xt:

$$\varphi : U \to U(t),\ x_\lambda \to x_\lambda t \qquad (x_\lambda \in X). \tag{1}$$

On U(t) we have the t-adic valuation v_o and we can define a valuation v on U by putting

$$v(p) = v_o(\varphi(p)),\ p \in U.$$

Here the value group is $\mathbb{Z}$, but the residue class field is very much larger than E; it contains as indeterminates xy^{-1} $(x \neq y,\ x,y \in X)$ and other similar elements. Similarly we can instead of (1) take the mapping $x_\lambda \to a_\lambda + (x_\lambda - a_\lambda)t$; this defines a different 'place' of U, again with value group $\mathbb{Z}$ and a similarly complicated residue class field. Let us call this the <u>simple valuation</u> at $x_\lambda = a_\lambda$.

This case can be generalized as follows. Let E have the centre C, where $|C|$ and $[E:C]$ are infinite, and write $F = E_C\langle X\rangle$, $U = E_C⦃X⦄$. Given any homomorphism α of F into a field L, the ring $G = L_C\langle X\rangle$ is again a tensor ring extending F. It is not hard to verify that U is a subfield of $L_C⦃X⦄$ (cf. [2], p. 114), and if v_o is the simple valuation on $L_C⦃X⦄$ at the point $x = x\alpha$, then the valuation on U induced by v_o is again $\mathbb{Z}$-valued, with residue class field an extension of L.

In all these cases the residue class field is quite complicated, while the value group is $\mathbb{Z}$. We can obtain a simpler residue class field, at the cost of complicating the value group, as follows. Let us index X by a well-ordered set I: $X = \{x_\lambda | \lambda \in I\}$, and denote by T the free abelian group on a set $\{t_\lambda | \lambda \in I\}$ (in multiplicative notation). We can regard T as an ordered group, using the lexicographic ordering, and form the field of fractions of the group algebra over U. This field U(t) may be regarded as a subfield of U((t)), the set of formal series over T with well-ordered support. This is the Mal'cev-Neumann construction mentioned in § 2, (ii) and it provides a valuation on U(T) with value group T. Now consider the embedding of U in U(T) which is E-linear and such that

$$x_\lambda \to a_\lambda + (x_\lambda - a_\lambda)t_\lambda .$$

This induces a T-valued valuation on U. But we note that the residue class field is still very much larger than E, since it contains elements such as $(x_\lambda - a_\lambda)^{-1}c(x_\lambda - a_\lambda)$ (for any $c \in E$). At the cost of complicating the value group further, e.g. taking it to be free on generators t_λ, we can ensure a residue class field isomorphic to E, but the construction of § 3 allows us to find a quasicommutative valuation.

Let E be a field with centre C such that $|C|$ and $[E:C]$ are infinite, and assume that we have a quasicommutative valuation v on E (cf. the example in § 3). We claim that v can be extended to a quasicommutative valuation on $U = E_C⦃X⦄$. If this were not so, we would by Theorem 3.2 have an equation

$$\Sigma a_i p_i + \Sigma b_j (q_j - 1) = 1, \tag{2}$$

where $a_i, b_j \in E$, $v(a_i) > 0$, $v(b_j) > 0$, $p_i, q_j \in U^c$. Let A be a common denominator for all the $p_i, p_i^{-1}, q_j, q_j^{-1}$, when expressed as products of commutators. This means that whenever A becomes invertible under a homomorphism, then p_i, q_j are all defined and non-zero and are still products of commutators. By the specialization lemma we can find a homomorphism $\alpha : E_C\langle X\rangle \to E$ such that $A\alpha$ is invertible over E. It follows that p_i, q_j become well-defined non-zero products of commutators when $x \in X$ is replaced by $x\alpha \in E$. Hence over E we have

$$v(\Sigma a_i \cdot p_i \alpha + \Sigma b_j (q_j \alpha - 1)) > 0.$$

But this contradicts (2). Hence no equation (2) is possible and it follows that the desired extension of v to U exists. In particular this provides a valuation on U with abelian value group and commutative residue class field.

REFERENCES

1. S.A. Amitsur, Generalized polynomial identities and pivotal monomials. Trans.Am.Math.Soc. 114(1965), 210-226.
2. P.M. Cohn, Skew field constructions. Lond.Math.Soc.Lect.Note Ser. No. 27. Cambridge University Press, Cambridge 1977.
3. P.M. Cohn, Universal Algebra (2nd ed.). D. Reidel, Dordrecht 1981.
4. P.M. Cohn, Free rings and their relations (2nd ed.). Lond.Math.Soc. Monographs No. 19. Academic Press, London 1985.
5. P.M. Cohn, The construction of valuations on skew fields. To appear.
6. P.M. Cohn and M. Mahdavi-Hezavehi, Extensions of valuations on skew fields. In "Ring Theory Antwerp 1980", ed. F. Van Oystaeyen. Lect. Notes Math. No. 825, Springer-Verlag, Berlin 1980, p. 28-41.
7. W. Dicks and E.D. Sontag, Sylvester domains. J.Pure Appl.Algebra 13(1978), 243-275.
8. K. Hensel, Theorie der algebraischen Zahlen. Teubner, Leipzig 1908.
9. M. Krasner, unpublished.
10. W. Krull, Allgemeine Bewertungstheorie. J.Reine Angew.Math. 167 (1932), 160-196.
11. A.I. Mal'cev, On the embedding of group algebras in division algebras. Dokl.Akad.Nauk SSSR 60(1948), 1499-1501 (= Izbrannye Trudy I, Nauka, Moscow, 1976, p. 211-213) (Russian).
12. B.H. Neumann, On ordered division rings. Trans.Am.Math.Soc. 66 (1948), 202-252.
13. O.F.G. Schilling, Noncommutative valuations. Bull.Am.Math.Soc. 31(1945), 297-304.

"This paper is in final form and no version of it will be submitted for publication elsewhere"

STANDARD BASES AND HOMOLOGY

E. S. GOLOD
Department of Mathematics and Mechanics
Moscow State University
Moscow 119 899, U.S.S.R

The purpose of this note is to give an interpretation of some standard bases criteria in terms of homology modules of certain complexes. Some known results are obtained as consequences, namely the confluence (diamond) lemma of A.I. Shirshov-G. Bergman [1, 2] and its counterparts concerning ideals in polynomial rings [3] and in universal enveloping algebras of Lie algebras [4].

1. Standard bases.

Let Γ be an ordered semigroup satisfying the descending chain condition for elements, let k be a commutative ring, and let R be a k-algebra endowed with an ascending Γ-filtration $\{F_\gamma R\}_{\gamma\in\Gamma}$ that is $R = \bigcup_{\gamma\in\Gamma} F_\gamma R$, where $F_\gamma R$ are k-submodules of R and $F_{\gamma'}R.F_{\gamma''}R \subseteq F_{\gamma'\gamma''}R$ for all $\gamma',\gamma''\in\Gamma$. We shall use the following notation: $\deg a = \min\{\gamma\in\Gamma : a\in F_\gamma R\}$, $F^*_\gamma R = \bigcup_{\gamma'<\gamma} F_{\gamma'}R$, $\bar{R}_\gamma = F_\gamma R/F^*_\gamma R$; then $\bar{R} = \bigoplus_{\gamma\in\Gamma} \bar{R}_\gamma$ is the associated Γ-graded ring. The canonical homomorphisms $\omega_\gamma : F_\gamma R \to \bar{R}_\gamma$ define a homomorphism of Γ-graded rings $\omega_* : \bigoplus_{\gamma\in\Gamma} F_\gamma R \to \bar{R}$. If $\gamma = \deg a$, then $\omega_\gamma(a)$ will be denoted by $\omega(a)$ and this element of $\bar{R}_\gamma$ will be called the leading term of a. Note that $\omega(a) \neq 0$ if $a \neq 0$, and that for $a_1, a_2 \in R$ either $\omega(a_1)\omega(a_2) = 0$ or $\omega(a_1 a_2) = \omega(a_1)\omega(a_2)$.

Let I be a (two-sided) ideal of R and let $\bar{I} = \bigoplus_{\gamma\in\Gamma} \bar{I}_\gamma$ denote the homogeneous ideal of $\bar{R}$ whose homogeneous elements are precisely the leading terms of elements of I. Let now $\underline{x} = \{x_\alpha\}_{\alpha\in A}$ be some fixed basis of I, $y_\alpha = \omega(x_\alpha)$ and let $\tilde{I} = \bigoplus_{\gamma\in\Gamma} I_\gamma$ be the homogeneous ideal of $\bar{R}$ generated by the set $\underline{y} = \{y_\alpha\}_{\alpha\in A}$. One always has $\tilde{I} \subseteq \bar{I}$.

Definition. The set $\underline{x} = \{x_\alpha\}_{\alpha\in A}$ is called a standard basis of I if the following equivalent conditions are satisfied:

1) $\tilde{I} = \bar{I}$;

2) each element $x \in I$ with $\gamma = \deg x$ has a presentation as a finite sum $x = \sum_i a_i x_{\alpha_i} b_i$ where $a_i, b_i \in R$ and $(\deg a_i)(\deg x_{\alpha_i})(\deg b_i) \leq \gamma$ for all i.

The implication 2) $\Rightarrow$ 1) is clear. The inverse implication is immediately proved by induction on deg x.

For an element $x \in I$ we shall call $\min\{\max_i\{(\deg a_i)(\deg x_{\alpha_i})(\deg b_i)\}\}$ the generative degree of x, where minimum is taken over all presentations $x = \sum_i a_i x_{\alpha_i} b_i$, and we shall denote this number by gdeg x. One has gdeg x $\geq$ deg x. Condition 2) in the definition above states that gdeg x = deg x for all $x \in I$.

For each $\gamma \in \Gamma$ choose a set-theoretical section $\tau_\gamma : \bar{R}_\gamma/\tilde{I}_\gamma \to F_\gamma R$ of the canonical surjection $\pi_\gamma : F_\gamma R \to \bar{R}_\gamma/\tilde{I}_\gamma$, subject to the mild restriction that $\tau_\gamma(0) = 0$. Denote $\operatorname{Im}\tau_\gamma$ by T_γ. An element $a \in R$ will be called reduced if it has an (evidently unique) presentation $a = \sum_{\gamma\in\Gamma} t_\gamma$, where $t_\gamma \in T_\gamma$, almost all t_γ are zero, and all indices are pairwise distinct. If an element $a \in R$ for some $\gamma_0 \in \Gamma$ can be written in the form $a = a_0 + b$ where $a_0 = \sum_{\gamma > \gamma_0} t_\gamma$ is reduced and $\deg b \leq \gamma_0$, then we shall say that a admits a reduction of degree γ_0; in a similar vein a reduction of a of degree γ_0 is any element $a' \in R$ such that $a' = a_0 + t_{\gamma_0} + b'$, where $t_{\gamma_0} \in T_{\gamma_0}$, $\deg b' < \gamma_0$, $a - a' \in I$ and $\operatorname{gdeg}(a-a') \leq \gamma_0$. Evidently, such a reduction really exists and $t_{\gamma_0} = \tau_{\gamma_0}(\pi_{\gamma_0} b)$ is uniquely determined; furthermore each element $a \in R$ admits reductions of degree equal to deg a, and by means of a finite sequence of reductions of decreasing degrees $\leq$ deg a one can pass from a given element a to some reduced element. Any reduced element obtained in this way will be called a complete reduction of a. The only complete reduction of zero is zero.

Proposition. The following conditions for the basis $\underline{x}$ of the ideal I are equivalent:

1) $\underline{x}$ is a standard basis;

2) two elements of R lie in the same residue class modulo I iff any of their complete reductions coincide;

3) zero is the unique complete reduction for any element of I;

4) any element of I has zero as its complete reduction.

Proof. 1) $\Rightarrow$ 2). Let a, a' be two elements of R such that $a - a' \in I$. Replacing a and a' by their complete reductions we may assume that a and a' are reduced, that is $a = \sum_{\gamma\in\Gamma} t_\gamma$, $a' = \sum_{\gamma\in\Gamma} t'_\gamma$. Let γ_0 be max {deg a, deg a'}. From $a - a' \in I$ and $\overline{I}_{\gamma_0} = \tilde{I}_{\gamma_0}$ it follows that $\pi_{\gamma_0}(a-a') = 0$, hence $\pi_{\gamma_0}(t_{\gamma_0}) = \pi_{\gamma_0}(t'_{\gamma_0})$ and $t_{\gamma_0} = t'_{\gamma_0}$. Passing to $a-t_{\gamma_0}$ and $a'-t'_{\gamma_0}$, proceeding by an obvious induction we obtain that $t_\gamma = t'_\gamma$ for all $\gamma\in\Gamma$.

The implications 2) $\Rightarrow$ 3) $\Rightarrow$ 4) are trivial and 4) $\Leftrightarrow$ 1) is a reformulation of the definition of standard basis.

2. The Shafarevich complex [5].

The Shafarevich complex Sh(x,R) of a k-algebra R relative to a set $\underline{x} = \{x_\alpha\}_{\alpha\in A}$ of elements of R is a complex of R-R-bimodules which is defined as the free product over k of R and the free k-algebra freely generated by the set $\{u_\alpha\}_{\alpha\in A}$ endowed with $\mathbb{Z}$-grading by degrees of monomials on u_α (each u_α being of degree one) and with differential d defined by the formula:

$$d(a_0 u_{\alpha_1} a_1 \ldots a_{n-1} u_{\alpha_n} a_n) = \sum_{i=1}^{n} (-1)^{i-1} a_0 u_{\alpha_1} a_1 \ldots a_{i-1} x_{\alpha_i} a_i \ldots a_{n-1} u_{\alpha_n} a_n .$$

For a Γ-filtered algebra R the complex $Sh(\underline{x},R)$ has a natural Γ-filtration $\{F_\gamma Sh(\underline{x},R)\}_{\gamma\in\Gamma}$ where $F_\gamma Sh(\underline{x},R)$ is the k-subcomplex of $Sh(\underline{x},R)$ generated as a k-module by the elements $a_0 u_{\alpha_1} a_1 \ldots a_{n-1} u_{\alpha_n} a_n$ with $(\deg a_0)(\deg x_{\alpha_1})(\deg a_1)\ldots(\deg a_{n-1})(\deg x_{\alpha_n})(\deg a_n) \leq \gamma$.

We shall consider also the Shafarevich complex $Sh(\underline{y},R)$ (where as above $\underline{y} = \{y_\alpha\}_{\alpha\in A}$, $y_\alpha = \omega(x_\alpha)$). This complex has a natural structure of a Γ-graded complex, that is $Sh(\underline{y},\overline{R}) = \bigoplus_{\gamma\in\Gamma} Sh(\underline{y},\overline{R})_\gamma$ where $Sh(\underline{y},\overline{R})_\gamma$ is the k-subcomplex of $Sh(\underline{y},\overline{R})$ generated as k-module by the elements $\overline{a}_0 u_{\alpha_1} \overline{a}_1 \ldots \overline{a}_{n-1} u_{\alpha_n} \overline{a}_n$ with $\overline{a}_i$ being homogeneous elements of $\overline{R}$ and $(\deg \overline{a}_0)(\deg y_{\alpha_1})(\deg \overline{a}_1)\ldots(\deg y_{\alpha_n})(\deg \overline{a}_n) = \gamma$. There is a natural surjective Γ-homogeneous morphism of Γ-graded complexes of $\overline{R}$-$\overline{R}$-bimodules $\varphi: Sh(\underline{y},\overline{R}) \to grSh(\underline{x},R)$, where $grSh(\underline{x},R)$ is the Γ-graded complex associated with the Γ-filtered complex $Sh(\underline{x},R)$: if $\overline{a}_0 u_{\alpha_1} \overline{a}_1 \ldots \overline{a}_{n-1} u_{\alpha_n} \overline{a}_n \in Sh(\underline{y},\overline{R})_\gamma$ then $\varphi(\overline{a}_0 u_{\alpha_1} \overline{a}_1 \ldots \overline{a}_{n-1} u_{\alpha_n} \overline{a}_n)$ is the class modulo $F^*_\gamma Sh(\underline{x},R)$ of any element $a_0 u_{\alpha_1} a_1 \ldots a_{n-1} u_{\alpha_n} a_n \in F_\gamma Sh(\underline{x},R)$ such that $\omega(a_i) = \overline{a}_i$,

$i = 0,\dots,n$. Furthermore, let Ψ be the canonical morphism of Γ-graded complexes $\bigoplus_{\gamma\in\Gamma} F_\gamma Sh(\underline{x},R) \to grSh(\underline{x},R)$. Let φ_* and Ψ_* denote the homology homomorphisms induced respectively by φ and Ψ. Denote the Γ-graded $\bar{R}$-$\bar{R}$-subbimodule $\varphi_*^{-1}(\mathrm{Im}\Psi_*)$ of $H_*(Sh(\underline{y},\bar{R})$ by $E_*(Sh(\underline{y},\bar{R}))$. We shall call the homogeneous elements of $E_*(Sh(\underline{y},\bar{R}))$ and the cycles representing them extendable classes and cycles. Note that a cycle $\bar{z} = \sum_i \omega(a_i)u_{\alpha_i}\omega(b_i)\in Sh_1(\underline{y},\bar{R})_\gamma$ is extendable iff there exists a cycle $z\in F_\gamma Sh_1(\underline{x},R)$ such that $z - \Sigma a_i u_{\alpha_i} b_i = dy + y^*$ where $y\in F_\gamma Sh_2(\underline{x},R)$, $y^*\in F_\gamma^* Sh_1(\underline{x},R)$, but this means that there exists a $y^*\in F_\gamma^* Sh_1(\underline{x},R)$ such that $\sum_i a_i u_{\alpha_i} b_i + y^*$ is a cycle.

Example. Let a' and a'' be two reductions of degree γ of the same element a, that is $a = a' + \sum_{i=1}^{k} a_i' x_{\alpha_i} b_i' = a'' + \sum_{j=1}^{l} a_j'' x_{\alpha_j} b_j''$ and $a' = t_\gamma + b'$, $a'' = t_\gamma + b''$, where $\deg b' < \gamma$, $\deg b'' < \gamma$ and we may assume without loss of generality that $(\deg a_i')(\deg x_{\alpha_i})(\deg b_i') = (\deg a_j'')(\deg x_{\alpha_j})(\deg b_j'') = \gamma$ for all $i = 1,\dots,k$ and $j = 1,\dots,l$. This situation gives rise to a cycle $\bar{z} = \sum_{i=1}^{k} \omega(a_i')u_{\alpha_i}\omega(b_i') - \sum_{j=1}^{l} \omega(a_j'')u_{\alpha_j}\omega(b_j'')\in Sh_1(\underline{y},\bar{R})_\gamma$. If a' and a'' have a common complete reduction or moreover if the element $x = a'' - a' = \sum_{i=1}^{k} a_i' x_{\alpha_i} b_i' - \sum_{j=1}^{l} a_j'' x_{\alpha_j} b_j'' \in I$ has zero complete reduction, then the cycle $\bar{z}$ is extendable. The converse is not true in general.

3. Criteria for standard bases.

Theorem 1. For a basis $\underline{x} = \{x_\alpha\}_{\alpha\in A}$ of the ideal I the following conditions are equivalent:

1) $\underline{x}$ is a standard basis;
2) $E_1(Sh(\underline{y},\bar{R})) = H_1(Sh(\underline{y},\bar{R}))$;
3) the $\bar{R}$-$\bar{R}$-bimodule $H_1(Sh(\underline{y},\bar{R}))$ is generated by extendable classes.

Proof. The equivalence 2) $\Leftrightarrow$ 3) is evident as $E_1(Sh(\underline{y},\bar{R}))$ is an $\bar{R}$-$\bar{R}$-subbimodule of $H_1(Sh(\underline{y},\bar{R}))$.

1) $\Rightarrow$ 2). Let $\bar{z} = \sum_i \omega(a_i)u_{\alpha_i}\omega(b_i)$ be any cycle of $Sh_1(\underline{y},\bar{R})_\gamma$. Then

$x = \sum_i a_i x_{\alpha_i} b_i \in I$ and $\deg x < \gamma$; hence $x = \sum_j c_j x_{\alpha_j} d_j$ with $(\deg c_j) \times (\deg x_{\alpha_j})(\deg d_j) < \gamma$ for all j. Therefore $z = \sum_i a_i u_{\alpha_i} b_i - \sum_j c_j u_{\alpha_j} d_j$ is a cycle of $F_\gamma Sh(\underline{x},R)$ extending $\bar{z}$.

2) → 1). For any $x \in I$ we have to show that $\mathrm{gdeg}\, x = \deg x$. Suppose $\mathrm{gdeg}\, x = \gamma > \deg x$, so that $x = \sum_{i=1}^{n} a_i x_{\alpha_i} b_i$ with $(\deg a_i)(\deg x_{\alpha_i}) \times (\deg b_i) = \gamma$ for $i = 1,\dots,k$ and $(\deg a_i)(\deg x_{\alpha_i})(\deg b_i) < \gamma$ for $i = k+1,\dots,n$. Then $\bar{z} = \sum_{i=1}^{k} \omega(a_i) u_{\alpha_i} \omega(b_i)$ is a cycle of $Sh_1(\underline{y},\bar{R})_\gamma$, and let $z = \sum_{i=1}^{k} a_i u_{\alpha_i} b_i + \sum_{j=1}^{l} c_j u_{\alpha_j} d_j$, $\sum_{j=1}^{l} c_j u_{\alpha_j} d_j \in F_\gamma^{*} Sh_1(\underline{x},R)$, be a cycle of $F_\gamma Sh_1(\underline{x},R)$ which extends $\bar{z}$. We have $x = \sum_{i=k+1}^{n} a_i x_{\alpha_i} b_i - \sum_{j=1}^{l} c_j x_{\alpha_j} d_j$, whence $\mathrm{gdeg}\, x < \gamma$. This is a contradiction which proves that $\mathrm{gdeg}\, x = \deg x$.

Corollary. Let $\{z_\beta = \sum_i \omega(a_{i\beta}) u_{\alpha_{i\beta}} \omega(b_{i\beta})\}_{\beta \in B}$ be a set of homogeneous cycles of $Sh_1(\underline{y},\bar{R})$ whose homology classes generate $H_1(Sh(\underline{y},\bar{R}))$. Then the basis $\underline{x}$ is standard iff all elements $x_\beta = \sum_i a_{i\beta} x_{\alpha_{i\beta}} b_{i\beta} \in I$, $\beta \in B$, have zero complete reduction.

Remark. If a basis $\underline{x}$ of the ideal I is not standard, add to $\underline{x}$ the complete reductions of all elements x_β from the corollary. Apply the same procedure to the resulting enlarged basis and iterate it: under proper assumptions of noetherity and effective computability of the 1-homology of the Shafarevich complex one obtains a standard bases of I after a finite number of steps.

Suppose now that the ring R and the semigroup Γ are commutative. Then the ring $\bar{R}$ is also commutative. Let us consider the Koszul complexes $K(\underline{x},R)$ and $K(\underline{y},\bar{R})$ endowed respectively with the natural Γ-filtration $\{F_\gamma K(\underline{x},R)\}_{\gamma\in\Gamma}$ and the natural Γ-grading $K(\underline{y},\bar{R}) = \bigoplus_{\gamma\in\Gamma} K(\underline{y},\bar{R})_\gamma$ defined analogously to the case of Shafarevich complexes. There is a natural isomorphism $K(\underline{y},\bar{R}) \cong \mathrm{gr}K(\underline{x},R)$ which allows us to define the morphism of Γ-graded complexes $\Psi: \bigoplus_{\gamma\in\Gamma} F_\gamma K(\underline{x},R) \to K(\underline{y},\bar{R})$. Denote by $E_*(K(\underline{y},\bar{R}))$ the Γ-graded $\bar{R}$-submodule $\mathrm{Im}\Psi_*$ of $H_*(K,(\underline{y},\bar{R}))$. We shall call the homogeneous elements of $E_*(K(\underline{y},\bar{R}))$ and the cycles representing them extendable classes and cycles. Note that a cycle $\bar{z} = \sum_\alpha \omega(a_\alpha) e_\alpha \in K_1(\underline{y},\bar{R})_\gamma$ is extendable iff there exists a $y^* \in F_\gamma^* K_1(\underline{x},R)$ such that $\sum_\alpha a_\alpha e_\alpha + y^*$ is a cycle.

Theorem 2. Let R be a commutative ring and let Γ be a commutative semigroup. For a basis $\underline{x} = \{x_\alpha\}_{\alpha \in A}$ of an ideal $I \subset R$ the following conditions are equivalent:

1) $\underline{x}$ is a standard basis;
2) $E_1(K(\underline{y},\bar{R}) = H_1(K(\underline{y},\bar{R}))$;
3) the R-module $H_1(K(\underline{y},\bar{R}))$ is generated by extendable classes.

The proof is completely similar to that of Theorem 1.

Corollary. Let R and Γ be as in Theorem 2 and let $\{z_\beta = \sum_\alpha \omega(a_{\alpha\beta})e_\alpha\}$ be a set of homogeneous cycles of $K_1(\underline{y},\bar{R})$ whose homology classes generate $H_1(K(\underline{y},\bar{R}))$. Then the basis $\underline{x}$ is standard iff all elements $x_\beta = \sum_\alpha a_{\alpha\beta} x_\alpha \in I$, $\beta \in B$, have zero complete reduction.

The remark on the construction of a standard basis holds in this situation as well.

4. Some applications.

The applications depend on the calculations of the 1-homology of the Shafarevich and Koszul complexes. We shall consider three examples.

Example 1. Let $\underline{y} = \{y_\alpha\}_{\alpha \in A}$ be a set of monomials in a free associative k-algebra R. By the expression "intersection between two monomials y_{α_1} and y_{α_2}" we describe one of the following situations: 1) there exist non-empty monomials h_1, p, h_2 such that $y_{\alpha_1} = h_1 p$, $y_{\alpha_2} = ph_2$; 2) $\alpha_1 \neq \alpha_2$ and there exist (possibly empty) monomials h_1, h_2 such that $y_{\alpha_1} = h_1 y_{\alpha_2} h_2$. With each intersection π of y_{α_1}, y_{α_2} we associate a 1-cycle z_π of the Shafarevich complex $Sh(\underline{y},R)$: $z_\pi = h_1 u_{\alpha_2} - u_{\alpha_1} h_2$ in the first case, and $z = h_1 u_{\alpha_2} h_2 - u_{\alpha_1}$ in the second one.

Claim: the set of homology classes of the cycles z_π over all intersections π generates the R-R-bimodule $H_1(Sh(\underline{y},R))$.

Let $z = \sum_{i=1}^{n} \lambda_i a_i u_{\alpha_i} b_i$ $(\lambda_i \in k)$ be any 1-cycle which may be assumed homogeneous relative to the grading of R by the semigroup of all monomials of R; this means that a_i, b_i are monomials and the monomials $a_i y_{\alpha_i} b_i$ are all equal to some monomial m. We use induction on the number n of summands in z to show that the homology class of z is contained in the R-R-subbimodule generated by the homology classes of z_π. The assertion is trivial for n = 0. If $n > 0$ then $n \geq 2$. If the subwords y_{α_1}, y_{α_2} of the monomial $m = a_1 y_{\alpha_1} b_1 = a_2 y_{\alpha_2} b_2$ do not overlap, for example $m = a_1 y_{\alpha_1} c y_{\alpha_2} b_2$ then the cycle $z - \lambda_1 d(a_1 u_{\alpha_1} c u_{\alpha_2} b_2)$ contains fewer than n summands and we can apply the induction. If y_{α_1}, y_{α_2} do overlap in m then one of the two situations described above takes place, that is $m = a_1 h_1 p h_2 b_2$ or $m = a_1 h_1 y_{\alpha_2} h_2 b_1$. We see that the cycle $z + \lambda_1 a_1 z_\pi b_2$ in the first case, or the cycle $z + \lambda_1 a_1 z_\pi b_1$ in the second one, contains fewer than n summands, and we can again apply the induction. Our claim is proved.

It should be noted that in the particular case where there are no intersections, the claim can be obtained from D. Anick's work in [6]. In Anick's notation, $(\tilde{A},d)$ is the Shafarevich complex, and his "combinatorially free" is precisely the absence of intersections. Now combine theorems 3.1 and 2.9 from [6].

Example 2. Let $\underline{y} = \{y_\alpha\}$ be a set of monomials in the polynomial algebra R over k. The expression "intersection between monomials y_{α_1} and y_{α_2}" describes the following situation: $\alpha_1 \neq \alpha_2$ and $y_{\alpha_1} = h_1 p$, $y_{\alpha_2} = h_2 p$ where $p = \text{g.c.d.}(y_{\alpha_1}, y_{\alpha_2}) \neq 1$. We associate with each intersection π of y_{α_1}, y_{α_2} the 1-cycle $z_\pi = h_1 e_{\alpha_2} - h_2 e_{\alpha_1}$ of the Koszul complex $K(\underline{y},R)$.

Claim: the set of homology classes of the cycles z_π over all intersections π generates the R-module $H_1(K(\underline{y},R))$.

The proof is completely similar to that in Example 1.

Example 3. Let R be a commutative k - algebra with a generating set $\{X_\beta\}_{\beta\in B}$, let $\underline{y} = \{Y_\alpha\}_{\alpha\in A}$ be a set of elements of R, and let $\{\sum_\alpha a_{\delta\alpha} e_\alpha\}_{\delta\in\Delta}$ be a set of 1-cycles of the Koszul complex $K(\underline{y},R)$ whose homology classes generate $H_1(K(\underline{y},R))$. We consider the following set of 1-cycles of the Shafarevich complex $Sh(\underline{y},R)$: $z_\delta = \sum_\alpha a_{\delta\alpha} u_\alpha$, $\delta\in\Delta$; $z_{\alpha\beta} = X_\beta u_\alpha - u_\alpha X_\beta = [X_\beta, u_\alpha]$, $\alpha\in A$, $\beta\in B$.

Claim: the set of homology classes of cycles z_δ, $\delta\in\Delta$, and $z_{\alpha\beta}$, $\alpha\in A$, $\beta\in B$, generates the R-R-bimodule $H_1(Sh(\underline{y},R))$.

Firstly, note that all 1-cycles of the form $[h, u_\alpha]$, $h\in R$, are contained in the R-R-subbimodule generated by $\{z_{\alpha\beta}\}_{\alpha\in A,\beta\in B}$. Hence the R-R-bimodule of the 1-cycles of the complex $Sh(\underline{y},R)$ is generated by $\{z_{\alpha\beta}\}_{\alpha\in A,\beta\in B}$ and by the cycles of the form $\sum_\alpha a_\alpha u_\alpha$. Since the homology classes of the cycles $\sum_\alpha a_{\delta\alpha}e_\alpha$, $\delta\in\Delta$, generate $H_1(K(\underline{y},R)$, each cycle $\sum_\alpha a_\alpha u_\alpha$ is contained in the left R-module generated by $\{z_\delta\}_{\delta\in\Delta}$ and by the cycles $y_{\alpha_1}u_{\alpha_2} - y_{\alpha_2}u_{\alpha_1}$, $\alpha_1, \alpha_2\in A$. But $y_{\alpha_1}u_{\alpha_2} - y_{\alpha_2}u_{\alpha_1} = y_{\alpha_1}u_{\alpha_2} - u_{\alpha_1}y_{\alpha_2} - [y_{\alpha_2},u_{\alpha_1}] = d(u_{\alpha_1}u_{\alpha_2}) - [y_{\alpha_2},u_{\alpha_1}]$. This completes the proof of our claim.

The confluence (or diamond) lemma of A.I. Shirshov - G. Bergman [1, 2] is an immediate consequence of the corollary of Theorem 1 and of the claim of Example 1. Similarly the corollary of Theorem 2 and the claim of Example 2 lead to the algorithm of B. Buchberger [3] for constructing Gröbner basis of polynomial ideals. Finally, from the corollary of Theorem 1 and the claim of Example 3 one can obtain the algorithm of V.N. Latyshev [4] for the construction of Gröbner bases of the ideals in universal enveloping algebras of Lie algebras.

References

1. A.I.Shirshov, Selected works. Nauka, Novosibirsk, 1984 (in Russian).
2. G.M.Bergman, The diamond lemma for ring theory, Adv. Math., 29, (1978), No 2, 178-218.
3. B.Buchberger, Gröbner bases: An algorithmic method in polynomial ideal theory. CAMP - Publ., No 83 - 29.0, Nov. 1983.
4. V.N.Latyshev, On the equality algorithm in Lie - nilpotent associative algebras, Vestn. Kiev. Univ., Math. Mech., 27, (1985), 67 (in Ukrainian).
5. E.S. Golod, I.R. Shafarevich, On the class - field tower. Izv. Akad. Nauk SSSR, Ser. Math., 1964, 28, No 2, 261-272 (in Russian).
6. D. Anick, Non-commutative graded algebras and their Hilbert series, J. Algebra 78 (1982), 120-140.

"This paper is in final form and no version of it will be submitted for publication elsewhere"

SEMISIMPLE SUPERALGEBRAS

Tadeusz Józefiak
Institute of Mathematics
Polish Academy of Sciences
Chopina 12, 87-100 Toruń, Poland

§1. Introduction

This paper grew out of an attempt to understand I.Schur's paper [9] on characters of projective representations of the symmetric and alternating groups, and to find a natural setting for an updated account of his theory that would explain and justify certain mysterious points in Schur's original exposition. This will be presented in detail in another article (see [5]).

We use here a fashionable terminology from supermathematics (see [6], [7],[8]) for what was earlier known as $\mathbb{Z}_2$-graded objects. It turned out that simple superalgebras were already classified by C.T.C.Wall in [10]. By using the formalism of the theory of semisimple algebras we get a class of semisimple superalgebras which enjoy nice properties. As an example we mention a relationship between simple supermodules over a semisimple superalgebra and simple modules over the underlying algebra which resembles familiar relations from A.H.Clifford's theory [2] applied to a finite group and its subgroup of index 2 (see Proposition (2.17)). General theory of semisimple superalgebras is presented in §2.

Two main classes of examples which should be kept in mind are : Clifford superalgebras (named after W.K.Clifford) and group superalgebras. We consider them in §3 and §4, respectively.

It turns out that a Clifford superalgebra of a nondegenerate quadratic form over an algebraically closed field is simple (see [10]). In the usual theory a Clifford algebra is either a matrix algebra or a product of two matrix algebras (hence not simple) depending on the rank of the quadratic form. This shows the advantage of the superalgebra approach when considering Clifford algebras. We give in §3 explicit isomorphisms using ideas of Schur from [9]. This is needed in [5] to describe basic spin characters of the symmetric group.

If G is a finite group and G_0 is a subgroup of index 2, then the group algebra of G can be equipped with a $\mathbb{Z}_2$-grading. The resulting superalgebra is semisimple if e.g. the coefficient field is of characteristic zero. The theory of group superalgebras developed in §4 is parallel to the classical representation theory of finite groups over $\mathbb{C}$.

If y is an element of order 2 contained in G_0 and in the center of G (we call it the distinguished central element of G), then we can study the so called negative supermodules over the superalgebra of G, i.e. supermodules on which y acts as multiplication by -1. This is exactly the context in which we can apply our results to get another outlook on the Schur's theory of characters of projective representations of the symmetric groups. There are some other applications of the theory presented here that we hope to employ elsewhere.

When writing §4B we were partly inspired by some ideas from a recent preprint of P.N.Hoffman and J.F.Humphreys [4].

§2. Structure theorems

Let K be a field. A superalgebra over K is a K-algebra A equipped with a grading $\{A_0, A_1\}$ such that $A_r A_s \subseteq A_{r+s}$ where $r,s \in \mathbb{Z}_2$ and r+s should be understood to be performed in $\mathbb{Z}_2 = \{0,1\}$ - a cyclic group of order 2.

Example (2.1) Let q be a quadratic form on a vector space W over K. Then the Clifford algebra C(q) can be endowed with a structure of a superalgebra over K by setting $C(q)_0$ to be a linear span over K of 1 and products $w_1 w_2 \dots w_p$, $w_r \in W$, p even, and $C(q)_1$ to be a linear span over K of products $w_1 w_2 \dots w_p$, $w_r \in W$, p odd. We call C(q) with this grading the Clifford superalgebra of q.

A supermodule over a superalgebra A (or an A-supermodule) is a K-vector space M equipped with a grading $\{M_0, M_1\}$ and a left action of A on M such that $A_r M_s \subset M_{r+s}$, $r,s \in \mathbb{Z}_2$.

Example (2.2) The field K with a grading $K_0=K$, $K_1=0$, can be treated as a superalgebra over K. A $\mathbb{Z}_2$-graded vector space M with the obvious action of K can be viewed as a supermodule over K.

If $x \in M_r$, then x is called a homogeneous element of M of degree r and we write d(x)= r. We denote by h(M) the set of all homogeneous elements of M, i.e. $h(M)= M_0 \cup M_1$. A map $f:M \to N$ between two A-supermodules is an A-homomorphism if

1) $f(M_r) \subset N_{r+p}$ for any $r \in \mathbb{Z}_2$ and for some $p \in \mathbb{Z}_2$ (p is called the degree of f in this case and we write d(f)= p),

2) f is A-linear, i.e. f is additive and

$$f(ax)=(-1)^{d(a)d(f)}af(x)$$

for $a \in h(A)$, $x \in M$.
We define $HOM_A(M,N)$ to be the K-supermodule whose set of homogeneous elements is the set of all A-homomorphisms from M to N. We have

$$HOM_A(M,N)_p = \text{the set of all A-homomorphisms of degree } p,\ p \in \mathbb{Z}_2.$$

We denote by A-MOD the category of all A-supermodules and their A-homomorphisms. Usual categorical notions from the category of modules over a ring carry over to A-MOD, e.g. that of a subsupermodule or a direct sum. A <u>left (right) superideal of A</u> is a subsupermodule of A viewed as an A-supermodule with respect to left (right) multiplications by elements of A. A subset of A which is both left and right superideal of A is called a <u>superideal of A</u>. We call an A-supermodule M <u>semisimple</u> if every A-subsupermodule of M is a direct summand and M is called <u>simple</u> if it has no nontrivial A-subsupermodules.

The proof of the following fact carries over almost verbatim from the ungraded case (see [3],p.87).

<u>Lemma (2.3)</u> The following statement about an A-supermodule M are equivalent :

(i) M is semisimple,

(ii) M is a direct sum of simple subsupermodules,

(iii) M is a sum (not necessarily direct) of simple subsupermodules.

A superalgebra A is called <u>simple</u> if it has no nontrivial superideals, i.e. different from zero and A.

<u>Proposition (2.4)</u> Let A be a superalgebra over a field K such that $\dim_K A < \infty$. The following are equivalent :

(i) every A-supermodule is semisimple,

(ii) A is a finite direct sum of left simple superideals,

(iii) A is a direct product of a finite number of simple superalgebras over K.

The proof in the ungraded case as presented in [3], §25, carries over with some adjustments to the superalgebra case. We point out the necessary changes. As in the ungraded case it follows from (ii) that $A = Ae_1 \oplus \ldots \oplus Ae_p$ where e_k are mutually orthogonal idempotents of degree zero and each Ae_k is a left simple superideal of A. When proving (ii)⇒(i) we write $M = \sum_{x \in h(M)} \sum_{k=1}^{p} Ae_k x$ for any A-supermodule M.

Next we define a mapping $f : Ae_k \to Ae_k x$ by

$$f(ae_k) = (-1)^{d(a)d(x)} ae_k x \qquad \text{for} \quad a \in h(A),\ x \in h(M)\ .$$

It is easily checked that f is an A-homomorphism of Ae_k onto $Ae_k x$. Since Ae_k is simple Ker f is either zero or Ae_k. Hence $Ae_k x$ is either simple or zero, i.e. M is a sum of simple A-supermodules and by Lemma (2.3) M is semisimple. To prove (ii)$\Rightarrow$(iii) we proceed as in [3], (25.15), to get a decomposition into a direct product $A = B^{(1)} \times \ldots \times B^{(r)}$ where $B^{(k)}$ is the superideal of A which is a sum of all simple left superideals of A isomorphic to a given one (degree one isomorphisms are allowed). One easily checks that $B^{(k)}$ is a simple superalgebra over K. A proof of (iii)$\Rightarrow$(ii) is standard as in the ungraded case.

A superalgebra A satisfying equivalent conditions of Proposition(2.4) will be called semisimple.

Before we pass to the structure theorem for simple superalgebras over an algebraically closed field we give two examples.

Examples (2.5) 1) We define the superalgebra $M(r|s)$ over K to be the K-algebra of square (r+s)-matrices with entries in K equipped with the following grading :

$M(r|s)_0$ = all square (r+s)-matrices for which D=E=0 in the (r,s)-block form $\begin{pmatrix} C & D \\ E & F \end{pmatrix}$,

$M(r|s)_1$ = all square (r+s)-matrices for which C=F=0 in the above block form.

Obviously $M(r|s)$ is a simple superalgebra over K since the underlying algebra is simple.

2) Another example is given by the superalgebra $Q(n)$ .The underlying algebra is the subalgebra of the K-algebra of square 2n-matrices consisting of matrices of the form $\begin{pmatrix} C & D \\ D & C \end{pmatrix}$ in the (n,n)-block form. $Q(n)_0$ consists of all matrices from $Q(n)$ with D=0, and $Q(n)_1$ of all matrices from $Q(n)$ with C=0. In other terms, every element of $Q(n)$ can be written uniquely in the form $C+Dt$ where C,D are square n-matrices over K, $t = i\begin{pmatrix} 0 & I \\ I & 0 \end{pmatrix}$, I is the identity n-matrix, $i^2=-1$; moreover

$$(C + Dt)(C' + D't) = (CC' - DD') + (CD' + DC')t\ .$$

Straightforward computation shows that $Q(n)$ is a simple superalgebra.

We denote by $|A|$ <u>the underlying algebra of the superalgebra A</u> and by $|M|$ <u>the underlying module of the A-supermodule M</u> ; obviously $|A|$ is a K-algebra and $|M|$ is an $|A|$-module. We write <u>z(A) for the center of the algebra $|A|$</u>; it inherits from A the structure of a K-supermodule. We have $z(M(r|s)) = \{K,0\}$, $z(Q(n)) = \{K,K\}$. A map <u>$f : A \to B$ is a homomorphism of superalgebras</u> if $f \in HOM_K(A,B)_0$ and f preserves multiplication.

<u>Theorem (2.6)</u> (C.T.C.Wall [10]) Let A be a simple superalgebra over an algebraically closed field K.

1) If $z(A)_1 = 0$, then A is isomorphic to $M(r|s)$ for some r,s.

2) If $z(A)_1 \neq 0$, then A is isomorphic to $Q(n)$ for some n.

<u>Remark (2.7)</u> C.T.C.Wall proved more general result without the assumption on the field K. We reproduce here, for the sake of completeness, the proof from [10] with some simplifications due to the hypothesis that K is algebraically closed.

For the proof of the Theorem we will need an auxiliary result.

<u>Lemma (2.8)</u> Let A be a simple superalgebra over K.

1) If J is a proper ideal of $|A|$ and $p_k : A \to A_k$, k=0,1, are natural projections, then they induce isomorphisms $p_k|J$ of vector spaces $p_k|J : J \to A_k$ and $J \cap A_k = 0$ for k=0,1.

2) If $z(A)=K$, i.e. $z(A)_0 = K$, $z(A)_1 = 0$, then $|A|$ is a simple K-algebra.

<u>Proof.</u> First notice that if I is a non-zero ideal of A_0, then

$$I + A_1 I A_1 = A_0 , \qquad A_1 I + I A_1 = A_1 \tag{2.1}$$

Indeed, the set $L = I + IA_1 + A_1 I + A_1 I A_1$ is closed under left and right multiplications by elements of A_0 and A_1, hence it is an ideal of $|A|$. Since L is graded it is a superideal of A. However A is simple and $I + A_1 I A_1 \subset A_0$, $A_1 I + I A_1 \subset A_1$ so that we get (2.1).

Passing to the proof of 1) we note that $J \cap A_0$ and $p_0(J)$ are ideals of A_0. If they are equal then J is a superideal of A contradicting simplicity ; hence $J \cap A_0 \subsetneqq p_0(J)$. Moreover $A_1 (J \cap A_0) A_1 \subset J \cap A_0$ and $A_1 p_0(J) A_1 \subset p_0(J)$ which shows in view of (2.1) that $J \cap A_0$ and $p_0(J)$ cannot be proper ideals of A_0 ; therefore $J \cap A_0 = 0$, $p_0(J) = A_0$.
We have $A_1^2 = A_0$ since otherwise $\{A_1^2, A_1\}$ would be a proper superideal of A. Hence

$$J \cap A_1 = A_0 (J \cap A_1) = A_1^2 (J \cap A_1) \subset A_1 (J \cap A_0) = 0.$$

Since $p_1(J) \supset A_1 p_0(J) = A_1 A_0 = A_1$ we have $p_1(J) = A_1$.

2) Assume that $|A|$ is not simple and let J be a proper ideal of $|A|$. Denote $w = p_0^{-1}(1) \in J$ and $u=p_1(w)\in A_1$, $u \neq 0$. We have $w=1+u$ and $xw=x+xu$, $wx=x+ux$ for $x\in h(A)$; hence $xu=ux$ by 1) and $u\in z(A)_1$. From $z(A)_1 =0$ we get $u=0$ - a contradiction.

Proof of Theorem (2.6) 1) By Lemma (2.8) $|A|$ is a simple algebra over K, i.e. $|A|$ is a matrix algebra over K since K is assumed to be algebraically closed. Consider an automorphism $f : |A| \rightarrow |A|$ defined by setting $f(a_k)= (-1)^k a_k$, $a_k\in A_k$, $k=0,1$. Since every automorphism of a matrix algebra is inner there exists $u\in|A|$ such that $f(x)=uxu^{-1}$ for any $x\in|A|$. Since $f(u)=u$ we infer that $u\in A_0$. Notice that $f^2=Id$, i.e. $u^2\in z(A)_0=K$. We can assume $u^2=1$ since K is algebraically closed. Then

$$A_0 = \{a\in|A|;\ au=ua\}\ , \qquad A_1 = \{a\in|A|;\ au=-ua\}$$

so that the grading is determined by the choice of u. Since by a change of basis we can take for u a diagonal matrix with r one's and s minus one's on the main diagonal the superalgebra A is isomorphic to $M(r|s)$.

2) Let $w\in z(A)_1$, $w\neq 0$; then $w^2=a\in z(A)_0=K$ and $a\neq 0$. Indeed $w^2=0$ would imply that the annihilator of w is a proper superideal of A - contrary to the assumption. Since K is algebraically closed we can assume that $a=-1$, i.e. $w^2=-1$. We have $z(A)_1=(z(A)_1w)w \subset z(A)_0w=Kw \subset z(A)_1$ hence $z(A)_1=Kw$, i.e. $z(A)=K+Kw$. Moreover $A_1=(A_1w)w \subset A_0w \subset A_1$, i.e. $A_1=A_0w$. We claim that A_0 is a simple algebra over K. Indeed if I is a nontrivial ideal of A_0 then $I+Iw$ is a nontrivial superideal of A - a contradiction. By the structure theorem for simple algebras A_0 is a matrix algebra over K, say A_0 is isomorphic to an algebra of square n-matrices over K. Since $A_1=A_0w$ and $w\in z(A)$ we infer that A is isomorphic to the superalgebra $Q(n)$.

By the center $Z(A)$ of the superalgebra A we mean the K-subsupermodule of A such that

$$Z(A)_p = \{a\in A_p\ ;\ ab = (-1)^{d(a)d(b)} ba \quad \text{for any} \quad b\in h(A)\}.$$

Obviously $K \subset Z(A)_0$. If $Z(A)= K$ the superalgebra A is called central.

A straightforward calculation and Theorem (2.6) imply

Corollary (2.9) Every simple superalgebra over an algebraically closed field is central.

If A and B are superalgebras over K, then their tensor product over K is a superalgebra C such that $C_0 = A_0\otimes B_0 + A_1\otimes B_1$, $C_1 = A_0\otimes B_1 + A_1\otimes B_0$ with multiplication

$$(a\otimes b)(a'\otimes b') = (-1)^{d(b)d(a')} aa'\otimes bb'$$

for $a,a' \in h(A)$, $b,b' \in h(B)$.

Wall proved in [10] that the tensor product of simple superalgebras over any field is again a simple superalgebra. In our situation we have

Proposition (2.10) There exist isomorphisms of superalgebras

1) $M(r|s) \otimes M(p|q) \approx M(rp+sq|rq+sp)$,
2) $M(r|s) \otimes Q(n) \approx Q(rn+sn)$,
3) $Q(m) \otimes Q(n) \approx M(mn|mn)$.

Proof. To prove 1) let us consider a K-supermodule N such that $\dim_K N_0 = r$, $\dim_K N_1 = s$ and write $END(N)=HOM_K(N,N)$. END(N) becomes a superalgebra over K by defining multiplication to be superposition of maps and $END(N) \approx M(r|s)$. Let P be another K-supermodule with $\dim_K P_0 = p$, $\dim_K P_1 = q$ so that $END(P) \approx M(p|q)$. We have a map

$$\alpha : END(N) \otimes END(P) \longrightarrow END(N\otimes P) ,$$

where $N\otimes P$ denotes tensor product of K-supermodules, defined by

$$\alpha(f\otimes g)(x\otimes y) = (-1)^{d(g)d(x)} f(x)\otimes g(y)$$

for homogeneous f,g,x,y. It is easily checked that α is a well-defined homomorphism of superalgebras over K and that it is in fact an isomorphism. This imply formula 1).

To prove 2) notice first that $Q(n) \approx M(n|0) \otimes Q(1)$ which follows directly from the definition of Q(n). Using this and 1) we get 2) by the associativity of the tensor product.

The formula 3) follows from 1), 2) and the fact that $Q(1)\otimes Q(1) \approx M(1|1)$. If $Q(1) = K+Kt$, $t^2=-1$, then the last isomorphism is established by the map

$$1\otimes 1 \mapsto \begin{pmatrix} 1 & 0 \\ 0 & 1 \end{pmatrix}, \quad t\otimes 1 \mapsto \begin{pmatrix} 0 & i \\ i & 0 \end{pmatrix}, \quad 1\otimes t \mapsto \begin{pmatrix} 0 & -1 \\ 1 & 0 \end{pmatrix}, \quad t\otimes t \mapsto \begin{pmatrix} i & 0 \\ 0 & -i \end{pmatrix}$$

where $i^2=-1$.

We record the following facts for future reference.

Proposition(2.11) 1) We have a decomposition of M(r|s) into a sum of left simple superideals

$$M(r|s) = I_1 \oplus \dots \oplus I_r \oplus I_{r+1} \oplus \dots \oplus I_{r+s} ,$$

where $I_k = M(r|s)E_{k,k}$ and $E_{k,k}$ is an (r+s)-matrix with 1 at the (k,k) place and zero elsewhere.

2)

$$HOM_{M(r|s)}(I_k,I_l) \approx \begin{cases} \{K,0\} & \begin{array}{c} 1 \le k,l \le r \\ \text{or} \\ r+1 \le k,l \le r+s \end{array} \\ \{0,K\} & \begin{array}{c} 1 \le k \le r \\ r+1 \le l \le r+s \\ \text{or} \\ r+1 \le k \le r+s \\ 1 \le l \le r \end{array} \end{cases}$$

3) Every simple supermodule over $M(r|s)$ is isomorphic to one of I_k.

4) We have a decomposition of $Q(n)$ into a sum of left simple superideals

$$Q(n) = J_1 \oplus \dots \oplus J_n$$

where $J_k=I_k+I_kt$, I_k is a simple superideal of $M(n|0)$ defined in 1), and we use the isomorphism $Q(n)\approx M(n|0)\otimes Q(1)$, $Q(1)=K+Kt$, $t^2=-1$.

5) $HOM_{Q(n)}(J_k,J_l)\approx \{K,K\}$.

6) Every simple supermodule over $Q(n)$ is isomorphic to one of J_k.

As a consequence of the preceding results we obtain the following facts concerning a semisimple superalgebra A over an algebraically closed field K.

Corollary (2.12) We have an isomorphism of superalgebras over K

$$A \approx \prod_{i=1}^{m} M(r_i|s_i) \times \prod_{j=1}^{q} Q(n_j) \tag{2.2}$$

and $m=m(A)$, $q=q(A)$ are invariants of A.

Corollary (2.13) Every supermodule over A is a direct sum of simple supermodules.

As in the ungraded case every simple supermodule P over A is annihilated by all but one factor in (2.2). We say that P is of type M if this factor is of the form $M(r_i|s_i)$ and of type Q if this factor is of the form $Q(n_j)$.

Corollary (2.14) Every simple A-supermodule is isomorphic to a simple superideal of A. Hence the number of non-isomorphic simple supermodules over A is equal to $m(A)+q(A)$.

Corollary (2.15) If N and P are simple A-supermodules then

$$\dim_K \mathrm{HOM}_A(N,P) = \begin{cases} 1 & \text{if } N\ P \text{ is of type } M, \\ 2 & \text{if } N\ P \text{ is of type } Q, \\ 0 & \text{if } N \not\approx P\ . \end{cases}$$

Corollary (2.16) The algebra $|A|$ is semisimple over K and the number of non-isomorphic simple $|A|$-modules is equal to $m(A)+2q(A)$.

Proof. This follows from the fact that the algebras of the form $|Q(n)|$ decompose into a product of two matrix algebras over K.

One can get a deeper insight into the relationship between Corollaries (2.14) and (2.16).

Consider an involution α on A defined by $\alpha(a_k)=(-1)^k a_k$, $a_k \in A_k, k=0,1$. It induces an operation on A-supermodules which assigns to a supermodule N a supermodule N' where N=N' as K-supermodules and the action of A on N' is defined by $a\ x=\alpha(a)x$. Obviously α induces an algebra involution on $|A|$ and for any $|A|$-module P we define P' exactly as stated above. Whereas $N\approx N'$ for any A-supermodule N, the map $N \to N'$, $x_k \mapsto (-1)^k x_k$, $x_k \in N_k, k=0,1$, is the required isomorphism, it is not so for all $|A|$-modules.

Proposition (2.17) 1) If P is a simple A-supermodule of type M, then $|P|$ is a simple $|A|$-module and $|P|' \approx |P|$ over $|A|$.

2) Let char $K \neq 2$. If P is a simple A-supermodule of type Q, then there exists a simple $|A|$-module N such that $|P| \approx N \oplus N'$ and $N \not\approx N'$ over $|A|$.

3) With the notation as in 2) there exists an isomorphism of A-supermodules $P \approx D(N)$ where

$$D(N)_0 = \{(x,x) \in N \oplus N' \ ;\ x \in N\} \quad , \quad D(N)_1 = \{(x,-x) \in N \oplus N' \ ;\ x \in N\}$$

with the action of A on D(N) induced from that of $|A|$ on $N \oplus N'$.

Proof. The assertions in 1) are obvious since $|A|$ is a matrix algebra over K. To prove 2) it is enough to assume that $P=J_k$ is one of superideals of a direct factor Q(n) of A. Recall (Proposition (2.11)) that $J_k=I_k+I_k t$ where I_k is the simple superideal of $M(n|0)$ and $Q(n) = M(n|0) + M(n|0)t$, $t^2=-1$. Since $|Q(n)|$ is a product of two matrix algebras and the corresponding central idempotents are $f_1=(1+it)/2$, $f_2=(1-it)/2$ we have $|J_k| \approx N_1 \oplus N_2$ where $N_1=I_k f_1$, $N_2=I_k f_2$. Notice that $N_2' \approx N_1$ over $|A|$ so that setting $N=N_1$ we get the required isomorphism $|J_k| \approx N \oplus N'$. Obviously $N \not\approx N$ over $|Q(n)|$, and hence over $|A|$, since N,N' are contained in distinct matrix algebra factors of $|Q(n)|$. The statement of 3) follows from the formulas $a=af_1+af_2$, $bit=bf_1-bf_2$ for $a \in I_k=(J_k)_0$, $b \in I_k$, $bit \in (J_k)_1$.

§3. Clifford superalgebras

In this section we return to Clifford superalgebras, see Example (2.1). Let K be an algebraically closed field and let q be a nondegenerate quadratic form on a vector space W of rank n over K. We write C(n) instead of C(q) for the Clifford superalgebra of q over K. The well-known structure theorem for Clifford algebras takes the following form in our setup and shows that it is the proper language.

Proposition (3.1) We have isomorphisms of superalgebras

1) $C(2k) \approx M(2^{k-1}|2^{k-1})$,
2) $C(2k+1) \approx Q(2^k)$.

onsequently the Clifford superalgebra C(n) is simple for every n.

Proof I. It is known (see e.g. [1]) that the Clifford superalgebra of an orthogonal direct sum of quadratic forms is isomorphic to the tensor product of Clifford superalgebras of the summands. This fact, Proposition (2.10) and an observation that $C(1) \approx Q(1)$ complete the proof.

Proof II. We are going to define suitable explicit isomorphisms which are used in [5] to compute the basic spin character of the symmetric group.

Since K is algebraically closed there exists a basis $e_1, e_2, \ldots, e_{2k}$ of W which is orthogonal with respect to q and such that $e_j^2=-1$, $j=1,2,\ldots,2k$, in C(2k). Let U be a K-supermodule such that $U_0=Ku_0$, $U_1=Ku_1$. We define an explicit isomorphism $\beta : C(2) \xrightarrow{\approx} END(U) \approx M(1|1)$. To this end notice that C(2) is generated as a superalgebra over K by e_1, e_2 which satisfy relations $e_1^2=e_2^2=-1$, $e_1e_2=-e_2e_1$. We define $D_1, D_2 \in END(U)$ by setting $D_1(u_0)=iu_1$, $D_1(u_1)=iu_0$, $D_2(u_0)=u_1$, $D_2(u_1)=-u_0$. A straightforward computation shows that the map β determined by $\beta(e_p)=D_p$, $p=1,2$, is an isomorphism.

For arbitrary k we have an isomorphism $\gamma : C(2k) \xrightarrow{\approx} \otimes^k C(2)$ whose existence follows from the fact that the Clifford superalgebra of the orthogonal direct sum is the tensor product of Clifford superalgebras of the summands. Explicitly

$$\gamma(e_{2j-1})=1\otimes \ldots \otimes e_1 \otimes \ldots \otimes 1, \quad \gamma(e_{2j})=1\otimes \ldots \otimes e_2 \otimes \ldots \otimes 1, \qquad j=1,\ldots,k,$$

and e_1, e_2 are located in the j-th place of the tensor product.

Next we consider the isomorphisms

$$\otimes^k \beta : \otimes^k C(2) \xrightarrow{\approx} \otimes^k END(U) \quad \text{and} \quad \alpha : \otimes^k END(U) \xrightarrow{\approx} END(\otimes^k U) ,$$

where the latter is an iteration of a map defined in the proof of Proposition (2.10). Explicitly

$$\alpha(f_1\otimes \dots \otimes f_k)(x_1\otimes \dots \otimes x_k)=(-1)^r f_1(x_1)\otimes \dots \otimes f_k(x_k)$$

where $x_j \in h(U)$, $f_j \in h(END(U))$ and $r = \sum_{s=2}^{k} d(f_s) \sum_{j=1}^{s-1} d(x_j)$.

The composition of the three isomorphisms gives the required isomorphism $C(2k) \approx END(\otimes^k U) \approx M(2^{k-1}|2^{k-1})$. Explicit action of the generators e_j on $\otimes^k U$ is as follows :

$$e_{2j-1}(x_1\otimes \dots \otimes x_k)=(-1)^{\sum_{s=1}^{j-1} d(x_s)} x_1\otimes \dots \otimes D_1(x_j)\otimes \dots \otimes x_k \ ,$$

$$e_{2j}(x_1\otimes \dots \otimes x_k)=(-1)^{\sum_{s=1}^{j-1} d(x_s)} x_1\otimes \dots \otimes D_2(x_j)\otimes \dots \otimes x_k \ , \qquad (3.1)$$

for j=1,...,k. This determines the structure of the only (up to an isomorphism) simple C(2k)-supermodule $\otimes^k U$ over K ; obviously $\dim_K \otimes^k U = 2^k$.

The formula 2) can easily be obtained from previous results : $C(2k+1) \approx C(2k)\otimes C(1) \approx M(2^{k-1}|2^{k-1})\otimes Q(1) \approx Q(2^k)$. To get the natural description of a simple C(2k+1)-supermodule we embed C(1) into END(U) by $1 \mapsto Id$, $e_1 \mapsto D_1$. Then we have the embedding

$$C(2k+1) \approx C(2k)\otimes C(1) \hookrightarrow END(\otimes^k U)\otimes END(U) \approx END(\otimes^{k+1} U)$$

which defines the structure of C(2k+1)-supermodule on $\otimes^{k+1} U$. Explicitly we have

$$e_{2j-1}(x_1\otimes \dots \otimes x_{k+1})=(-1)^{\sum_{s=1}^{j-1} d(x_s)} x_1\otimes \dots \otimes D_1(x_j)\otimes \dots \otimes x_{k+1}$$

for j=1,...,k+1,

$$e_{2j}(x_1\otimes \dots \otimes x_{k+1})=(-1)^{\sum_{s=1}^{j-1} d(x_s)} x_1\otimes \dots \otimes D_2(x_j)\otimes \dots \otimes x_{k+1} \qquad (3.2)$$

for j=1,...,k.

The K-supermodule $\otimes^{k+1} U$ is a simple C(2k+1)-supermodule since $\dim_K \otimes^{k+1} U = 2^{k+1}$ and this is the smallest possible dimension of non-zero supermodules over $C(2k+1) \approx Q(2^k)$.

It is now obvious in view of Proposition (3.1) and definitions of simple superalgebras of type M and Q that

Corollary (3.2) We have $C(n)_0 \approx |C(n-1)|$.

§4. Semisimple group superalgebras

4A. Arbitrary supermodules

Let G be a finite group and G_0 a subgroup of G of index 2 ; we denote $G_1 = G \setminus G_0$. The group algebra of G over a field K can be endowed with a $\mathbb{Z}_2$-grading by setting

$$\deg \left(\sum_{g \in G} n_g g \right) = \begin{cases} 0 & \text{if } n_g = 0 \text{ for } g \in G_1 , \\ 1 & \text{if } n_g = 0 \text{ for } g \in G_0 . \end{cases}$$

We call the resulting object the group superalgebra of the pair G, G_0 over K and denote it by $K[G,G_0]$. If $A = K[G,G_0]$, then $\dim_K A_0 = \dim_K A_1 = \#(G_0)$ where $\#(X)$ means the cardinality of a finite set X.

We have the counterpart of Maschke's Theorem for group superalgebras.

Theorem (4.1) If the characteristic of K does not divide the order of G, then the group superalgebra $K[G,G_0]$ is a semisimple superalgebra.

The proof consists in showing that every $K[G,G_0]$-supermodule is semisimple and the arguments given in [3],p.41, go over unchanged.

From Corollary (2.12) and Theorem (4.1) we get

Corollary (4.2) If K is an algebraically closed field and the characteristic of K does not divide the order of a finite group G, then we have an isomorphism of superalgebras over K

$$K[G,G_0] \approx \prod_{i=1}^{m} M(r_i|s_i) \times \prod_{j=1}^{q} Q(n_j) \qquad (4.1)$$

and $m = m(G,G_0)$, $q = q(G,G_0)$ are invariants of the pair G, G_0.

From now on we assume that K satisfies the hypotheses of Corollary (4.2).

Corollary (4.3) We have a decomposition

$$K[G,G_0] \qquad \bigoplus_{i=1}^{m} (\dim_K N_i) N_i \oplus \bigoplus_{j=1}^{q} (\tfrac{1}{2}\dim_K P_j) P_j \qquad (4.2)$$

of the regular supermodule of the superalgebra $K[G,G_0]$ as a sum of simple supermodules where N_i are of type M and P_j are of type Q. Every simple supermodule over $K[G,G_0]$ is isomorphic to a simple component appearing in (4.2).

Proof. The proof follows from Corollary (4.2), Proposition (2.11) and Corollary (2.14).

Corollary (4.4) If a superalgebra $M(r|s)$ occurs as a factor in (4.1), then $r=s$.

Proof. Let $A=K[G,G_0]$; if $g\in G_1$, then $G_1=G_0g$ and multiplication by g defines a K-vector space isomorphism $A_0\approx A_1$ for A as well as for any factor in (4.1). Hence $\dim_K M(r|s)_0 = \dim_K M(r|s)_1$,i.e. $r^2+s^2=2rs$ and $r=s$.

Corollary (4.5) If $A=K[G,G_0]$, then

1) the number of non-isomorphic simple A-supermodules is equal to $m+q$,
2) the number of non-isomorphic simple $|A|$-modules is equal to $m+2q$,
3) the number of non-isomorphic simple A_0-modules is equal to $2m+q$.

Proof. 1) and 2) follow from Corollaries (2.14) and (2.16). 3) follows from (4.1), Corollary (4.4) and the fact that $M(r|r)_0$ is a product of two matrix algebras whereas $Q(n)_0$ is a matrix algebra.

We are going to interpret the invariants m and q in terms of conjugacy classes.

Definition (4.6) We set c_p to be the number of conjugacy classes of G contained in G_p, $p=0,1$.

Proposition (4.7) We have $c_0=m+q$, $c_1=q$ so that $c_0 \geq c_1$.

Proof. We write $A=K[G,G_0]$ and consider the K-supermodule $z(A)$ - the center of the algebra $|A|$. By (4.1) we get $\dim_K z(A)_0=m+q$, $\dim_K z(A)_1=q$ since $z(M(r|s))=\{K,0\}$, $z(Q(n))=\{K,K\}$. On the other hand $z(A)_p$ is freely generated over K by elements $\sum_{x\in C} x$ for all conjugacy classes C of G contained in G_p, $p=0,1$. Hence $\dim_K z(A)_p=c_p$, $p=0,1$, and the required equalities follow.

Proposition (4.8) A conjugacy class of G contained in G_0 can either become a conjugacy class of G_0 or split into the union of two conjugacy classes of G_0 of the same cardinality.

Proof. If $x\in G_0$, then we write $C(x)$ for the conjugacy class of x with respect to G, $C_0(x)$ the conjugacy class of x with respect to G_0, $z(x)$ the centralizer of x in G, $z_0(x)$ the centralizer of x in G_0. We have $C_0(x) \subset C(x)$, $z_0(x) \subset z(x)$ and

$$\#(C(x))\#(z(x))=\#(G) \quad , \quad \#(C_0(x))\#(z_0(x))=\#(G_0)=\frac{1}{2}\#(G).$$

If $z(x)=z_0(x)$, then $\#(C(x))=2\#(C_0(x))$ and $C(x)$ splits into two G_0-conjugacy classes of the same cardinality. If $z(x) \neq z_0(x)$, then there exists

$u\in z(x)\cap G_1$. If $h=gxg^{-1}$, then $(gu)x(gu)^{-1}=gxg^{-1}=h$ so that either g or gu is in G_0, i.e. $C(x)=C_0(x)$.

Corollary (4.9) The number d_0 of conjugacy classes of G contained in G_0 which split into two conjugacy classes with respect to G_0 is equal to m.

Proof. A_0 is the group algebra of G_0 over K so that we have $c_0+d_0 = 2m+q$ by Corollary (4.5). Since by Proposition (4.7) $c_0=m+q$ we get $d_0=m$.

Let $R(G,G_0)$ be the Grothendieck group of the category $K[G,G_0]$-MOD of supermodules over $K[G,G_0]$. It is a free abelian group with a basis consisting of classes of simple supermodules. The rank of $R(G,G_0)$ is $m+q$ by 1) of Corollary (4.5). The group $R(G,G_0)$ is equipped with an involution. It is induced by the mapping $N\longmapsto N^*$ where $N^*=HOM_K(N,K)$, N is an A-supermodule and K is treated here as K-supermodule $\{K,0\}$. N^* is made into A-supermodule by setting $(g\phi)(x)=\phi(g^{-1}x)$ for $g\in G$, $\phi\in N^*$, $x\in N$.

From now on we assume that $K=\mathbb{C}$ is the field of complex numbers and we write $A=\mathbb{C}[G,G_0]$ for short.

We define the pairing

$$[N,P] = \dim_{\mathbb{C}} HOM_A(N,P) \qquad (4.3)$$

for $N,P \in R(G,G_0)$.

We get another interpretation of (4.3) by considering the action of G on $HOM_{\mathbb{C}}(N,P)_r$, $r=0,1$. We set

$$(g\phi)(x) = (-1)^{d(g)d(\phi)} g\phi(g^{-1}x)$$

for $\phi\in HOM_{\mathbb{C}}(N,P)_r$, $g\in G$, $x\in N$.

In this way each $HOM_{\mathbb{C}}(N,P)_r$, $r=0,1$, becomes a G-module and it follows just from definitions that

$$HOM_A(N,P)_r = HOM_{\mathbb{C}}(N,P)_r^G = \{\ \phi\in HOM_{\mathbb{C}}(N,P)_r\ ;\ g\phi=(-1)^{d(g)d(\phi)}\phi,\ \forall\ g\in G\},$$

i.e.

$$[N,P] = \dim_{\mathbb{C}} HOM_{\mathbb{C}}(N,P)^G \qquad (4.4)$$

As in the usual (non-graded) case one can easily check that $[\ ,\ ]$ is a scalar product on $R(G,G_0)$.

From Corollary (2.15) it follows

Corollary (4.10) If N,P are simple A-supermodules, then

$$[N,P] = \begin{cases} 1 & \text{if } N\approx P \text{ is of type } M, \\ 2 & \text{if } N\approx P \text{ is of type } Q, \\ 0 & \text{if } N\not\approx P\ . \end{cases}$$

We call <u>the canonical basis</u> of $R(G,G_0)$ the basis consisting of classes of all simple A-supermodules . Here is its characterization in terms of the scalar product.

<u>Proposition (4.11)</u> Let Z be a basis of $R(G,G_0)$ consisting of pairwise orthogonal elements of $R(G,G_0)$ such that $Z=X\cup Y$ and $[x,x]=1$ for $x\in X$, $[y,y]=2$ for $y\in Y$. Then Z is (up to a sign of its elements) the canonical basis of $R(G,G_0)$.

<u>Proof.</u> We have to show that $[y,y]=2$ implies that $\pm y$ is the class of a simple supermodule ; the rest is standard. Assume that $y=u_1+u_2$ for some $u_1,u_2\in R(G,G_0)$, $[u_1,u_1]=1$, $[u_1,u_2]=0$, and write u_1 in the basis Z : $u_1=ay+\sum_{z\in Z\setminus\{y\}} bz$, $a,b\in\mathbb{Z}$. Then $1=[u_1,y]=a[y,y]=2a$ - a contradiction.

For any A-supermodule N one can consider the function χ^N ; $\chi^N(g)$ is the trace of the left multiplication by g on N. Obviously

$$\chi^N(g) = \chi^N(hgh^{-1}) \qquad \text{for } g,h\in G ;$$

moreover

$$\chi^N(g) = 0 \qquad \text{for } g\in G_1,$$

since $gN_r \subset N_{r+1}$ for $g\in G_1$.

Inspired by this we consider the space $CF(G,G_0)$ of functions $f:G\longrightarrow\mathbb{C}$ such that

$$f(g) = f(hgh^{-1}) \qquad \text{for } g,h\in G,$$

$$f(g) = 0 \qquad \text{for } g\in G_1 .$$

Notice that $CF(G,G_0)$ is a $\mathbb{C}$-vector space of dimension c_0 (see Definition (4.6)) and by Corollary (4.7) c_0 is equal to m+q, i.e. to the rank of $R(G,G_0)$. We have an involution on $CF(G,G_0)$ which sends f to $\bar{f}$ where $\bar{f}(g)=\overline{f(g)}$ and the bar denotes the complex conjugation. We define

$$[f_1,f_2] = \frac{1}{\#(G)}\sum_{x\in G}\bar{f}_1(x)f_2(x) = \frac{1}{\#(G)}\sum_{x\in G_0}\bar{f}_1(x)f_2(x)$$

for $f_1,f_2\in CF(G,G_0)$ and it is easily checked that this determines a scalar product on $CF(G,G_0)$. Here is the basic relationship between $R(G,G_0)$ and $CF(G,G_0)$.

<u>Theorem (4.12)</u> The mapping $\chi : R(G,G_0)\longrightarrow CF(G,G_0)$ defined by $\chi(N)=\chi^N$ is a homomorphism of groups preserving involution and scalar product. The map χ is an injection so that it identifies $R(G,G_0)$ with the image of χ. This image is called <u>the character group of G,G_0</u>. The images by χ of simple supermodules form an orthogonal basis of $CF(G,G_0)$.

Proof. We indicate some arguments leading to the proof that χ preserves scalar product ; the other assertions are straightforward.

We recall that for a G-module V we have $\dim_{\mathbb{C}} V^G = \frac{1}{\#(G)} \sum_{g\in G} \operatorname{tr}(g)$ where $V^G = \{ v\in V \; ; \; gv=v$ for every $g\in G \}$. We apply this formula to the G-module $V = \mathrm{HOM}_{\mathbb{C}}(N,P)_0 + \mathrm{HOM}_{\mathbb{C}}(N,P)_1$. Since $V \simeq N^*\otimes P$ where the G-module structure on $N^*\otimes P$ is given by $g(x\otimes y)=gx\otimes gy$, $x\in N^*$, $y\in P$, we have

$$[N,P] = \dim_{\mathbb{C}} \mathrm{HOM}_{\mathbb{C}}(N,P)^G = \frac{1}{\#(G)} \sum_{g\in G} \chi^V(g) = \frac{1}{\#(G)} \sum_{g\in G} \chi^{N^*\otimes P}(g) =$$

$$\frac{1}{\#(G)} \sum_{g\in G_0} \chi^{N^*}(g)\chi^P(g) = \frac{1}{\#(G)} \sum_{g\in G_0} \bar{\chi}^N(g)\chi^P(g) = [\chi^N,\chi^P].$$

4B. Negative supermodules

Let G be a finite group, G_0 a subgroup of index 2 of G and y an element of G_0 of order 2 belonging to the center of G. We call y the distinguished central element of G.

An important example for which the resulting statements apply (see [5]) is the representation group $\tilde{S}_n$ of the symmetric group S_n . There exists a nontrivial double covering $\theta_n : \tilde{S}_n \longrightarrow S_n$, $(\tilde{S}_n)_0$ is the preimage of the alternating group and the distinguished central element y is the nontrivial element from $\theta_n^{-1}(1)$.

Let K be an algebraically closed field such that the superalgebra $A=K[G,G_0]$ is semisimple (see Theorem (4.1)).

Definition (4.13) A supermodule N over A is called negative if y acts on N as multiplication by -1. In the same way we define negative $|A|$-modules and A_0-modules.

We define $\bar{m}=\bar{m}(G,G_0,y)$ to be the number of simple negative A-supermodules of type M and similarly $\bar{q}=\bar{q}(G,G_0,y)$ to be the number of simple negative A-supermodules of type Q. Our next task is to express these invariants intrinsically in terms of G,G_0 and y.

If D is a conjugacy class in G, then so is yD. We have two possibilities : either $D=yD$ or $D\cap yD=\emptyset$. We define $\bar{c}_p$ to be the number of conjugacy classes D of G contained in G_p such that $D\cap yD=\emptyset$, p=0,1.

Proposition (4.14) We have $\bar{c}_0=\bar{m}+\bar{q}$, $\bar{c}_1=\bar{q}$, hence $\bar{c}_0 \geq \bar{c}_1$.

Proof. The element y acts on the center $z(A)$ of the algebra $|A|$.Let $\bar{z}(A)$ be a K-subsupermodule of $z(A)$ of all elements on which y acts as multiplication by -1. We have $\dim_K \bar{z}(A)_0=\bar{m}+\bar{q}$, $\dim_K \bar{z}(A)_1=\bar{q}$ directly from definitions and (4.1). On the other hand, we write down all conjugacy classes of G as follows : D_1, yD_1, D_2, yD_2, ... , D_r, yD_r, D_{r+1} ... , D_s , where $D_k \cap yD_k = \emptyset$ for $1 \le k \le r$, $yD_k = D_k$ for $r+1 \le k \le s$. Let $d = \sum_{x \in D} x \in z(A)$ for each conjugacy class D. Then the elements d_1, yd_1, ... , d_r, yd_r, d_{r+1}, ... , d_s form a basis of z(A) over K. A matrix of multiplication by y in this basis has a diagonal block form with $\begin{pmatrix} 0 & 1 \\ 1 & 0 \end{pmatrix}$ corresponding to d_k, $1 \le k \le r$, and (1) corresponding to d_k, $r+1 \le k \le s$. Since $\begin{pmatrix} 0 & 1 \\ 1 & 0 \end{pmatrix}$ is similar to $\begin{pmatrix} 1 & 0 \\ 0 & -1 \end{pmatrix}$ we infer that $\bar{c}_p = \dim_K \bar{z}(A)_p$, p=0,1, which completes the proof.

<u>Corollary (4.15)</u> If $A=K[G,G_0]$ and y is distinguished central element of G_0, then

1) the number $\bar{m}+\bar{q}$ of non-isomorphic simple negative A-supermodules is equal to the number of conjugacy classes D of G contained in G_0 such that $D \cap yD = \emptyset$,

2) the number of non-isomorphic simple negative $|A|$-modules is equal to $\bar{m}+2\bar{q}$,

3) the number of non-isomorphic simple negative A_0-modules is equal to $2\bar{m}+\bar{q}$.

<u>Proof.</u> 1) is a reformulation of Proposition (4.14). 2) and 3) follow from (4.1) and Proposition (2.17).

In order to state the next result we define $H=G/\{1,y\}$ and consider the natural surjection $\theta : G \to H$. Let $h \mapsto h'$ be any section of θ, i.e. $\theta^{-1}(h)=\{h',yh'\}$. For $a,b \in G$ we write $a \sim b$ ($a \not\sim b$) if a and b are conjugate (are not conjugate) in G.

<u>Proposition (4.16)</u> Let C be a conjugacy class in H.

1) If for some $h \in C$ we have $h' \sim yh'$, then for any $g \in C$ we have $g' \sim yg'$ and $D=\theta^{-1}(C)$ is a conjugacy class in G; moreover $D=yD$ and $\#(D)=2\#(C)$.

2) If for some $h \in C$ we have $h' \not\sim yh'$, then for any $g \in G$ we have $g' \not\sim yg'$ and $\theta^{-1}(C)$ is a union of two conjugacy classes in G. If one of them is termed D, then the other one is yD so that $\theta^{-1}(C)=D \cup yD$ and $D \cap yD = \emptyset$; moreover $\#(D)=\#(yD)=\#(C)$.

<u>Proof.</u> 1) Let $g \in G$ and $g=uhu^{-1}$ for some $u \in H$. Then $\theta(u'h'u'^{-1})=uhu^{-1}=g$, i.e. $u'h'u'^{-1}$ equals either g' or yg'. Consequently, $g' \sim h' \sim yh' \sim yg'$ in the first case and $yg' \sim h' \sim yh' \sim g'$ in the second case. Since $h' \sim g'$ or $h' \sim yg'$ $D=\theta^{-1}(C)$ forms a conjugacy class in G and yD=D. Obviously $\#(D)=2\#(C)$.

2) Take $h \in C$ such that $h' \not\sim yh'$ and $g=uhu^{-1}$, $u \in H$; then $u'h'u'^{-1} \in \theta^{-1}(g)$.

Write $D=\{u'h'u'^{-1};\ u\in H\}$, $yD=\{yv\ ;\ v\in D\}$. Then $D\cup yD=\theta^{-1}(C)$ and $D\cap yD=\phi$; indeed, $u'h'u'^{-1}=yw'h'w'^{-1}$ implies $(w'^{-1}u')h'(w'^{-1}u')^{-1}=yh'$ thus contradicting the hypothesis $h'\not\sim yh'$. It is clear from the definition that D and yD are conjugacy classes in G ; the formula for $\#(D)$ is also obvious.

References

[1] M.Atiyah, R.Bott, A.Shapiro, Clifford modules, Topology 3(1964), 3-38

[2] A.H.Clifford, Representations induced in an invariant **subgroup,** Ann. Math. 38(1937), 533-550

[3] C.W.Curtis, I.Reiner, Representation theory of finite groups and associative algebras, Interscience Publishers, New York, 1962

[4] P.N.Hoffman, J.F.Humphreys, Hopf algebras and projective representations of $G\wr S_n$ and $G\wr A_n$, Can.J.Math. 27(1986), 1380-1458.

[5] T.Józefiak, Characters of projective representations of symmetric groups, Preprint, Toruń, 1987

[6] D.A.Leites, Introduction to supermanifolds, Uspekhi Mat. Nauk 35(1980), 3-57

[7] D.A.Leites, Theory of supermanifolds, Petrozavodsk, 1983 (in Russian)

[8] Yu.I.Manin, Gauge fields and complex geometry, Moscow, 1984 , (in Russian)

[9] I.Schur, Über die Darstellung der symmetrischen und der alternierenden Gruppe durch gebrochene lineare Substitutionen, J.Reine Angew.Math., 139(1911), 155-250

[10] C.T.C.Wall, Graded Brauer groups, J.Reine Angew.Math., 213(1964), 187-199

"This paper is is final form and no version of it
will be submitted for publication elsewhere"

ON THE LAWS OF FINITE DIMENSIONAL REPRESENTATIONS OF SOLVABLE LIE ALGEBRAS AND GROUPS

A.N. Krasil'nikov
Department of Mathematics and Physics
Moscow Regional Pedagogical Institute
Moscow 107 005, USSR

A.L. Šmel'kin
Department of Mathematics and Mechanics
Moscow State University
Moscow 119 899, USSR

1. INTRODUCTION

1.1. Representations of Lie algebras and their laws.

Let K be a field and V a vector space over K. Let $\underline{G}$ be a Lie algebra over K and $\rho: \underline{G} \to gl(V)$ be a representation of $\underline{G}$ on V. Suppose that $f = f(x_1,..,x_m)$ is an associative polynomial with coefficients in K (i.e. f is an element of the free associative algebra over K freely generated by $\{x_1,x_2,...\}$). We say that $f \equiv 0$ is a law of ρ (or a weak law in terms of [1]) if for all $g_1,..,g_m$ in $\underline{G}$ we have

$$f(\rho(g_1),..,\rho(g_m)) = 0.$$

In this case we also say that f itself is a law of ρ.

For example, if ρ is a nilpotent representation of dimension n of a certain Lie algebra, then by Engel's Theorem

$$x_1.x_2.\dots.x_n \equiv 0$$

is a law of ρ. Let now ρ be a triangular representation of dimension n and set $[x_1,x_2]_{(a)} = x_1x_2 - x_2x_1$; then

$$[x_1,x_2]_{(a)}[x_3,x_4]_{(a)} \cdots [x_{2n-1},x_{2n}]_{(a)} \equiv 0 \qquad (1)$$

is a law of ρ.

All terminology and basic facts relating to the laws of algebras can be found in [2] or [3]. The following questions are the most interesting versions of the Finite Basis Problem (see [2] or [3]) for representations of Lie algebras:

1. Are the laws of any representation of a Lie algebra over K finitely based?

2. Are the laws of any finite dimensional representation of a Lie algebra over K finitely based?

In the case of a field K of characteristic zero both these questions remain open. If the field K is of non-zero characteristic then the results of [4] and [5] imply that for any such K there exists an infinite dimensional representation of a solvable Lie algebra over K whose laws are not finitely based. Thus the first question has a negative answer in this case. But the results of [4] and [5] do not imply any answer to the second question.

The main result of the first part of our paper is that over any zero characteristic field any finite dimensional representation of a solvable Lie algebra has a finite basis for its laws. This provides a positive answer to the second question in the case of a solvable Lie algebra and of a zero characteristic field. In fact, we shall prove the following

THEOREM 1.1. Let ρ be a representation of a Lie algebra over a zero characteristic field, and let (1) be a law of ρ. Then ρ has a finite basis for its laws.

Note without proof that the answer to the second question for a field K of non-zero characteristic is generally 'no' even for triangular representations of Lie algebras. For let K be an infinite field of characteristic 2, $\underline{G}$ the Lie subalgebra of gl(4,K) generated by matrices $(E_{22}+E_{33})$, $(E_{12}+E_{34})$, $(E_{13}+E_{24})$, where E_{ij} is a matrix unit. Let ρ be the natural representation of $\underline{G}$ on the vector space of dimension 4 over K. Then the laws of the representation ρ are not finitely based. The Lie algebra theoretical origins of this result may be found in [4] and [5].

However, the answer to the second question in the case of a triangular representation and of a finite field K is 'yes'.

PROPOSITION 1.2. Let ρ be a finite dimensional triangular representation of a Lie algebra over a finite field. Then ρ has a finite basis for its laws.

The proof of this proposition uses the techniques of Cross varieties (see [3]).

1.2. Laws of group representations.

Let K be a field, F the free group freely generated by $\{x_1, x_2, \ldots\}$ and KF its group algebra over K. Let V be a vector space over K, G a group, $\rho: G \to \mathrm{Aut}(V)$ a representation of G on V. Suppose that $u = u(x_1, \ldots, x_m)$ is an element of KF. We say that $u \equiv 0$ is a law of ρ, if for all $g_1, \ldots, g_m$ in G we have

$$u(\rho(g_1), \ldots, \rho(g_m)) = 0.$$

Occasionally we shall also say that u itself is a law of ρ.

For example, if ρ is a unipotent representation of a group of dimension n then

$$(1-x_1)(1-x_2)\ldots(1-x_n) \equiv 0$$

is a law of ρ. Let now ρ be a triangular representation of dimension n and set $[x_1, x_2]_{(m)} = x_1^{-1}x_2^{-1}x_1x_2$; then

$$(1-[x_1,x_2]_{(m)})(1-[x_3,x_4]_{(m)})\ldots(1-[x_{2n-1},x_{2n}]_{(m)}) \equiv 0 \qquad (2)$$

is a law of ρ.

For basic facts and terminology relating to the laws of group representations see [6] and [7]. The results of [8 - 10] imply the existence over any field of infinite dimensional representations of groups such that their laws are not finitely based. Over infinite fields the existence problem for finite dimensional group representations with the same property is not yet solved.

The main result of the second part of our paper is that over any zero characteristic field laws of any finite dimensional triangular group representation are finitely based (we say that a representation ρ of a group G over K is triangular if it is equivalent to a representation of G by triangular matrices over the algebraic closure of K).

In fact, we shall prove the following

THEOREM 1.3. *Suppose that a group representation ρ over a field K satisfies (2). If K is of characteristic 0 or p, $p \geq n$, then the laws of ρ are finitely based.*

The proof of the following proposition is similar to the proof of Theorem 1.1 from [11].

PROPOSITION 1.4. *Let ρ be a group representation over some Noetherian associative and commutative ring K, $1 \in K$. Suppose that ρ satisfies the law*

$$(1-[x_1,x_2]_{(m)})(1-[x_3,x_4]_{(m)})(1-[x_5,x_6]_{(m)}) \equiv 0.$$

Then the laws of ρ are finitely based.

In particular, the laws of any triangular group representation of dimension two or three are finitely based.

2. PROOF OF THEOREM 1.1.

In this section K is an arbitrary field of characteristic zero, N is the set of all positive integers, and $[x_1,x_2] = [x_1,x_2]_{(a)} = x_1x_2 - x_2x_1$.

Let A be the free associative algebra over K freely generated by $\{x_1,x_2,...\}$ and let L be the free Lie algebra freely generated by the same set, so that $L \subseteq A$. We say that an ideal U is a weakly fully-invariant ideal or a weakly verbal ideal of A if U is closed under all endomorphisms α of A such that $\alpha(x_i) \in L$ for all $i \in N$. We say that an ideal J is a fully-invariant ideal or a verbal ideal (or a T-ideal) of A if J is closed under all endomorphisms of A.

If ρ is a representation of a Lie algebra over K then the set of all laws of ρ is a weakly verbal ideal of A. On the other hand, if U is a weakly verbal ideal of A, then there exists a representation ρ such that the laws of ρ are precisely the elements of U. If ρ is a representation of a Lie algebra over K and U is the weakly verbal ideal of its laws then the set of polynomials $\{f_1,f_2,...\}$ of A is a basis of laws of ρ if and only if the polynomials $f_1,f_2,...$ generate U as a weakly verbal ideal.

Let U_n be the weakly verbal ideal of A generated by

$$[x_1,x_2][x_3,x_4]...[x_{2n-1},x_{2n}].$$

We shall prove Theorem 1.1 by showing that any weakly verbal ideal U of A containing U_n is finitely generated as a weakly verbal ideal or, equivalently, that A satisfies the ascending chain condition on weakly verbal ideals U such that $U_n \subseteq U$.

It is easy to prove the following

LEMMA 2.1. $U_n = [A,A]^n$ for all $n \in N$.

Thus we have

COROLLARY 2.2. U_n is a verbal ideal generated by the polynomial

$$[x_1,x_2][x_3,x_4]\cdots[x_{2n-1},x_{2n}].$$

Let T_n be the Lie algebra of upper triangular n×n matrices over K, ρ be its natural representation on a vector space of dimension n over K. The set of all laws of ρ is precisely the set of all laws of the associative algebra of upper triangular matrices, also denoted by T_n. However, the set of all laws of the associative algebra T_n is $[A,A]^n$ (see [12]), so it is a finite dimensional representation, namely the representation ρ, which generates the variety of Lie algebra representations determined by the law (1). Now, repeating the proof of Proposition 2 from [13], we obtain the following lemma.

LEMMA 2.3. Suppose that V is a variety of representations of Lie algebras over K, (1) is a law of V and $r = \frac{n(n+1)}{2}$ is the dimension of the algebra T_n over K. Then V can be determined by the law (1) and laws involving only the variables $x_1,x_2,\ldots,x_r$.

Let A_r be the free associative algebra over K freely generated by $\{x_1,x_2,\ldots,x_r\}$ so that $A_r \subseteq A$. Let B_r be a free algebra of the variety of associative algebras over K determined by the law (1), $\{y_1,y_2,\ldots,y_r\}$ a set of free generators of B_r. If V_1 and V_2 are weakly verbal ideals of A containing U_n, then by Lemma 2.3 $A_r \cap V_1 = A_r \cap V_2$ implies $V_1 = V_2$. Thus to show that A has the ascending chain condition on weakly verbal ideals containing U_n it suffices to show that B_r has the ascending chain condition on weakly verbal ideals (i.e. ideals of B_r closed under all endomorphisms β of B_r such that $\beta(y_i) \in L_r$ $(1 \leq i \leq r)$ where L_r is the Lie subalgebra of B_r generated by $y_1,y_2,\ldots,y_r$). By induction arguments it is sufficient to consider only chains of ideals contained in the (weakly) verbal ideal of B_r generated by the polynomial

$$[x_1,x_2][x_3,x_4]\cdots[x_{2n-3},x_{2n-2}].$$

We shall denoted this verbal ideal by V_{n-1}.

By Lemma 2.1 the ideal V_{n-1} is a subspace of the vector space B_r spanned by elements of the form

$$g_1[y_{i_1},y_{i_2}]g_2\cdots g_{n-1}[y_{i_{2n-3}},y_{i_{2n-2}}]g_n \qquad (3)$$

where

$$g_j = y_{k_{j1}}^{m_{j1}}\, y_{k_{j2}}^{m_{j2}} \cdots y_{k_{js}}^{m_{js}},$$

$m_{jl} \geq 0$, $1 \leq k_{jl} \leq r$, $1 \leq l \leq s$, $1 \leq j \leq n$.

It is easy to prove the following

LEMMA 2.4. <u>Let</u> $g_j, g_j' \in B_r$ <u>and</u> $g_j = g_j'$ <u>modulo</u> $[B_r, B_r]$ $(1 \leq j \leq n)$. <u>Then</u>

$$g_1[y_{i_1}, y_{i_2}]g_2 \ldots g_{n-1}[y_{i_{2n-3}}, y_{i_{2n-2}}]g_n =$$
$$= g_1'[y_{i_1}, y_{i_2}]g_2' \ldots g_{n-1}'[y_{i_{2n-3}}, y_{i_{2n-2}}]g_n'.$$

Lemma 2.4 shows that, in fact, the monomials g_j in (3) are commutative $(1 \leq j \leq n)$.

Now we shall use an idea from [14]. Let R be the algebra of polynomials in the variables t_{ij} $(1 \leq i \leq n,\ 1 \leq j \leq r)$ over K. Let M be the free module over R freely generated by

$$w_{i_1 i_2 \ldots i_{2n-2}}, \quad 1 \leq i_l \leq r.$$

Define a K-linear map δ from M to V_{n-1} by

$$\delta(w_{i_1 i_2 \ldots i_{2n-2}} t_{11}^{l_{11}} \ldots t_{1r}^{l_{1r}} t_{21}^{l_{21}} \ldots t_{nr}^{l_{nr}}) =$$

$$= y_1^{l_{11}} \ldots y_r^{l_{1r}} [y_{i_1}, y_{i_2}] y_1^{l_{21}} \ldots y_r^{l_{2r}} \ldots [y_{i_{2n-3}}, y_{i_{2n-2}}] y_1^{l_{n1}} \ldots y_r^{l_{nr}}.$$

Let $\Psi_{\alpha i}$ $(\alpha \in K,\ 1 \leq i \leq r)$ denote the endomorphism of B_r given by

$$\Psi_{\alpha i}(y_j) = y_j + \alpha[y_j, y_i] \quad (1 \leq j \leq r).$$

Let $f = t_{11}^{l_1} \ldots t_{1r}^{l_r}$, $y = y_1^{l_1} \ldots y_r^{l_r}$,

$$q_{\alpha i} = (1 + \alpha t_{2i}')(1 + \alpha t_{3i}') \ldots (1 + \alpha t_{ni}'),$$

$\alpha \in K$, $1 \leq i \leq r$, $t_{jl}' = t_{jl} - t_{(j-1)l}$.

Define endomorphisms of the module M, denoted by f^* and $q_{\alpha i}^*$, by

$$f^*(m) = m.f, \quad q_{\alpha i}^*(m) = m.q_{\alpha i}$$

for all $m \in M$, and an endomorphism of a vector space V_{n-1}, denoted by g^*, by

$$g^*(v) = g.v$$

for all $v \in V_{n-1}$.

LEMMA 2.5. The following diagrams are commutative:

(1) $\begin{array}{ccc} M & \xrightarrow{f^*} & M \\ \downarrow\delta & & \downarrow\delta \\ V_{n-1} & \xrightarrow{g^*} & V_{n-1} \end{array}$ (2) $\begin{array}{ccc} M & \xrightarrow{q^*_{\alpha i}} & M \\ \downarrow\delta & & \downarrow\delta \\ V_{n-1} & \xrightarrow{\Psi_{\alpha i}} & V_{n-1} \end{array}$

The proof of Lemma 2.5 is straightforward.

Let R_1 be the subalgebra of R generated by the polynomials $q_{\alpha i}$ ($\alpha \in K$, $1 \le i \le r$) and the variables t_{1i} ($1 \le i \le r$). If W is a weakly verbal ideal of B_r contained in V_{n-1} then by Lemma 2.5 its inverse image $\delta^{-1}(W)$ is a module over R_1. Now we can prove that B_r has the ascending chain condition on weakly verbal ideals contained in V_{n-1}. Suppose otherwise, and let

$$W_1 \subsetneqq W_2 \subsetneqq \dots$$

be an infinite strictly ascending chain of weakly verbal ideals of B_r such that $W_i \subseteq V_{n-1}$ for all $i \in N$. Then there is an infinite strictly ascending chain

$$\delta^{-1}(W_1) \subsetneqq \delta^{-1}(W_2) \subsetneqq \dots$$

of R_1-submodules in M.But it is easy to check that M is a Noetherian R_1-module (see [14]). This contradiction completes the proof of Theorem 1.1.

3. PROOF OF THEOREM 1.3.

3.1. This theorem was proved by the first author. Its proof required improvements of the techniques due to D.E. Cohen [15] and G. Higman [16]. Using these techniques, it can be proved that any subgroup of the triangular n×n matrix group over a field on non-zero characteristic p, where $p \ge n$, has a finite basis for its (group) laws. Furthermore, the proof of Theorem 1.1 from [11] can be simplified by using these techniques.

In this section K is an arbitrary field, N is the set of all positive integers and $[x_1,x_2] = [x_1,x_2]_{(m)} = x_1^{-1}x_2^{-1}x_1x_2$.

Let F be the free group freely generated by $\{x_1,x_2,\dots\}$, KF the group algebra of F over K. We say that an ideal U is a verbal ideal if U is closed under all endomorphisms α of KF such that $\alpha(x_i) \in F$ for all $i \in N$. If ρ is a group representation over K then the set of all laws of ρ is a verbal ideal of KF. On the other hand, if U is a verbal ideal of KF, then there exists a group representation ρ such that the laws of ρ are precisely the elements of U.

Let U_k be the verbal ideal of KF generated by

$$(1-[x_1,x_2])(1-[x_3,x_4]) \ldots (1-[x_{2k-1},x_{2k}]).$$

As in §2, to prove Theorem 1.3 it suffices to show that KF has the ascending chain condition on verbal ideals U such that $U_{n-1} \supseteq U \supseteq U_n$ or, equivalently, that KF/U_n has the ascending chain condition on verbal ideals contained in the image of U_{n-1} (denoted by V_{n-1}) under the natural epimorphism from KF onto KF/U_n.

It is easy to check that V_{n-1} is a vector subspace of KF/U_n spanned by elements of the form

$$g_1(1-[x_{i_1},x_{i_2}])^{g_2}(1-[x_{i_3},x_{i_4}])^{g_3}\ldots(1-[x_{i_{2n-3}},x_{i_{2n-2}}])^{g_n} \quad (4)$$

where $g_i \in F$ for all i,

$$(1-[x_1,x_m])^{g_i} = g_i^{-1}(1-[x_1,x_m])g_i.$$

Note that if $g_i = g_i'$ modulo F', $1 \le i \le n$, then the element

$$g_1'(1-[x_{i_1},x_{i_2}])^{g_2'}(1-[x_{i_3},x_{i_4}])^{g_3'} \ldots (1-[x_{i_{2n-3}},x_{i_{2n-2}}])^{g_n'}$$

is equal to (4), so, in fact, in (4) $g_i \in F/F'$ for all i.

Let $R = K[t_{ij}: 1 \le i \le n,\ j \in N]$ be an algebra of polynomials over K, M the free module over R freely generated by

$$w_{i_1 i_2 \ldots i_{2n-2}},\ i_l \in N,\ 1 \le l \le 2n-2.$$

Let Φ be the set of all one-to-one order-preserving maps of the set of positive integers N into itself. Let Φ^* denote both the set of all endomorphisms φ^* of the algebra R such that

$$\varphi^*(t_{ij}) = t_{i(\varphi j)} \quad (1 \le i \le n;\ j \in N),$$

and the set of all K-linear mappings of M into itself such that

$$\varphi^*(w_{i_1 i_2 \ldots i_{2n-2}} r) = w_{(\varphi i_1)(\varphi i_2)\ldots(\varphi i_{2n-2})}(\varphi^*(r)) \quad (i_l \in N,\ r \in R).$$

Let Θ^* be the set of all endomorphisms Θ^*_{ij} $(i < j)$ of R such that

$$\Theta^*_{ij}(t_{kl}) = t_{kl}\ (l \ne j),\quad \Theta^*_{ij}(t_{kj}) = t_{ki}t_{kj}.$$

Let Θ^* denote also the set of all K-linear mappings Θ_{ij} $(i < j)$ of M into itself which are defined by

$$\Theta^*_{ij}(w_{i_1 i_2 \ldots i_{2n-2}} r) = w_{i_1 i_2 \ldots i_{2n-2}}(\Theta^*_{ij}(r)) +$$

$$+ \sum_k w_{i_{k1} i_{k2} \ldots i_{k(2n-2)}} r_k ,$$

where for all k $(i_{k1}, i_{k2}, \ldots, i_{k(2n-2)})$ is less than $(i_1, i_2, \ldots, i_{2n-2})$ under the lexicographic order on J^{2n-2}. Let $\Omega^* = \Phi^* \cup \Theta^*$.

Let $\sigma_1^{(j)}, \sigma_2^{(j)}, \ldots, \sigma_n^{(j)}$ be the elementary symmetric polynomials of $t_{1j}, t_{2j}, \ldots, t_{nj}$ and let R_1 be the subalgebra of R generated by all polynomials $\sigma_i^{(j)}$ $(1 \le i \le n,\ j \in N)$. If M_1 is a vector subspace of M, we say that M_1 is an Ω^*-module, if M_1 is an R_1-module which is closed under all endomorphisms $\varphi^* \in \Phi^*$ and $\Theta^*_{ij} \in \Theta^*$. Define a K-linear map δ from M to V_{n-1} by

$$\delta(w_{i_1 \ldots i_{2n-2}} t_{11}^{l_{11}} t_{12}^{l_{12}} \ldots t_{21}^{l_{21}} t_{22}^{l_{22}} \ldots t_{n1}^{l_{n1}} t_{n2}^{l_{n2}} \ldots) =$$

$$= g_1 (1 - [x_{i_1}, x_{i_2}])^{g_2} (1 - [x_{i_3}, x_{i_4}])^{g_3} \ldots (1 - [x_{i_{2n-3}}, x_{i_{2n-2}}])^{g_n}$$

where $g_i = x_1^{l_{i1}} x_2^{l_{i2}} \ldots$ for all i. As in the proof of Theorem 1.1, it can be shown that if the field K is of characteristic 0 or p, where $p \ge n$, then to prove Theorem 1.3 it suffices to establish the following lemma.

LEMMA 3.1. The module M has the ascending chain condition on Ω^* - submodules.

3.2. Proof of Lemma 3.1. Let J be the set of all non-negative integers, and J^n be the set of all ordered n-tuples of elements of J. Let $\le$ denote the lexicographic order on J^l $(l \in N)$, i.e.

$$(k_1, k_2, \ldots, k_l) < (k_1', k_2', \ldots, k_l')$$

if and only if there exists an i, $1 \le i \le l$, such that $k_i < k_i'$ but $k_j = k_j'$ for all $j < i$. We define partial orders $\le_1$, $\le_3$ and a good order $\le_2$ on the set J^n. If $k, k' \in J^n$, $k = (k_1, k_2, \ldots, k_n)$, $k' = (k_1', k_2', \ldots, k_n')$, $k_1 \ge k_2 \ge \ldots \ge k_n$, $k_1' \ge k_2' \ge \ldots \ge k_n'$ then we write $k <_1 k'$ if and only if $k < k'$ (under the lexicographic order). Let now l, l' be arbitrary n-tuples and let k, k' be non-increasing n-tuples, obtained by permutation of members of l and l' respectively. Then we write $l <_1 l'$ if and only if

$<_1 k'$. We shall write $k<_2 k'$, $k,k' \in J^n$ if and only if either

i) $k<_1 k'$;

or

ii) k, k' are incomparable under $\leq_1$ and $k < k'$ (under the lexicographic order).

Note that $(J^n, \leq_2)$ is a well ordered set.

Let $1^{(k)} \in J^n$, $1^{(k)} = (1_{k1},\dots,1_{kn})$, $(k = 1,2)$. We write $1^{(1)} \leq_{(3)} 1^{(2)}$ if and only if the following conditions are satisfied:

1) there exists a permutation τ on the set $\{1,2,\dots,n\}$ such that

$$1_{k(\tau 1)} = \dots = 1_{k(\tau(r_1))} > 1_{k(\tau(r_1+1))} = \dots = 1_{k(\tau(r_2))} > \dots > 1_{k(\tau(r_q+1))} = \dots =$$

$$= 1_{k(\tau(n))} \quad (q = 1,2,..,n; \quad k = 1,2);$$

2) $1_{1(\tau(n))} \leq 1_{2(\tau(n))}$ and for all j, $1 \leq j \leq q$,

$$1_{1(\tau(r_j))} - 1_{1(\tau(r_j+1))} \leq 1_{2(\tau(r_j))} - 1_{2(\tau(r_j+1))}.$$

Let C_i be the semigroup of primitive monomials of $K[t_{i1},t_{i2},\dots,t_{in}]$,

$$C_i = \{t_{i1}^{m_1} t_{i2}^{m_2} \dots t_{in}^{m_n} : m_1,m_2,\dots,m_n \in J\}.$$

Suppose that $m^{(j)} \in C_i$ $(j = 1,2)$,

$$m^{(j)} = t_{i1}^{1_{j1}} t_{i2}^{1_{j2}} \dots t_{in}^{1_{jn}} \quad (j = 1,2).$$

We define a good order $\leq_2$ and a partial order $\leq_3$ on C_i by $m^{(1)} \leq_{(k)} m^{(2)}$ if and only if

$$(1_{11},\dots,1_{1n}) \leq_{(k)} (1_{21},\dots,1_{2n}) \quad (k = 2,3).$$

Let C be a semigroup of monomials of R, $C = \{ \prod_{i,j} t_{ij}^{m_{ij}} : i \in N,\ 1 \leq j \leq n,$ $m_{ij} \in J,\ m_{ij} = 0$ for almost all $i\}$,

and let W be the set of all elements of M of the form

$$w_{i_1 i_2 \dots i_{2n-2}} m,\ m \in C,\ i_1 \in N\ (1 \leq 1 \leq 2n-2).$$

Note that W is a basis of the vector space M. We define a good order $\leq_2$ and a partial order $\leq_3$ on C and W. Suppose that $m^{(j)} \in C$, $m^{(j)} = m_1^{(j)} m_2^{(j)} \dots$ $(j = 1,2)$, where $m_i^{(j)} \in C_i$ $(i \in N)$. Then we write $m^{(1)} <_2 m^{(2)}$ if and only if there exists $q \in N$ such that $m_q^{(1)} <_2 m_q^{(2)}$ but $m_r^{(1)} = m_r^{(2)}$ for all $r > q$, and we write $m^{(1)} \leq_3 m^{(2)}$ if and only if $m_i^{(1)} \leq_3 m_i^{(2)}$ for all $i \in N$.

We write

$$w_{i_1 i_2 \ldots i_{2n-2}}{}^{m} <_2 w_{i'_1 i'_2 \ldots i'_{2n-2}}{}^{m'}$$

if either

1) $(i_1, i_2, \ldots, i_{2n-2}) < (i'_1, i'_2, \ldots, i'_{2n-2})$,

or

2) $i_l = i'_l$ $(1 \le l \le 2n-2)$, $m <_2 m'$.

We also write

$$w_{i_1 i_2 \ldots i_{2n-2}}{}^{m} \le_3 w_{i'_1 i'_2 \ldots i'_{2n-2}}{}^{m'}$$

if and only if $i_l = i'_l$ $(1 \le l \le 2n-2)$ and $m \le_3 m'$.

Let $h \in M$, $h = \sum_{i=0}^{s} a_i w^{(i)}$, $a_i \in K$, $w^{(i)} \in W$ $(0 \le i \le s)$. We say that $a_o w^{(o)}$ is leading term of h, denoted by l.t.(h), if $a_o \ne 0$ and $w^{(i)} <_2 w^{(o)}$ $(1 \le i \le s)$. Now we define the order $\le_4$ on W by $w \le_4 w'$ if and only if there exists $\varphi^* \in \Phi^*$ and $\Theta^*_{j_1 j_2}, \Theta^*_{j_3 j_4}, \ldots, \Theta^*_{j_{2s-1} j_{2s}} \in \Theta^*$ such that $j_1, j_3, \ldots, j_{2s-1} \notin \operatorname{Im}\varphi$ and

$$\text{l.t.}(\Theta^*_{j_1 j_2} \Theta^*_{j_3 j_4} \cdots \Theta^*_{j_{2s-1} j_{2s}} \varphi^*(w)) \le_3 w'.$$

LEMMA 3.2. <u>Let</u> $h \in M$, $\text{l.t.}(h) = w$, $w \in W$. <u>Suppose that</u> $w' \in W$, $w \le_4 w'$. <u>Then there exist</u> $\varphi^* \in \Phi^*$, $\Theta^*_{j_1 j_2}, \Theta^*_{j_3 j_4}, \ldots, \Theta^*_{j_{2s-1} j_{2s}} \in \Theta^*$ $(j_1, j_3, \ldots, j_{2s-1} \notin \operatorname{Im}\varphi)$ <u>and</u> $f \in R_1$ <u>such that</u>

$$\text{l.t.}((\Theta^*_{j_1 j_2} \Theta^*_{j_3 j_4} \cdots \Theta^*_{j_{2s-1} j_{2s}} \varphi^*(h)) f) = w'.$$

The proof of Lemma 3.2 is straightforward.

A partially ordered set S is said to be partially well ordered if every infinite sequence of elements of S has an infinite (nonstrictly-) increasing subsequence or, equivalently, if there exist neither an infinite strictly decreasing sequence in S, nor an infinity of mutually incomparable elements of S (for other equivalent definitions see [16]). Using Lemma 2.1 from [11] and Lemma 3.2 it is easy to check that if $(W, \le_4)$ is partially well ordered then M has the ascending chain condition on Ω^*-submodules. Thus it suffices to prove the following lemma.

LEMMA 3.3. $(W, \le_4)$ <u>is partially well ordered</u>.

3.3. <u>Partially ordered sets</u>. If S is a set with a distinguished element 0, we shall write $V(S) = V(S,0)$ for the set of all sequences

$(s_i : i\in N) = (s_i)$ of elements of S in which $\{i : s_i \neq 0\}$ is finite. Suppose that $\varphi\in\Phi$; then φ also denotes the mapping of V(S) into itself such that $\varphi((s_i)) = (s_i')$, where

$$s_j' = \begin{cases} s_i, & \text{if } j = i\varphi; \\ 0, & \text{if } j \neq i\varphi \text{ for all } i\in N; \end{cases}$$

and Φ denotes the set of all such mappings.

Let $V^{(i)}$ be the set of all elements (s_i) of V(S) in which $s_i = 0$. Suppose that $s_l' = s_l$ for all $l \neq i$ and $s_i' = s_j$. Define a mapping Θ_{ij} from $V^{(i)}$ to V(S) by $\Theta_{ij}((s_l)) = (s_l')$. Let Θ denote the set of all mappings Θ_{ij} such that $i < j$. Let $\Omega = \Phi\cup\Theta$.

Suppose that S has a partial order $\leq$. Then we can define a partial order $\leq_\Phi$ on V(S). Let Imφ be the image of N under φ. We write $(s_i) \leq_\Phi (s_i')$ if and only if there exists an element φ of Φ such that if $(s_i'') = \varphi(s_i)$ then $s_j'' \leq s_j'$ for all $j\in$ Imφ. From [11], Lemma 3.1, or from [16], Theorem .3, it is easy to deduce

LEMMA 3.4. _If_ $(S,\leq)$ _is partially well ordered, then so is_ $(V(S),\leq_\Phi)$.

Now we define a partial order $\leq_\Omega$ on V(S). We write $(s_i) \leq_\Omega (s_i')$ if and only if there exist $\varphi\in\Phi$, $q\in J$, and $\Theta_{j_1 j_2}, \Theta_{j_3 j_4}, \ldots, \Theta_{j_{2q-1} j_{2q}} \in\Theta$ such that the following conditions are satisfied:

i) $j_1, j_3, \ldots, j_{2q-i} \notin$ Imφ;

ii) if $(s_i'') = \Theta_{j_1 j_2}\Theta_{j_3 j_4} \ldots \Theta_{j_{2q-1} j_{2q}}\varphi((s_i))$ then $s_i'' \leq s_i'$ for all $i\in N$.

LEMMA 3.5. _Let_ (s_i), (s_i') _be elements of V(S). Suppose the following conditions are satisfied:_

1) _there exists a_ $\varphi\in\Phi$ _such that if_ $(s_i'') = \varphi((s_i))$, _then_ $s_j'' \leq s_j$ _for all_ $j\in$ Imφ (i.e. $(s_i) \leq_\Phi (s_i')$);

2) _for all_ $j\in N$ _such that_ $j\notin$ Imφ _and_ $s_j' \neq 0$ _there exists_ a $k(j)\in N$ _such that_ $k(j) > j$ _and_ $s_{k(j)}'' \leq s_j'$.

Then $(s_i) \leq_\Omega (s_i')$.

Proof. Let $(s_i''') = \Theta_{j_1 k(j_1)}\Theta_{j_2 k(j_2)} \ldots \Theta_{j_q k(j_q)}\varphi((s_i))$, where $\{j_1, j_2, \ldots, j_q\}$ is the set of all $j\in N$ such that $j\notin$ Imφ and $s_j' \neq 0$. It is easy to check that $s_i''' \leq s_i'$ for all $i\in N$, so that $(s_i) \leq_\Omega (s_i')$, as required. Lemma 3.5 is proved.

The cardinal product of two partially ordered sets $(S,\leq)$ and $(T,\leq)$

is the set of all ordered pairs (s,t), with $s \in S$, $t \in T$, ordered by $(s_1,t_1) \leq (s_2,t_2)$ if and only if $s_1 \leq s_2$, $t_1 \leq t_2$. The following lemma was proved in [16], Theorem 2.3.

LEMMA 3.6. <u>Let $(S,\leq)$ and $(T,\leq)$ be partially well ordered, and let $(U,\leq)$ be their cardinal product. Then $(U,\leq)$ is partially well ordered.</u>

Let E be a finite set containing 0. Let $S = J^n \times E$ be a set with distinguished element $0 = (0,0,\ldots,0)$ ordered by

$$(j_1,j_2,\ldots,j_n,e) \leq (j_1',j_2',\ldots,j_n',e')$$

if and only if $j_l \leq j_l'$ for all l, $1 \leq l \leq n$, and $e = e'$. Note that $(S,\leq)$ is partially well ordered. We need the following result.

PROPOSITION 3.7. <u>Let the set $S = J^n \times E$ be ordered as above. Then $(V(S),\leq_\Omega)$ is partially well ordered.</u>

<u>Proof</u>. Let $(U,\leq)$ be a partially ordered set, and let U' be a subset of U. The closure of U' is the set $cl(U') = \{u : u \in U,\ u \geq u'$ for some $u' \in U'\}$. We shall say that the subset U' is closed, if

$$U' = cl(U').$$

Define the order $\leq_5$ on J^n by $(j_1,j_2,\ldots,j_n) \leq_5 (j_1',j_2',\ldots,j_n')$ if and only if $j_l \leq j_l'$ for all l, $1 \leq l \leq r$. Let T_n be the set of all closed subsets of $(J^n,\leq_5)$ ordered by $t \leq t'$ if and only if $t \supseteq t'$. Using induction on n, it can be proved that there are no infinite strictly descending chains and no infinite sets of mutually incomparable elements in $(T_n,\leq)$. So $(T_n,\leq)$ is partially well ordered.

Now suppose that $S = J^n \times E$ is ordered as above. Let T be the set of all closed subsets of $(S,\leq)$ ordered by $t \leq t'$ if and only if $t \supseteq t'$. The set S is the union of subsets $S_1,S_2,\ldots,S_m$ $(m = |E|)$ such that $(S_i,\leq)$ is isomorphic to $(J^n,\leq_5)$ for all i, $1 \leq i \leq m$, and s_i, s_j $(s_i \in S_i,\ s_j \in S_j)$ are incomparable is $i \neq j$. So $(T,\leq)$ is isomorphic to the set of all m-tuples of elements of T_n ordered by $(t_1,\ldots,t_m) \leq$ $\leq (t_1',\ldots,t_m')$ if and only if $t_i \leq t_i'$ for all i, $1 \leq i \leq m$. So, by Lemma 3.6, $(T,\leq)$ is partially well ordered.

Let $Q = S \times T$ be the cardinal product of the partially ordered sets $(S,\leq)$ and $(T,\leq)$. Set $V(Q) = V(Q,(0,\emptyset))$ and let $P = T \times V(Q)$ be the cardinal product of $(T,\leq)$ and $(V(Q),\leq_\Phi)$ so that $(t,v) \leq (t',v')$ if and only if $t \leq t'$, $v \leq_\Phi v'$. Note that, by Lemmas 3.6 and 3.4, $(P,\leq)$ is partially well ordered.

Let $v \in V(S)$, $v = (s_i)$ and let $t_i = cl\{s_j : j > i\}$ for all $i \in J$. We define a one-one mapping γ from $V(S)$ to P by

$$\gamma((s_i)) = (t_o, ((s_j,t_j): j\in N))).$$

Since P is partially well ordered, it suffices to show that $\gamma(v) \le \gamma(v')$ implies $v \le_\Omega v'$ for any $v,v' \in V(S)$.

Let $(s_i),(s_i') \in V(S)$, $\gamma((s_i)) = (t_o,((s_i,t_i)))$, $\gamma((s_i')) = (t_o',((s_i',t_i')))$. Suppose that $\gamma((s_i)) \le \gamma((s_i'))$, then $t_o \le t_o'$ and there exists a $\varphi\in\Phi$ such that if $(s_i'') = \varphi((s_i))$, $\gamma((s_i'')) = (t_o'',((s_i'',t_i'')))$, then $s_j'' \le s_j'$, $t_j'' \le t_j'$ for all $j\in \mathrm{Im}\varphi$. In particular, the mapping φ satisfies condition 1) of Lemma 3.5. We shall check that condition 2) of this lemma is satisfied as well.

Let j be a positive integer such that $j\notin \mathrm{Im}\varphi$ and $s_j' \ne 0$. Let

$$l = \begin{cases} \max(i: i\in \mathrm{Im}\varphi,\ i<j), & \text{if there exists an } i \text{ such that } i\in \mathrm{Im}\varphi,\ i<j; \\ 0 & \text{otherwise} \end{cases}$$

Since $l < j$, $s_j' \in t_l'$, so $s_j' \in cl(s_k''|k>l)$ since $t_l' \subseteq t_{\underline{l}}'' = cl(s_k'': k>l)$. Thus there exists a positive integer $k(j) > l$ such that $s_{k(j)}'' \le s_j'$. However $s_k'' = 0$ for all k, $l < k \leqq j$, by definition of l, so we may assume $k(j) > j$. Thus for any $j\in \mathrm{Im}\varphi$ such that $s_j' \ne 0$, there exists a $k(j)$ such that $k(j) > j$ and $s_{k(j)}'' \le s_j'$, i.e. condition 2) of Lemma 3.5 is satisfied. So by Lemma 3.5 $(s_i) \le_\Omega (s_i')$. This completes the proof of Proposition 3.7.

3.4. Proof of Lemma 3.3. Let i be an arbitrary positive integer, $m = t_{i1}^{l_1} t_{i2}^{l_2} \dots t_{in}^{l_n}$, $m' = t_{i1}^{l_1'} t_{i2}^{l_2'} \dots t_{in}^{l_n'}$ be monomials from C_i, and let τ, τ' be permutations of $\{1,2,..,n\}$ such that

$$l_{\tau 1} = \dots = l_{\tau(s_1)} > l_{\tau(s_1+1)} = \dots = l_{\tau(s_2)} > \dots > l_{\tau(s_q+1)} = \dots = l_{\tau n},$$

$$l_{\tau' 1}' = \dots = l'_{\tau'(r_1)} > l'_{\tau'(r_1+1)} = \dots = l'_{\tau'(r_2)} > \dots > l'_{\tau'(r_{q'}+1)} = \dots = l'_{\tau' n}.$$

Suppose that $s_o = r_o = 1$, $s_{q+1} = r_{q'+1} = n$. We say that the monomials m and m' are equivalent if $q = q'$, $s_k = r_k$ $(1 \le k \le q)$ and $\{\tau(s_k+1),\dots,\tau(s_{k+1})\} = \{\tau'(r_k+1),\dots,\tau'(r_{k+1})\}$ for all k, $1 \le k \le q$. Note that there is only a finite number of equivalence classes of monomials in C_i. Let E' denote the set of all equivalence classes of monomials from C_i and 0 denote the equivalence class including the unit element. Suppose that $J_2 = \{0,1\}$, $E = E' \times J_2^{2n-2}$, $S = J^n \times E$, $S' = J^n \times E'$. Let S, S' be ordered as defined in §3.3.

Let $m\in C_i$, $m = t_{i1}^{l_1} t_{i2}^{l_2} \dots t_{in}^{l_n}$ and let τ be a permutation of $\{1,2,..,n\}$ sich that

$$l_{\tau 1}=\ldots=l_{\tau(s_1)}>l_{\tau(s_1+1)}=\ldots=l_{\tau(s_2)}>\ldots>l_{\tau(s_q+1)}=\ldots=l_{\tau n}.$$

Suppose that

$$k_j=\begin{cases} l_{\tau(s_1)}-l_{\tau(s_1+1)}, & \text{if } j\geq s_1,\\ l_{\tau(s_r)}-l_{\tau(s_r+1)}, & \text{if } s_{r-1}>j\geq s_r,\\ l_{\tau n} & , \text{if } j>s_q.\end{cases}$$

Let $d(m)$ be the equivalence class including the monomial m. Define a mapping ν from C_i to S' by

$$\nu(m) = (k\ ,\ldots\ k_n,\ d(m)).$$

Let $w\in W$, $w = w_{i_1\ldots i_{2n-2}}m$, $m = m_1m_2\ldots,m_i\in C_i$ for all $i\in N$. Suppose that

$$k_{lj}=\begin{cases}1 \text{ if } i_l = j,\\ 0 \quad \text{otherwise},\end{cases}$$

where $1\leq l\leq 2n-2$, $j\in N$. Now we define a mapping ν from W to V(S) by $\nu(w) = (s_j)$, where $s_j = (\nu(m_j),k_{1j},\ldots,k_{(2n-2)j})$. Note that ν is injective.

Using the definition of $\leq_4$, the following lemma can be proved.

LEMMA 3.8. <u>Let</u> $w, w'\in W$, <u>and</u> $\nu(w)\leq_\Omega \nu(w')$. <u>Then</u> $w\leq_4 w'$.

Since $(V(S), \leq_\Omega)$ is partially well ordered by Proposition 3.7, Lemma 3.8 implies that $(W, \leq_4)$ is partially well ordered also. So Lemma 3.3 is proved.

This completes the proof of Theorem 1.3.

REFERENCES

1. Yu.P. Razmyslov. Finite basing of the identities of a matrix algebra of second order over a field of characteristic zero (in Russian). Algebra Logika 12 (1973), 83-113. (English translation: Algebra Logic, 12 (1973), 47-63 (1974)).
2. Hanna Neumann. Varieties of groups. Ergebnisse der Mathematik und ihrer Grenzgebiete, Band 37. Springer-Verlag, 1967.
3. Yu.A. Bahturin. Lectures on Lie algebras. Studien zur Algebra und ihre Anwendungen, Band 4. Akademic-Verlag, Berlin, 1978.
4. M.R. Vaughan-Lee. Varieties of Lie algebras. Q.J. Math., Oxf. II Ser., 21 (1970), 297-308.
5. V.S. Drenski. On identities in Lie algebras (in Russian). Algebra Logika 13(1974), 265-290. (English translation: Algebra Logic 13 (1974), 150-165 (1975)).

6. B.I. Plotkin. Varieties of group representations (in Russian). Usp. Mat. Nauk, 32 (1977), No. 5(197), 3-68. (English translation: Russ. Math. Surveys, 32 (1977), No. 5, 1-72).

7. B.I. Plotkin. Varieties in representations of finite groups. Locally stable varieties. Matrix groups and varieties of representations (in Russian). Usp. Mat. Nauk, 34 (1979), No. 4(208), 65-95. (English translation: Russ. Math. Surveys, 34 (1979), No. 4, 69-103).

8. A.Yu. Ol'šanskii. On the problem of a finite basis of identities in groups (in Russian). Izv. Akad. Nauk SSSR, Ser. Mat., 34 (1970), 376-384. (English translation: Math. USSR, Izv., 4 (1970), 381-389 (1971)).

9. S.I. Adian. Infinite irreducible systems of group identities (in Russian). Dokl. Akad. Nauk SSSR, 190 (1970), 499-501. (English translation: Soviet Math., Dokl., 11 (1970), 113-115).

0. M.R. Vaughan-Lee. Uncountably many varieties of groups. Bull. Lond. Math. Soc., 2 (1970), 280-286.

1. R.M. Bryant and M.F. Newman. Some finitely based varieties of groups. Proc. Lond. Math. Soc., III Ser., 28 (1974), 237-252.

2. Yu.N. Mal'cev. Basis for identities of the algebra of upper triangular matrices (in Russian). Algebra Logika, 10 (1971), 393-400. (English translation: Algebra Logic, 10 (1971), 242-247 (1973)).

3. Yu.P. Razmyslov. The Jacobson radical in PI-algebras (in Russian). Algebra Logika, 13 (1974), 337-360. (English translation: Algebra Logic, 13 (1974), 192-204 (1975)).

4. A.N. Krasil'nikov. The finite basis property for certain varieties of Lie algebras (in Russian). Vestn. Mosk. Univ., Ser. 1, No. 2 (1982), 34-38. (English translation: Mosc. Univ. Math. Bull., 37, No. 2 (1982), 44-48).

5. D.E. Cohen. On the laws of a metabelian variety. J. Algebra 5 (1967), 267-273.

6. G. Higman. Ordering by divisibility in abstract algebras. Proc. Lond. Math. Soc., III Ser., 2 (1952), 326-336.

"This paper is in final form and no version of it will be submitted for publication elsewhere"

LIE GROUPS AND ERGODIC THEORY

G.A. Margulis
Institute for Problems of Information Transmission
Moscow 103051, USSR, Ermolovoi 19

In these notes, some relations between Lie group theory and ergodic theory will be presented. Some applications to number theory will also be provided. Because the notes are mainly written for algebraists, a sufficiently detailed survey of elementary ergodic theory is given.

As usual $\mathbb{R}$, $\mathbb{Q}$, $\mathbb{Z}$ and $\mathbb{N}^+$ denote, respectively, the sets of reals, rationals, integers and positive integers.

§1. Preliminaries from ergodic theory

1.1. The subject of ergodic theory is the study of groups (or, more generally, semigroups) of transformations of measure spaces. By a measure on a set X we mean a non-negative countably additive function μ defined on a σ-algebra Ω of subsets of X and taking finite or infinite values. The subsets of X belonging to Ω are called μ-measurable or simply measurable. A measure μ on X is said to be normalized if $\mu(X) = 1$, finite if $\mu(X) < \infty$, and σ-finite if X is representable in the form of a countable union of measurable subsets of finite measure. In what follows all measures are supposed to be σ-finite.

By a Borel measure on a locally compact space X we mean a measure defined on the σ-algebra of Borel subsets of X. A Borel measure is called locally finite if the measure of any compact subset is finite. If X is a separable metric space then any locally finite Borel measure μ on X is σ-finite and regular (i.e., for every closed subset $Y \subset X$ and for each $\varepsilon > 0$, there exists on open set $U \subset X$ such that $U \supset Y$ and $\mu(U-Y) < \varepsilon$).

In measure theory the expressions "almost everywhere" and "for almost all" mean respectively "everywhere but on a null set" and "for all but belonging to a null set" (a null set is by definition a set of measure 0).

1.2. Let X be a measure space. A map from X to a measure space Y s called measurable if the preimage of any measurable subset of Y is easurable. A measurable map $T:X \to X$ is said to be an endomorphism of he measure space X provided it preserves μ, i.e., if $\mu(A) = \mu(T^{-1}A)$ or every measurable $A \subset X$. An endomorphism of X is said to be an auto- orphism if it is bijective.

Let G be a locally compact group. We say that an action of G on X s given if $gx \in X$ is defined for every $g \in G$ and $x \in X$ so that the following onditions are satisfied:

(a) $g_1(g_2x) = (g_1g_2)x$ and $ex = x$ for all $g_1, g_2 \in G$ and $x \in X$;

(b) the map $(g,x) \to gx$ is measurable, i.e. for every measurable $A \subset X$ he set $\{(g,x) \in G \times X \mid gx \in A\}$ belongs to the natural σ-algebra of subsets of $\times X$;

(c) the measure μ is G-invariant, i.e. $\mu(gA) = \mu(A)$ for every $g \in G$ nd every measurable $A \subset X$ (or, in other words, for every $g \in G$, the map $\to gx$, $x \in X$, is an automorphism of X).

If an action of G on X is given, we say that X is a G-space. In G- paces, G acts on X on the left. Therefore G-spaces are also called eft G-spaces. Sometimes it is more convenient to consider right G-spa- es in which $xg \in X$ is defined for every $g \in G$ and every $x \in X$ and conditions imilar to (a), (b) and (c) are satistied.

We say that a G-space X is ergodic or that G acts on X ergodically f the following condition is satisfied: if $A \subset X$ is measurable and $(A \Delta (gA)) = 0$ for every $g \in G$, then either $\mu(A) = 0$ or $\mu(X-A) = 0$. In his case we also say that the measure μ is G-ergodic or simply ergodic. or right G-spaces, the ergodicity can be defined analogously. For a -compact G, the ergodicity of an action of G on X is equivalent to the ollowing condition: if $A \subset X$ is measurable and $GA = A$, then either $(A) = 0$ or $\mu(X-A) = 0$. Thus if G is σ-compact and G acts transitively n X then G acts ergodically on X (for arbitrary G this assertion is ot always true).

An automorphism T of X is said to be ergodic if the group $\{T^m \mid m \in \mathbb{Z}\}$ cts ergodically on X. It is easily seen that the ergodicity of the utomorphism T is equivalent to the foloowing condition: if $A \subset X$ is easurable and $T(A) = A$ then either $\mu(A) = 0$ or $\mu(X-A) = 0$.

The study of arbitrary G-spaces can in a sense be reduced to the tudy of ergodic G-spaces. Namely, under sufficiently general conditions in particular if X is a locally compact metrizable G-space, μ is a lo- ally finite Borel measure, and G is metrizable and separable), there xist a partition of X into G-invariant subsets X_y, $y \in Y$, measures μ_y on

X_y and a measure ν on Y such that:

1) for any measurable $A \subset X$, the set $A \cap X_y$ is μ_y-measurable for almost all $y \in Y$ and $\mu(A) = \int_Y \mu_y(A \cap X_y) d\nu(y)$;

2) for almost all $y \in Y$, the restriction to X_y of the action of G is ergodic relative to the measure μ_y.

In case that μ is finite, we say that an automorphism T of X has the <u>mixing property</u> if for every measurable $A, B \subset X$

$$\lim_{n \to \infty} \tilde{\mu}((T^n A) \cap B) = \tilde{\mu}(A) \tilde{\mu}(B)$$

where $\tilde{\mu}(C) = \mu(C)/\mu(X)$, $C \subset X$. One can easily see that if T has the mixing property, then T is ergodic. But the converse is in general false.

1.3. Let an action of a locally compact group G on a space X with a measure μ be given. The space of complex-valued power p integrable functions on X will be denoted by $L_p(X,\mu)$ and the norm on $L_p(X,\mu)$ by $\| \ \|_p$. For every $p \geq 1$, let us define the representation of G on the space $L_p(X,\mu)$ by the formula

$$(\rho(g)f)(x) = f(g^{-1}x), \quad f \in L_p(X,\mu), \quad g \in G.$$

Since the measure μ is G-invariant, one has that $\|\rho(g)f\|_p = \|f\|_p$ for every $g \in G$ and $f \in L_p(X,\mu)$. In particular, ρ defines an unitary representation of G on $L_2(X,\mu)$. It is not difficult to prove that this representation is continuous, i.e. that $\rho(g)f$ depends continuously on $(g,f) \in G \times L_2(X,\mu)$. One can easily prove the following assertions.

(I) Let $p \geq 1$ and let the measure μ be finite. Then the G-space X is ergodic if and only if every $\rho(G)$-invariant function $f \in L_p(X,\mu)$ is constant.

(II) Let us denote by $L_2^0(X,\mu)$ the subspace of $L_2(X,\mu)$ consisting of functions with integral 0. Let $g \in G$ and let $T(x) = gx$, $x \in X$. If the measure μ is finite and the restriction of the unitary operator $\rho(g)$ to $L_2^0(X,\mu)$ has absolutely continuous spectrum, then T has the mixing property.

1.4. The following simple but important theorem holds.

Poincaré recurrence theorem. Let T be an endomorphism of a space X with a measure μ. Let us assume that μ is finite. Then for every measurable $A \subset X$ and almost all (with respect to μ) $x \in A$, the set $\{n \in \mathbb{N}^+ \mid T^n x \in A\}$ is infinite.

Sketch of the proof. For $m \in \mathbb{N}^+$, set $B_m = \{x \in A \mid T^n x \notin A \text{ for all } n \geq m\}$. Then $T^{-mi}B_m \cap T^{-mj}B_m = \phi$ for all $i, j \in \mathbb{N}^+$, $i \neq j$. But T preserves μ, and μ is finite. Thus $\mu(B_m) = 0$ and, hence, the measure μ of the set $\{x \in A \mid$ the set $\{n \in \mathbb{N}^+ \mid T^n x \in A\}$ is finite$\} = \bigcup_{m \in \mathbb{N}^+} B_m$ is equal to 0.

An analogue for G-spaces of the Poincaré recurrence theorem is the following

Theorem. Let G be a locally compact σ-compact group and let a G-space X with a measure μ be given. Then for every measurable $A \subset X$ and for almost all $x \in A$, the set $\{g \in G \mid gx \in A\}$ is not relatively compact in G.

1.5. The central fact of ergodic theory is the

Birkhoff individual ergodic theorem. Let T be an automorphism of a space X with a finite measure μ and let $f \in L_1(X, \mu)$. Set

$$f_n(x) = \frac{1}{n} \sum_{i=1}^{n} f(T^i x) \text{ if } n > 0, \text{ and}$$

$$f_n(x) = \frac{1}{-n} \sum_{i=1}^{-n} f(T^{-i} x) \text{ if } n < 0.$$

Then there exists a function $\tilde{f} \in L_1(X, \mu)$ such that for almost all $x \in X$

$$\tilde{f}(Tx) = \tilde{f}(x), \quad \int_X \tilde{f}(x) d\mu(x) = \int_X f(x) d\mu(x)$$

and $\lim_{|n| \to \infty} f_n(x) = \tilde{f}(x)$ a.e.

If the automorphism T is ergodic then

$$\tilde{f}(x) = \frac{1}{\mu(X)} \int_X f(x) d\mu(x).$$

The proof of this theorem, which can be found in any textbook on ergodic theory (see for example [2], [11]), is rather complicated. It is easier to prove the so called statistical ergodic theorem which asserts that f_n tends to f with respect to the norm of the space $L_1(X, \mu)$.

There are analogues of the individual and statistical ergodic theorems for G-spaces (see [20]); for this one takes f_n to be the functions of the form

$$f_n(x) = \frac{1}{\mu_G(K_n)} \int f(g^{-1}x)\,d\mu_G(x)$$

where $\{K_n\}$ is a suitable exhaustive sequence of compact sets in G and μ_G is a left-invariant Haar measure on G. It should be noted that such generalizations do not exist always, but only for some classes of groups G (in particular for compactly generated nilpotent groups).

1.6. Now let X be a locally compact separable metric space, let G be a separable metrizable locally compact group, and let a continunous action of G on X be given (i.e. $gx \in X$ is defined for all $g \in G$ and $x \in X$ and (a) $g_1(g_2x) = (g_1g_2)x$ and $ex = x$; (b) gx depends continuously on $(g,x) \in G\times X$). Let us denote by Ω the set of all non-zero G-invariant locally finite Borel measures on X and by Ω_o the set of G-ergodic measures $\mu \in \Omega$. The set Ω is a cone in the space of measures and Ω_o coincides with the set of extreme points of this cone ($\mu \in \Omega$ is called extreme if the equality $\mu = \mu_1 + \mu_2$, $\mu_1 \in \Omega$, $\mu_2 \in \Omega$ implies that μ_1 and μ_2 are proportional to μ). Every measure $\mu \in \Omega$ can bc decomposed into an integral of measures from Ω_o, i.e. there exists a measure ν_μ on Ω_o such that

$$\mu(A) = \int_{\Omega_o} \omega(A)\,d\nu_\mu(\omega)$$

for every μ-measurable $A \subset X$.

If G is amenable and X is compact, then $\Omega \neq \phi$. (The group G is called <u>amenable</u> if there is an invariant mean on the space of bounded continuous functions of G; all solvable groups are amenable). This property of amenable groups is characteristic, i.e. if G is not amenable then there exists a continuous action of G on a compact metric space Y such that there is no non-zero G-invariant Borel measure on Y.

We say that the action of G on X is <u>uniquely ergodic</u> if $\Omega = \Omega_o$ or, equivalently, if every two measures from Ω_o are proportional. A homeomorphism $T: X \to X$ is called <u>uniquely ergodic</u> if the action on X of the group $\{T^n | n \in \mathbb{Z}\}$ is uniquely ergodic. For compact X, the homeomorphism T is uniquely ergodic if and only if, for every continuous function f on X, one can find a constant c(f) such that $\lim_{n\to\infty} f_n(x) = c(f)$ for all $x \in X$ where the f_n are defined in the formulation of the Birkhoff theorem (moreover: a) $c(f) = \int_X f(x)\,d\mu(x)$ where μ is the unique normalized T-invariant Borel measure on X; b) f_n converges to c(f) uniformly on X).

We say that a measure μ on X is strictly positive if $\mu(U) > 0$ for very non-empty open $U \subset X$. If $\mu \in \Omega$ and μ is strictly positive, then the -ergodicity of μ implies that the orbit G_x is dense in X for almost ll (with respect to μ) $x \in X$; the converse is in general false. If X is ompact, the homeomorphism $T: X \to X$ is uniquely ergodic, and the unique ormalized T-invariant Borel measure on X is strictly positive, then ne orbit $\{T^n x \mid n \in \mathbb{Z}\}$ is dense in X for all $x \in X$ (but there exist homeo- orphisms $T: X \to X$ of compact spaces which are not uniquely ergodic and ich that each orbit $\{T^n x \mid n \in \mathbb{Z}\}$ is dense in X).

§2. Ergodic properties of actions on homogeneous spaces

2.1. In the study of ergodic properties of actions on homogeneous aces, an essential role is played by results of the theory of infi- ite dimensional unitary group representations. In 2.2-2.4 and 2.7-2.9 ome of these results will be presented.

2.2. Generalized Mautner lemma. Let H be a topological group and t $x, y \in H$ be elements such that the sequence $\{x^n y x^{-n}\}$ converges to e as $\to +\infty$. If ρ is a continuous unitary representation of the group H on Hilbert space W, $w \in W$, and $\rho(x)w = w$ then $\rho(y)w = w$.

Proof. Since $\rho(x)w = w$ and ρ is unitary, we have for each $n \in \mathbb{Z}$ that

$$\|\rho(y)w - w\| = \|\rho(y)\rho(x^{-n})w - \rho(x^{-n})w\| = \|\rho(x^n y x^{-n})w - w\|.$$

t $\{x^n y x^{-n}\} \to e$ as $n \to +\infty$, and ρ is continuous. Therefore $\|\rho(y)w - w\| = 0$, d hence $\rho(y)w = w$.

2.3. Corollary. Let ρ be a continuous unitary representation of the oup $SL_2(\mathbb{R})$ on a Hilbert space W, $w \in W$ and $d = \begin{pmatrix} a & 0 \\ 0 & a^{-1} \end{pmatrix} \in SL_2(\mathbb{R})$ where $\neq \pm 1$. If $\rho(d)w = w$ then $\rho(SL_2(\mathbb{R}))w = w$.

Proof. We set

$$U = \left\{ \begin{pmatrix} 1 & x \\ 0 & 1 \end{pmatrix} \middle| x \in \mathbb{R} \right\} \text{ and } U^- = \left\{ \begin{pmatrix} 1 & 0 \\ x & 1 \end{pmatrix} \middle| x \in \mathbb{R} \right\}.$$

placing if necessary d by d^{-1}, one can assume that $|a| < 1$. Then it n be easily checked that, for every $u \in U$, $\{d^n u d^{-n}\} \to e$ as $n \to +\infty$. The- fore and in view of Lemma 2.2, $\rho(U)w = w$. Analogously replacing d by ^{1} and U by U^-, we obtain that $\rho(U^-)w = w$. But the subgroups U and U^- nerate $SL_2(\mathbb{R})$. Thus $\rho(SL_2(\mathbb{R}))w = w$.

2.4. Proposition. Let ρ be a continuous unitary representation of the group $SL_2(\mathbb{R})$ on a Hilbert space W and let $U = \{\begin{pmatrix}1 & x\\ 0 & 1\end{pmatrix} | x \in \mathbb{R}\}$. If w is an element of W such that $\rho(U)w = w$, then $\rho(SL_2(\mathbb{R}))w = w$.

Proof. We may assume that $||w|| = 1$. Let us consider the continuous function $\varphi(g) = \langle\rho(g)w,w\rangle$, $g \in SL_2(\mathbb{R})$ where $\langle , \rangle$ denotes the scalar product. Since ρ is initary and $\rho(U)w = w$, the function φ is constant on a double coset modulo U. But if $\begin{pmatrix}a & b\\ d & d\end{pmatrix} \in SL_2(\mathbb{R})$ and $c \neq 0$, then we have that

$$\begin{pmatrix}1 & c^{-1}(1-a)\\ 0 & 1\end{pmatrix}\begin{pmatrix}a & b\\ c & d\end{pmatrix}\begin{pmatrix}1 & c^{-1}(1-d)\\ 0 & 1\end{pmatrix} = \begin{pmatrix}1 & 0\\ c & 1\end{pmatrix}.$$

Therefore, one has for $c \neq 0$

$$\varphi\begin{pmatrix}a & 0\\ c & a^{-1}\end{pmatrix} = \varphi\begin{pmatrix}1 & 0\\ c & 1\end{pmatrix}.$$

Passing in this equality to the limit as $c \to 0$, we obtain that $\varphi\begin{pmatrix}a & 0\\ 0 & a^{-1}\end{pmatrix} = 1$. It then follows from the property of ρ being unitary that $\rho(g)w = w$ for all $g = \begin{pmatrix}a & 0\\ 0 & a^{-1}\end{pmatrix}$. Applying now Corollary 2.3, we obtain that $\rho(SL_2(\mathbb{R}))w = w$.

2.5. Let G be a locally compact σ-compact group. Let us denote a right-invariant Haar measure on G by μ_G. For any discrete subgroup $\Gamma \subset G$, the measure μ_G induces a measure on G/Γ which will also be denoted by μ_G. A discrete subgroup $\Gamma \subset G$ is said to be a lattice if $\mu_G(G/\Gamma) < \infty$. A lattice Γ is called uniform if G/Γ is compact, and non-uniform otherwise. If G contains a lattice, then G is unimodular, i.e., the measure μ_G is left-invariant.

The group G acts on G/Γ by left translations. The measure μ_G is invariant under this action iff G is unimodular. Since the group G is σ-compact and acts transitively on G/Γ, we have the following:

Lemma. Let G be unimodular. For any discrete subgroup $\Gamma \subset G$, the measure μ_G on G/Γ is G-ergodic.

In view of this lemma and the assertion (I) in 1.3, Corollary 2.3 and Proposition 2.4 imply the following.

2.6. Theorem. Let $G = SL_2(\mathbb{R})$ and let Γ be a lattice in G.

(a) If $d = \begin{pmatrix}a & 0\\ 0 & a^{-1}\end{pmatrix}$ G where $a \neq \pm 1$, then the automorphism $x \to dx$, $x \in G/\Gamma$, of the space G/Γ (with the measure μ_G) is ergodic.

(b) Set $U = \{\begin{pmatrix} 1 & x \\ 0 & 1 \end{pmatrix} | x \in \mathbb{R}\}$. Then U acts ergodically by left translaions on G/Γ.

2.7. Corollary 2.3 and Proposition 2.4 are particular cases of heorem 2.9 stated below. Before the formulation of this theorem we ive the following:

Definition. Let G be a topological group, $H \subset G$ and $F \subset G$. We say hat the triple (G,H,F) has the Mautner property if the following condiion is satisfied: $\rho(H)w = w$ for every continuous unitary representaion ρ of G on a Hilbert space W and every $w \in W$ such that $\rho(F)w = w$.

Lemma 2.2, Corollary 2.3 and Proposition 2.4 asset respectively hat the triples $(H,\{y\},\{x\})$, $(SL_2(\mathbb{R}), SL_2(\mathbb{R}), \{d\})$ and $(SL_2(\mathbb{R}), SL_2(\mathbb{R}), U)$ ave the Mautner property.

2.8. Let G be a connected Lie group, $\mathfrak{g}$ be its Lie algebra, and Ad e the adjoint representation of G. We say that the subgroup $F \subset G$ is d-compact if the subgroup Ad(F) is relatively compact in the group of inear transformations of $\mathfrak{g}$. If H_1 and H_2 are two connected normal subroups of G such that the image of F in G/H_i $(i = 1,2)$ is Ad-compact or $i = 1,2$ then the image of F in $G/(H_1 \cap H_2)^o$ is Ad-compact as well, $(H_1 \cap H_2)^o$ denoting the connected component of the indentity of $H_1 \cap H_2$. onsequently there is a unique smallest connected normal subgroup H_F f G such that the image of F in G/H_F is Ad-compact. If G is a connected semisimple Lie group with trivial center, then H_F coincides with he product of all simple factors G_i of G for which the subgroup $\pi_i(F)$ s not relatively compact in G_i, where $\pi_i : G \to G_i$ is the natural proection.

2.9. Theorem (see [3], [16]). Let G be a connected Lie group, F be connected Lie group, F be a subgroup of G and H_F be defined as in .8. Let us assume that either G is semisimple or F is a one-parameter ubgroup. Then the triple (G, H_F, F) has the Mautner property.

2.10. It is not difficult to deduce from Theorem 2.9 the theorem n ergodicity of actions on homogeneous spaces. For this one should nly use the assertion (I) in 1.3 and the following simple

Lemma. Let G be an unimodular locally compact σ-compact group, let be a normal subgroup of G and let $\Gamma \subset G$ be a discrete subgroup. Suppose hat the subgroup $H.\Gamma$ is dense in G. Then H acts ergodically by left ranslations of G/Γ.

2.11. Theorem (see [3]). Let G be a connected Lie group, Γ be a lattice in G, F be a subgroup of G, and let H_F be defined as in 2.8. Suppose the following conditions are satisfied:

(a) either G is semisiple or F is a one-parameter subgroup;

(b) the subgroup $H_F.\Gamma$ is dense in G.

Then F acts ergodically by left translations of G/Γ.

2.12. Let us state a result on the mixing property for actions on homogeneous spaces.

Theorem (see [3]). Let G be a connected semisimple Lie group, Γ be a lattice in G, $F \subset G$ be a one-parameter subgroup subgroup and let H_F denote the same as in 2.8. Suppose that the subgroup $H_F.\Gamma$ is dense in G. Then for every $g \in F$, $g \neq e$, the automorphism $x \to gx$, $x \in G/\Gamma$, of the space G/Γ with the measure μ_G has the mixing property.

This theorem can be easily deduced from the assertions (I) and (II) in 1.3, Lemma 2.10, and the following result of representation theory:

2.13. Theorem (see [16]). Let G be a connected semisimple Lie group, $F \subset G$ be a one-parameter subgroup, H_F be defined as in 2.8 and ρ be a continuous representation of G on a Hilbert space W. Suppose that there are no non-zero $\rho(H_F)$-invariant vectors in W. Then the unitary operator $\rho(g)$ has an absolutely continuous spectrum for every $g \in F$, $g \neq e$.

§3. Closure of orbits and invariant measures for actions on homogeneous spaces

3.1. Some of the results of the previous section can be interpreted as assertions on the behaviour of typical orbits. In particular, these results imply that, under sufficiently general conditions, "almost every" orbit is dense in the homogeneous space. But the structure of an "individual" orbit can be in general very complicated. Indeed, if $G = SL_2(\mathbb{R})$, Γ is a lattice in G and $D \subset G$ is the cyclic subgroup generated by a nontrivial diagonal matrix, then there exists a $x \in G/\Gamma$ such that the closure in G/Γ of the orbit Dx is a Cantor set. But if we replace D by a unipotent subgroup then the situation becomes quite different. Namely, the following theorem holds.

3.2. Theorem (see [5]). Let $G = SL_2(\mathbb{R})$, $u = \begin{pmatrix} 1 & 1 \\ 0 & 1 \end{pmatrix}$ and let Γ be a uniform lattice in G. Then the homeomorphism $x \to ux$, $x \in G/\Gamma$, of the space G/Γ is uniquely ergodic, and hence, for every $x \in G/\Gamma$, the orbit $\{u^n x \mid n \in \mathbb{Z}\}$ is dense in G/Γ.

3.3. Looking at Theorem 3.2, one may expect that if G is a connected semisimple Lie group and Γ is a uniform lattice, then, under some natural conditions on G and Γ, every unipotent subgroup of G acts uniquely ergodically by left translations on G/Γ. But this is not so. In particular, even in the group $G = SL_3(\mathbb{R})$, there exist a uniform lattice Γ, an element $x \in G/\Gamma$ and a one-parameter unipotent subgroup U, such that the closure of the orbit $\overline{Ux}$ is a submanifold of strictly positive codimension. Nevertheless, the following two conjectures are quite plausible.

<u>Conjecture 1</u>. Let G be a connected Lie group, let Γ be a lattice in G, and let U be a subgroup of G. Suppose that U is unipotent, i.e., the transformation $\mathrm{Ad}\,u$ is unipotent for every $u \in U$. Then for every U-invariant U-ergodic locally finite Borel measure σ on G/Γ, there exist a closed subgroup $P \subset G$ containing U and a $x \in X$ such that the set Px is closed in G/Γ, and σ is a finite P-invariant measure supported on Px.

<u>Conjecture 2.</u> Let, as in Conjecture 1, G be a connected Lie group, Γ a lattice in G, and U a unipotent subgroup of G. Then for every $x \in G/\Gamma$, there exists a closed subgroup $P \subset G$ containing U such that the closure of the orbit Ux coincides with Px.

Let us note that if the orbit Px is closed then the natural map $P/P \cap G_x \to Px$ is a homeomorphism where $G_x = \{g \in G \mid gx = x\}$ is the stabilizer of x.

Conjectures 1 and 2 were stated in [4] (for the case when G is reductive and U is a one-parameter subgroup). It is also noted there that Conjecture 1 is due to M.S. Raghunathan. Raghunathan also noted the connection of his conjecture with Davenport's conjecture (see Theorem 4.4 of the present paper).

Since the group U is nilpotent, then, for any action of this group on a compact space X, there exists a finite U-invariant Borel measure on X. Therefore, in case the closure Ux of the orbit Ux is a relatively compact U-minimal subset, the validity of Conjecture 2 follows from that of Conjecture 1 (U-minimality means the orbit Uy is dense in Ux for every $y \in Ux$).

Conjecture 1 was proved in [4] in the case where G is reductive and U is a maximal unipotent subgroup, and Conjecture 2 was proved in the case where G is reductive and U is horospherical (a subgroup W is called horospherical if there exists a $g \in G$ such that $W = \{w \in G \mid g^j w g^{-j} \to e \text{ as } j \to +\infty\}$; every horospherical subgroup is unipotent and every maximal unipotent subgroup of a reductive Lie group is horospherical). For the case $G = SL_2(\mathbb{R})$, Conjectures 1 and 2 were proved in [5] (with $\Gamma = SL_2(\mathbb{Z})$) and in [8] (for arbitrary Γ); the main difference of the case $G = SL_2(\mathbb{R})$ from the general one is that every connected nontrivial unipotent subgroup of $SL_2(\mathbb{R})$ is horospherical. The following theorem was also proved in [5] and [8].

3.4. Theorem. Let $G = SL_2(\mathbb{R})$, let Γ be a lattice in G, and let $x \in G/\Gamma$. We set

$$u_t = \begin{pmatrix} 1 & t \\ 0 & 1 \end{pmatrix} \text{ and } u = u_1 = \begin{pmatrix} 1 & 1 \\ 0 & 1 \end{pmatrix}.$$

Suppose that the orbit $\{u_t x \mid t \in \mathbb{R}\}$ is not periodic, i.e., that $u_t x \neq x$ for every $t \neq 0$. Then:

(I) the orbit $\{u_t x \mid t \in \mathbb{R}\}$ is uniformly distributed with respect to the measure μ_G (defined in 2.5), i.e., for any bounded continuous function f on G/Γ one has

$$\frac{1}{T}\int_0^T f(u_t x)\,dt \to \int_{G/\Gamma} f\,d\mu_G \text{ as } T \to \infty;$$

(II) the sequence $\{u^n x \mid n \in \mathbb{Z}\}$ is uniformly distributed with respect to the measure μ_G, i.e., for any bounded continuous function f on G/Γ one has

$$\frac{1}{N}\sum_0^{N-1} f(u^n x) \to \int_{G/\Gamma} f\,d\mu_G \text{ as } N \to \infty.$$

Let us note that, in the assertions (I) and (II), one can take instead of bounded continuous functions the characterististic functions of open sets whose boundary has measure zero.

3.5. Let φ be a homeomorphism of a locally compact metric space X. Then (I) a point $x \in X$ is said to be recurrent if there exists a sequence $\{n_k\}$ such that $\varphi^{n_k} x \to x$; (II) $x \in X$ is said to be generic if there exists a finite Borel measure μ_x on X such that

$$\lim_{n\to\infty} \frac{1}{n}\sum_{i=0}^{n-1} f(\varphi^i x) = \int_X f\,d\mu_x$$

for all bounded continuous functions f on X.

The Birkhoff ergodic theorem implies that if μ is a finite Borel φ-invariant measure on X then almost all (with respect to the measure

a) points $x \in X$ are generic. If the measure μ is strictly positive as well, then almost all points $x \in X$ are recurrent.

The following conjecture was stated in [5] (for the case $G = SL_n(\mathbb{R})$ and $\Gamma = SL_n(\mathbb{Z})$).

Conjecture 3. Let G be a connected Lie group, Γ be a lattice in G and $u \in G$. Suppose that u is unipotent, i.e., that the transformation Adu is unipotent. Then every point of G/Γ is both recurrent and generic with respect to the homeomorphism $x \to ux$, $x \in G/\Gamma$, of the space G/Γ.

Theorem 3.4 implies that Conjecture 3 is valid in the case that $G = SL_2(\mathbb{R})$. Attention should be paid to the close connection between Conjectures 1 and 3. Namely if one succeeds in proving Conjecture 1 then, most likely, he would succeed in proving Conjecture 3 (apparently, the converse is true as well).

3.6. Let a continuous action of a locally compact group G on a locally compact space X be given. We say that a subgroup H of G has property (D) with respect to X if, for every H-invariant locally finite Borel measure μ on X, there exist Borel subsets X_i, $i \in \mathbb{N}^+$, such that 1) $\mu(X_i) < \infty$ for all i; 2) $\mu(X_i \Delta (hX_i)) = 0$ for all $h \in H$ and $i \in \mathbb{N}^+$; 3) $X = \bigcup_{i \in \mathbb{N}^+} X_i$. If H has property (D) with respect to X, then, as one can easily see, every H-ergodic H-invariant locally finite Borel measure on X is finite (when G and X are separable and metrizable, the converse is also true).

The following theorem should be mentioned in connection with Conjecture 1.

3.7. Theorem (see [6]). Let G be a connected Lie group and let Γ be a lattice in G. Then any unipotent subgroup U of G has property (D) with respect to G/Γ.

3.8. Lemma. Let a continuous action of a locally compact group G on a locally compact space X be given and let H and F be subgroups of X. Suppose that the triple (G,H,F) has the Mautner property. Then

(I) if μ is a Borel measure on X, $A \subset X$ is a Borel subset, $\mu(A) < \infty$, and $\mu(A \Delta (fA)) = 0$ for all $f \in F$, then $\mu(A\Delta(hA)) = 0$ for all $h \in H$;

(II) if F has property (D) with respect to X, then H has property (D) with respect to X as well;

(III) if $F \subset H$ and F has property (D) with respect to X, then every H-ergodic H-invariant locally finite Borel measure on X is F-ergodic.

Assertion (I) follows from the equivalence of the conditions "$\rho(g)\chi_A = \chi_A$" and "$\mu(A\Delta(gA)) = 0$", where ρ is the unitary representation of G in the space $L_2(G,\mu_G)$ defined in 1.3, and χ_A is the characteristic function of the set A. Assertions (II) and (III) immediately follow from assertion (I) and the definitions of the property (D) and the Mautner property.

3.9. If H is a connected semisimple Lie group without compact factors, and U is a maximal unipotent subgroup of H, then, as follows from Theorem 2.9, the triple (G,H,U) has the Mautner property when G is an arbitrary Lie group containing H. This, Theorem 3.7 and Lemma 3.8 (II) imply:

3.10. Theorem. Let G be a connected Lie group, let Γ be a lattice in G, and let $H \subset G$ be a connected semisimple Lie group without compact factors. Then H has property (D) with respect to G/Γ.

A particular case of theorem 3.10 is

3.11. Theorem. Let G, Γ and H be the same as in Theorem 3.10 and let $x \in G/\Gamma$. Suppose that the orbit Hx is closed in H/Γ. Then $H \cap G_x$ is a lattice in H, where $G_x = \{g \in G \mid gx = x\}$ is the stabilizer of x.

3.12. Remarks. (I) In Theorems 3.10 and 3.11, the condition "H is a connected semisimple Lie group without compact factors" can be replaced by the condition" H is a connected Lie group such that the factor of the radical of H by the nilpotent radical of H is compact".

(II) Using remark (I) and choosing $G = SL(n, \mathbb{R})$ and $\Gamma = SL(n, \mathbb{Z})$, it is not difficult to deduce from Theorem 3.11 the Borel-Harish-Chandra Theorem on finiteness of volumes of quotient spaces of Lie groups by their arithmetic subgroups.

(III) Using Theorem 2.9 and Lemma 3.8 (III), it is not difficult to show that if we replace in Conjecture 1 the condition "U is unipotent" by the hypothesis "U is generated by unipotent elements", then we obtain an equivalent conjecture. Let us note that if the factor of a connected Lie group H by its nilpotent radical is semisimple and has no compact factors, then H is generated by unipotent elements.

3.13. In view of the last remark and in view of some other considerations, it seems reasonable to generalize Conjecture 2 in the following way.

Conjecture 2'. Let G be a connected Lie group, let Γ be a lattice n G, and let H be a subgroup of G. Suppose that H is generated by unipotent elements. Then, for every $x \in G/\Gamma$, there exists a closed subgroup $\subset G$ containing H, such that the closure of the orbit Hx coincides with x.

§4. Applications to number theory and concluding remarks

4.1. For any $t \in \mathbb{R}$ let $[t]$ denote the largest integer not exceeding and set $\{t\} = t - [t]$. For any two positive integers m and n, let (m,n) enote the g.c.d. of m and n. In [5], with the help of Theorem 3.4 (II), he following was established:

4.2. Theorem. For any irrational number Θ

$$\lim_{T\to\infty} \frac{1}{T} \sum_{\substack{0<m\le T\{m\Theta\} \\ (m,[m\Theta])=1}} \{m\Theta\}^{-1} = \frac{1}{\zeta(2)} = \frac{6}{\pi^2},$$

here ζ stands for the Riemann zeta function.

4.3. Let B be a real nondegenerate indefinite quadratic form in n ariables. It is well known that if $n \ge 5$ and the coefficients of B are ational, then B represents zero nontrivially, i.e., there exist integers $x_1,\ldots,x_n$, not all equal to 0, such that $B(x_1,\ldots,x_n) = 0$. Theorem .4 stated below can be considered as an analogue of this assertion in he case when B is not proportional to a form with rational coefficients. ote that in Theorem 4.4 the condition "$n \ge 5$" is replaced by the condition "$n \ge 3$".

4.4. Theorem (see [15]). Suppose that $n \ge 3$ and that B is not proortional to a form with rational coefficients. Then for any $\varepsilon > 0$, here exist integers $x_1,\ldots,x_n$, not all equal to 0, such that $|B(x_1,\ldots,x_n)| < \varepsilon$.

4.5. Theorem 4.4 easily implies that, under the conditions of this heorem, the set of values taken by B on integer points is dense in the et of reals.

One can easily understand that if Theorem 4.4 is proved for some n_0, then it is proved for all $n \ge n_0$. So it is enough to prove this heorem for n = 3. Let us note that if n = 2 then the analogous assertion is false; for this one can consider the form $x_1^2-\lambda x_2^2$, λ an irraional positive number such that $\sqrt{\lambda}$ has a continued fraction development with bounded partial quotients.

Theorem 4.4 answers Davenport's conjecture (see [9]). It has been proved earlier in the following cases: (a) $n \geq 21$ (see [10]); (b) $n = 5$ and B is of the form $B(x_1,\ldots,x_5) = \lambda_1 x_1^2+\ldots+\lambda_5 x_5^2$ (see [9]). The proofs given in [9] and [10] are based on the use of methods from analytic number theory.

In [15], Theorem 4.4 is deduced from Theorem 4.6 stated below. Theorem 4.6 answers Conjecture 2' in 3.10 in a particular case. The proof of this theorem given in [15] is based on the use of methods of algebraic group theory and ergodic theory (more exactly, the topological theory of dynamical system).

4.6. <u>Theorem</u>. Let $G = SL_3(\mathbb{R})$, $\Gamma = SL_3(\mathbb{Z})$ and H denote the group of elements of G preserving the form $2x_1x_3 - x_2^2$. Let $G_x = \{g \in G \mid gx = x\}$ denote the stablizer of $x \in G/\Gamma$. If $x \in G/\Gamma$ and the orbit Hx is relatively compact in G/Γ, then the quotient space $H/H \cap G_x)$ is compact.

4.7. Let us give the reduction of Theorem 4.4 to Theorem 4.6. Let H_B denote the group of elements of G preserving B. As explained in 4.5, it is enough to prove Theorem 4.4 for $n = 3$. In this case $H = g_B H_B g_B^{-1}$ for some $g_B \in G$. Let us suppose that the assertion of Theorem 4.4 fails. Then one can easily show using Mahler's compactness criterion that the set $H_B \mathbb{Z}^3$ is relatively compact in G/Γ (we identify G/Γ with the space of lattices in $\mathbb{R}^3$). Now we apply Theorem 4.6 for $x = g_B \mathbb{Z}^3$ and get that the quotient space $H/H \cap G_x$, and consequently the quotient space $H_B/H_B \cap \Gamma$, are compact. Then in view of Borel's density theorem (see [1]), $H_B \cap \Gamma$ is Zariski dense in H_B. But $\Gamma = SL_3(\mathbb{Z})$. So H_B is a $\mathbb{Q}$-subgroup of G and hence B is proportional to a form with rational coefficients. Contradiction.

Using the methods of [15], one can prove Raghunathan's conjecture (i.e. Conjecture 2 in 3.3) in case $G = SL_3(\mathbb{R})$ and the orbit Ux is relatively compact in G/Γ. This makes it possible to prove the following theorem, which can be considered as a generalization of Theorem 4.4.

4.8. <u>Theorem.</u> Let B_1 and B_2 be two real quadratic forms in 3 variables. Suppose that

1) there exists a basis of the space $\mathbb{R}^3$ in which B_1 and B_2 have the form $2x_1x_3-x_2^2$ and x_1^2, respectively;

2) every non-zero linear combination of B_1 and B_2 is not proportional to a form with rational coefficients.

Then for any $\varepsilon > 0$, there exist integers x_1, x_2, x_3, not all equal to 0, such that $|B_1(x_1,x_2,x_3)| < \varepsilon$ and $|B_2(x_1,x_2,x_3)| < \varepsilon$.

4.9. Concluding remarks. In connection with the content of §2, let us note that a detailed survey of ergodic properties of actions on homogeneous spaces is presented in [3]. As for the decomposition of such actions into ergodic components, see [19].

Many questions concerning connections between Lie group theory and ergodic theory remained untouched in this paper. In particular, results on rigidity for discrete subgroups and ergodic actions (see [14], [21], [22]), on the connection between finiteness of factor groups of discrete subgroups and invariant algebras of measurable sets (see [12], [13]), and on rigidity of horocycle flows (see [17], [18]) have been omitted.

References

1. Borel A. Density properties for certain subgroups of semisimple groups without compact components. Ann.Math. 72, (1960), 179-188.
2. Billingsley P. Ergodic Theory and Information. John Wiley and Sons, Inc. New York, London, Sydney, 1965.
3. Brezin J., Moore C.C. Flows on homogeneous spaces: a new look. Am. J.Math. 103, (1981), 571-613.
4. Dani S.G. Invariant measures and minimal sets of horospherical flows. Invent.Math. 64, (1981), 357-385.
5. Dani S.G. On uniformly distributed orbits of certain horocycle flows. Ergod.Th. and Dynam.Syst. 2, (1981), 139-158.
6. Dani S.G. On orbits of unipotent flows on homogeneous spaces. Ergod. Th. and Dynam.Syst. 4, (1984), 25-34.
7. Dani S.G. Orbits of horospherical flows. Duke Math.J. 53, (1986), 177-188.
8. Dani S.G., Smillie J. Uniform distribution of horocycle orbits for Fuchsian groups. Duke Math.J. 51, (1984), 185-194.
9. Davenport H., Heilbronn H. On indefinite quadratic forms in five variables. J.Lond.Math.Soc., II ser., 21, (1946), 185-193.
10. Davenport H., Ridout H. Indefinite quadratic forms. Proc.Lond.Math. Soc., III Ser., 9, (1959), 544-555.
11. Cornfel'd I.P., Sinai Ya.G., Fomin S.V. Ergodic theory. Nauka 1980, Moscow (in Russian; English Translation: Springer-Verlag, Berlin, Heidelberg, New-York 1982).
12. Margulis G.A. Quotient groups of discrete subgroups and measure theory. Funkts.Anal.Prilozh. 12, No 4, (1978), 64-76 (in Russian ; English translation: Funct.Anal.Appl. 12, (1978), 295-305).
13. Margulis G.A. Finiteness of quotient groups of discrete subgroups. (in Russian: Funkts.Anal.Prilozh. 13, No 3, (1979), 28-39; English translation: Funct.Anal.Appl. 13, (1979), 178-187).
14. Margulis G.A. Arithmeticity of irreducible lattices in semisimple groups of rank greater than 1 (in Russian). Appendix to the Russian translation of: Raghunathan M.S. Discrete subgroups of Lie groups. Mir, Moscow, 1977. (English translation: Invent.Math. 76, (1984), 93-120).

15. Margulis G.A. Formes quadratiques indéfinies et flots unipotents sur les espaces homogènes. C.R.Acad.Sci.Paris, Ser.I, 304, (1987), 249-253.

16. Moore C.C. The Mautner phenomenon for general unitary representations. Pac.J.Math. 86, (1980), 155-169.

17. Ratner M. Rigidity of horocycle flows. Ann.Math. 115, (1982), 597-614.

18. Ratner M. Horocycle flows, joinings and rigidity of products. Ann. Math. 118, (1983), 277-313.

19. Starkov A.N. The ergodic behavior of flows on homogeneous spaces. Dokl.Akad.Nauk SSSR, 273, (1983), 538-540 (in Russian; English translation: Sov.Math.Dokl., 28 (1983), 675-676).

20. Tempelman A.A. Ergodic theorems on groups. Mokslas, Vil'nius, 1986 (in Russian).

21. Zimmer R.J. Strong rigidity for ergodic actions of semisimple Lie groups. Ann.Math. 112, (1980), 511-529.

22. Zimmer R.J. Ergodic theory and semisimple groups. Birkhäuser Verlag, Boston, 1984.

GALOIS THEORY OF DATABASES

B.I. Plotkin
al. Vidzemes 8, kv. 35
226024 Riga, USSR

1. Introduction

Database theory is a large applied science, which uses various mathematical methods. The relational approach, discovered by E.F. Codd [1, 2], gives wide possibilities for applying universal algebra and algebraic logic in database theory. It is possible to define a database algebraic model and to create a suitable constructive model on this foundation.

Here are a few words about what the model can give for applications.

First of all, we hope that the algebraic model serves as a guide in understanding the nature of databases. This model allows us to consider a natural concept of isomorphism of two databases, and to give a precise definition of the informational equivalence of databases.

The model allows us to speak about algebraic structure of databases, to define different constructions of database composition and decomposition on an abstract level, to consider the problem of decomposition of databases, to define complexity of databases in these terms.

Due to this model it has become possible to construct for databases a Galois theory, which we shall consider in the article.

Finally, the model helps organizing calculations and programming. Different problems connected with complexity of calculations are formulated in a natural way. It is expected that the model in question will be enriched by additional structures connected with numeration and algorithms.

In other words, a constructive database model is to be created.

The aim of the article is to introduce an adequate algebraic database model, and to use it in order to provide an approach to the classification of databases by means of symmetries.

2. Database scheme

The construction of a database begins from its scheme. This scheme includes, first of all, a set of variables X, a set of sorts of variables Γ, and a map $n:X \to \Gamma$, which defines a stratification of the set X into components X_i, $i \in \Gamma$, consisting of variables of the same sort i.

Further, the scheme includes a variety of data algebras Θ, which plays the role of type of data. In general, algebras from Θ are multi-sort and are denoted $D = (D_i, i \in \Gamma)$. Thus, the set Γ is simultaneously a list of names of domains which are included in data algebras. To give a variety Θ means to specify, first of all, a definite set Ω of the main operations symbols connected with Γ. To each symbol $\omega \in \Omega$ corresponds a definite type $\tau = (i_1,\ldots,i_n)$, consisting from sorts in Γ, and for each $D = (D_i, i \in \Gamma)$ one has an operation

$$\omega: D_{i_1} \times \ldots \times D_{i_n} \to D_j .$$

The identities which link the operations from Ω define the variety Θ.

The set Φ of symbols of the main relations is then also included in the scheme, together with the set of symbols of operations Ω. Each $\varphi \in \Phi$ also has a type $\tau = (i_1,\ldots,i_n)$: with it φ is realized in an algebra $D \in \Theta$ as a relation, i.e. a subset of the Cartesian product $D_{i_1} \times \ldots \times D_{i_n}$. Realization is carried out by some function f, defined on the set Φ, and this function is treated as a database state.

Finally, the set of axioms A, which should be satisfied by all states considered, may be included in the scheme.

A database in the considered scheme is defined for various data algebras $D \in \Theta$. The aim of this article - to give an approach to database classification - is fulfilled in case the algebra $D = (D_i, i \in \Gamma)$ is finite, i.e. all D_i are finite and the set of variables in the scheme is sufficient to differentiate data. The latter means that one always has the inequality $|X_i| \geq |D_i|$ of cardinal numbers.

The next two sections are devoted to the definition of databases. A database is represented as an algebraic structure, which takes into account the scheme above.

3. Halmos algebras

In the first approximation, a database is an algebraic automaton of type (F,Q,R), where F is the set of states, Q is the algebra of reuests, and R is the algebra of replies. There is an operation $*:F\times Q\to R$. ere $f*q = r$ is a reply to the request q in the state f.

It is supposed that the algebras Q and R are of the same type, and t is asked that for each $f\in F$ the map $\hat{f}:Q\to R$, defined by $\hat{f}(q) = f*q$, hould be a homomorphism of algebras. The latter means that the strucure of the reply is coordinated with the structure of the request. uch an automaton we shall call a $*$-automaton. But this is not yet he database, because the connection with definite data algebra has ot been pointed out. Moreover, the concepts of request and reply need ore detailed definitions. The algebras Q and R are polyadic (we call them almos algebras, and we go over to define such algebras.

The requests of relational databases are usually recorded by means f the language of first-order logic.

We consider a specialized language for the variety Θ. Proceed from he scheme described above. Let $X = (X_i, i\in\Gamma)$ be the system of sets for he scheme, and based on it, take a free algebra $W = (W_i, i\in\Gamma)$ in Θ. urther on, define elementary formulas as formulas of the form $(w_1,\dots,w_k)$, where $\varphi\in\Phi$, the type of φ is $\tau = (i_1,\dots,i_n)$, and $w_s\in W_{i_s}$.

From the elementary formulas, as usual, construct all formulas, sing the boolean operations $\vee, \wedge, \neg$, and the existential quantifier $x, x\in X$. Denote the set of all formulas by $\tilde{\Phi}$. Axioms of predicate calcuus and rules of inference, connected with the chosen Θ, are singled out.

Two formulas u and v are called equivalent, if one has $\vdash(u\to v)\wedge(v\to u)$.

Denote this equivalence by ρ. If $u\rho v$ holds, then to the formulas u nd v corresponds one and the same request, and in any state f - one nd the same reply.

In general, the request is a class of equivalent formulas. Let $U = \tilde{\Phi}/\rho$. he set of requests U must be converted into an algebra. As a matter of act, here is an algebraization of predicate calculus.

It is known that an algebraization of propositional calculus leads o boolean algebras.

There are three approaches to predicate calculus: Tarski's cylinric algebras [5], Halmos' polyadic algebras [4], and the categorial pproach, based on Lawvere's ideas [8,9]. All these approaches have een considered in the pure case, that is when the variety Θ is absent rom the scheme. We can also consider the empty set of operations Ω. hese three approaches can be realized for any Θ.

We shall define the Halmos algebras now.

Definition (We adhere to the scheme from the previous section.) A almos algebra H in this scheme is, first of all, a boolean algebra. oreover, for each subset $Y\subset X$ an existential quantifier $\exists(Y)$ acts. By definition, a map $\exists:H\to H$ is an existential quantifier of the boolean algebra H, if the following three conditions hold:

1. $\exists 0 = 0$
2. $h < \exists h$
3. $\exists(h_1 \wedge \exists h_2) = \exists h_1 \wedge \exists h_2$.

Here 0 denotes zero in H, and h, h_1, h_2 are arbitrary elements from H.

For quantifiers $\exists(Y)$, $Y\subset X$, the following axioms hold:

1. $\exists(\emptyset)h = h$
2. $\exists(Y_1 \cup Y_2)h = \exists(Y_1)\exists(Y_2)h$, $h\in H$.

Suppose further, that the semigroup S = EndW acts on H as a semigroup of boolean endomorphisms. This action is connected with the quantifiers. We shall give the corresponding axioms.

Let σ and s be two elements from S, which act in the same way on those variables from X, which do not belong to the subset Y. Then

3. $\sigma\exists(Y)h = s\exists(Y)h$, $h\in H$.

Let, further, σ be an element of S, and let Y be a subset of X, which satisfy the following conditions:

if $\sigma(x) = \sigma(y)\in Y$, then $x = y$, and if $\sigma(x)\notin Y$, then all the variables from X, included in the record of the element $\sigma(x)$, do not belong to Y either. Then:

4. $\exists(Y)\sigma h = \sigma\exists(\sigma^{-1}Y)h$, $h\in H$.

All the enumerated axioms are the identities and, finally, the scheme defines the variety of Halmos algebras.

The set of requests $U = \tilde{\Phi}/\rho$ is naturally provided with a structure of Halmos algebra and thus U belongs to the variety described above. It should be emphasized that the algebra U depends only on the database scheme.

We now go over to the examples of Halmos algebras, defined by algebras $D\in\Theta$.

For each $D\in\Theta$ we shall consider the set of homomorphisms Hom(W,D), and let $\underline{M}_D$ be the set of all subsets of Hom(W,D). Then $\underline{M}_D$ is a boolean algebra. For $Y\subset X$ define the action of the existential quantifier $\exists(Y)$ in $\underline{M}_D$ by the rule: if $A\in\underline{M}_D$ and $\mu\in \mathrm{Hom}(W,D)$, then $\mu\in\exists(Y)A$ if one can find a $\nu\in A$, such that $\mu(x) = \nu(x)$ for each $x\in X$, $x\notin Y$.

Further, define the action of the semigroup S on the set Hom(W,D). For $s\in S$ and $\mu\in \mathrm{Hom}(W,D)$ set $(\mu s)(x) = \mu(s(x))$. Then $\mu\in sA \Leftrightarrow \mu s\in A$, $A\in\underline{M}_D$.

It is easy to verify that for $\underline{M}_D$ all axioms of Halmos algebra in the scheme in question hold.

Note further, that for each state f of signature Φ in an algebra D there corresponds canonically a homomorphism of Halmos algebras $\hat{f}:U\to\underline{M}_D$.

This homomorphism is constructed in the following way. First of ll, we construct the map $\tilde{f}:\tilde{\Phi} \to \underline{M}_D$. Let $\varphi(w_1,\ldots,w_n)$ be an elementary formula. Define $\tilde{f}(\varphi(w_1,\ldots,w_n))$ as the set of all $\mu\in\mathrm{Hom}(W,D)$, for which $(w_1^\mu,\ldots,w_n^\mu)\in f(\varphi)$ holds. Further, $\hat{f}$ extends inductively on the whole set , and then $\tilde{f}$ defines $\hat{f}$.

Separately we also consider the pure case. In this case $D = (D_i,\ i\in\Gamma)$ s the system of sets D_i and there are no algebraic operations in it. he algebra W coincides with the system $X = (X_i,\ i\in\Gamma)$, and the semigroup is a semigroup of transformations of the system X. It is clear that he set Hom(X,D) can now be identified with the Cartesian product of all $D_i^{X_i}$ $(i\in\Gamma)$. Thus, the algebra $\underline{M}_D$ consists of subsets - relations in this artesian product.

Really, it is sufficient to restrict to subsets, which have finite upport, i.e. subsets, which can be recognized by a finite set of variables. Having this in mind, we have in addition some definitions.

<u>Definition.</u> Let H be any Halmos algebra with arbitrary scheme, $h\in H$ nd $x\in X$. The variable x is said to belong to the support of the element (h depends on x), if $\exists(x)h \neq h$ holds. The set of all these x is designated as Δh. An element h has finite support, if Δh is a finite set.

<u>Definition</u>. An algebra H is called locally finite, if all its elements have finite support.

The algebra of requests U is always locally finite. In an arbitrary , all elements with finite support form a subalgebra - the locally inite part of U.

Denote the locally finite part of $\underline{M}_D$ by V_D. Then, for each state f e have a homomorphism $\hat{f}:U\to V_D$.

. <u>Database model</u>

For a given signature Φ and an algebra $D\in\Theta$, by $\underline{F}_D = \underline{F}_D(\Phi)$ we shall enote the set of all states of the set Φ in D. With the scheme and the lgebra D we form a triplet $(\underline{F}_D,\ U,\ V_D)$. To each $f\in\underline{F}_D$ corresponds a omomorphism $\hat{f}:U\to V_D$. Define then the operation $*:\underline{F}_D\times U\to V_D$ by the rule:

$$f*u = \hat{f}(u),\quad f\in\underline{F}_D,\ u\in U.$$

hus, we have a *-automaton $(\underline{F}_D, U, V_D) = \mathrm{Atm}D$.

For an arbitrary abstract *-automaton $\underline{A} = (F,Q,R)$ we shall consider representations

$$\rho:\underline{A}\to \mathrm{Atm}D.$$

ere, $\rho = (\alpha,\beta,\gamma)$, where:

$\alpha:F\to\underline{F}_D$ is the map, which transforms abstract states into real states;

$\beta: U \to Q$ is the homomorphism of Halmos algebras, connecting the requests with their records by language formulas;

$\gamma: R \to V_D$ is the homomorphism which assigns realizations of relations (i.e. subsets of Hom(W,D)) to abstract replies.

At last, the following equality is to hold:

$$(f_* u^{\beta})^{\gamma} = f^{\alpha}{}_* u, \ f \in F, \ u \in U.$$

<u>Definition</u> [12] A database is a $*$-automaton $\underline{A} = (F,Q,R)$, considered together with a representation $\rho: \underline{A} \to \mathrm{AtmD}$.

The representation ρ gives the connection with the data algebra. A database can also be denoted by $\underline{A}$, and finally $\underline{A} = (F,Q,R;U,D,\rho)$.

Here the universal algebra of requests U represents the scheme.

We shall consider an important special case.

Let (F,U,R) be a subautomaton of AtmD = $(\underline{F}_D, U, V_D)$, i.e. F is a subset of $\underline{F}_D$ and R is a subalgebra in V_D, and assume that for each $f \in F$ and each $u \in U$ one has $f * u \in R$.

Let us take a congruence τ in U, defined by the condition:

$$u_1 \tau u_2 \Rightarrow f * u_1 = f * u_2, \ \forall f \in F.$$

Passing to $Q = U/\tau$, we obtain a $*$-automaton (F,Q,R). The natural homomorphism $U \to Q$, together with the identity maps $F \to \underline{F}_D$ and $R \to V_D$, defines the required representation.

Such databases we shall call concrete, as distinct from general abstract databases. In the concrete case, (F,Q,R) may be interpreted as a subautomaton of an universal automaton - the database AtmD = $= (\underline{F}_D, U, V_D)$.

Note, further, that homomorphisms of Halmos algebras are well realized by ideals and filters. An ideal is a full inverse image of zero, and a filter is a full inverse image of the unit.

Homomorphisms of databases can also be considered. We can speak of homomorphisms in a given scheme, and of homomorphisms which change the scheme. We shall return to such homomorphisms below.

Some further remarks may be made. First of all, note that dynamical databases can also be considered. Here we have a semigroup Σ, which regulates the changes of states in F. Simultaneously this semigroup acts on the algebra of requests Q. Thus, Q is the Halmos dynamical algebra, and we may start from dynamical requests, i.e. requests that take into account changes of the states.

In the definitions we started from Halmos polyadic algebras as algebraic equivalents of predicate calculi.

On desire, we can start from cylindric algebras or from the cateorial approach, leading to relational algebras [9,10]. The categorial pproach is suitable because it allows to turn to databases with indisinct information. In this case the category of sets is replaced by ome topos of fuzzy sets, and categorial algebra is used.

. Galois theory of algebras of relations

Let a homomorphism $\delta: D' \to D$ of algebras in Θ be given. This homoorphism naturally induces a map $\tilde{\delta}: \mathrm{Hom}(W,D') \to \mathrm{Hom}(W,D)$.

If a homomorphism $\mu: W \to D'$ is given, then, applying $\delta: D' \to D$, we btain $\mu\tilde{\delta}$. If, later on, $A \in \underline{M}_D$, then we define $\delta_* A$ by $\mu \in \delta_* A$, if $\mu\tilde{\delta} \in A$. his gives a map $\delta_*: \underline{M}_D \to \underline{M}_{D'}$.

The map is coordinated with the boolean structure, and if the iniial $\delta: D' \to D$ is a surjection, then $\delta_*: \underline{M}_D \to \underline{M}_{D'}$ is an injection of Halos algebras. Here we have a monomorphism $\delta_*: V_D \to V_{D'}$.

Let us define $\delta_*: \underline{F}_D \to \underline{F}_{D'}$. As we know, an $\hat{f} \in \mathrm{Hom}(U,V_D)$ corresponds o each f from $\underline{F}_D$. It is easy to check, that the correspondence $:\underline{F}_D \to \mathrm{Hom}(U,V_D)$ is a bijection. We shall use this bijection now.

Starting from $\delta_*: V_D \to V_{D'}$, set $\delta^*: \mathrm{Hom}(U,V_D) \to \mathrm{Hom}(U,V_{D'})$, defining $^*(\mu) = \delta_* \mu$ for each $\mu: U \to V_D$. Then, the map $\delta_*: \underline{F}_D \to \underline{F}_{D'}$ we define by he commutative diagram:

$$\begin{array}{ccc} \underline{F}_D & \xrightarrow{\delta_*} & \underline{F}_{D'} \\ \wedge\downarrow & & \downarrow\wedge \\ \mathrm{Hom}(U,V_D) & \xrightarrow{\delta^*} & \mathrm{Hom}(U,V_{D'}) \end{array}$$

This diagram means that for any $f \in \underline{F}_D$ one has $\widehat{\delta_* f} = \delta_* \hat{f}$.

Note that $\underline{F}_D$ actually always is a boolean algebra, and the corresonding δ_* preserves this structure.

If follows from the definitions that for any $f \in \underline{F}_D$ and any $u \in U$, $_*(f*u) = (\delta_* f)*u$ holds. This formula means that two different δ_* give njection for *-automata $\delta_*: \mathrm{AtmD} \to \mathrm{AtmD}'$. Fix then an algebra $D \in \Theta$ and ake a group of automorphisms $G = \mathrm{AutD}$. In accordance with the corresondance $g \to g_*$ for each $g \in G$ we have a canonical representation

$$\mathrm{AutD} \to \mathrm{Aut}\underline{M}_D .$$

Theorem 1. The given representation is an isomorphism of groups.

(Notes about proofs of this and other results will be given later n.)

Simultaneously, we have a representation $G \to \mathrm{AutV}_D$, and on the base f this representation we can construct the Galois theory of the algebras of relations. In this case we proceed from the concept of Halmos algebra with equality.

Definition. Equality in a Halmos algebra H is a function d, which is connected with the free algebra $W = (W_i, i\in\Gamma)$ in Θ, and assigns to each pair of elements w and w' from W_i an element $d(w,w')\in H$. This function has to satisfy some conditions, which imitate the axioms of equality, namely:

1. $sd(w,w') = d(sw,sw')$, $s\in \mathrm{End}W$;
2. $d(w,w) = 1$ for each w;
3. $d(w_1,w_1')\wedge\ldots\wedge d(w_n,w_n') < d(w_1,\ldots,w_n\omega, w_1',\ldots,w_n'\omega)$,

it ω is an operation in Θ of the corresponding type;

4. $s_w^x h\wedge d(w,w') < s_{w'}^x h$, $h\in H$, $x\in X$, and $w,w'\in W_i$, if $n(x) = i$, where $s_w^x\in \mathrm{End}W$, $s_w^x(x) = w$, and $s_w^x(y) = y$ when $y\neq x$.

It can be proved, that if equality can be defined in H, then this can be done in a unique way. In an algebra of type $\underline{M}_D$ equality is defined as follows: an element $\mu\in \mathrm{Hom}(W,D)$ belongs to $d(w,w')$, if $w^\mu = w'^\mu$. For an algebra of requests U, it is supposed that the set of symbols of relations Φ is expanded by the symbols of equality $\equiv_i$ for each sort $i\in\Gamma$, and by the corresponding axioms of equality.

In the sequel we shall always consider Halmos algebras with equality, and each $d(w,w')$ will be considered as a nullary operation. Specifically, in questions dealing with subalgebras, all of these have to include all $d(w,w')$, and the $d(w,w')$ have to be invariant with respect to automorphisms. The elements of form $d(w,w')$ are also called diagonals.

In an algebra $\underline{M}_D$ the condition to be invariant by the automorphisms from AutD holds, and V_D is an algebra with equality as well.

We now go over to the Galois correspondence.

For any subset $R\subset V_D$, $H = R'$ denotes the subgroup in $G = \mathrm{Aut}D$, consisting of all $g\in G$, which induce the identity on R. If, further, H is a subset of G, then $R = H'$ denotes the set of all elements of V_D, which are invariant for each $g\in H$. This set R always is a subalgebra in the algebra with equality V_D.

Theorem 2. If an algebra D is finite and the condition of sufficiency of the set of variables holds, then the Galois correspondence described above, gives one-to-one correspondence between the subgroups of G and the subalgebras of V_D.

A similar theorem for pure algebras of relations was proved long ago by M.I. Krasner [6,7]. In the context of databases it was proved by E.M.

eniaminov [10] too. The transition to pure one-sort Halmos algebras as done by A. Daigneault [3]. For multisort Halmos algebras the theoem is proved by S.N. Boyko [11]. The generalization to Halmos algebas, specialized in some Θ, was done by the author in cooperation with .S. Maphtsir.

The following theorem on conjugacy holds:

Theorem 3. Under the assumptions of Theorem 2, let R_1 and R_2 be wo subalgebras of V_D. They are then isomorphic if and only if they re conjugate by some $g \in G$: $R_1 = gR_2$.

Hence it follows, that under the conditions we have imposed, the roup G really is the group of all automorphisms of an algebra V_D by he canonical representation $\mathrm{Aut}D \to \mathrm{Aut}V_D$.

We shall now make some remarks on the proofs of these theorems in he case of specialized Halmos algebras.

Suppose that the set Γ is finite, $\Gamma = \{1,\dots,k\}$. Denote $\tilde{D} = D_1^{x_1} \times \dots \times D_k^{x_k}$. ach $a \in \tilde{D}$ corresponds to a unique element $\bar{a} \in \mathrm{Hom}(W,D)$. This gives a bijecion $\nu: \tilde{D} \to \mathrm{Hom}(W,D)$. For each $x \in X$ we have $a(x) = a^{\nu}(x)$.

The semigroup $S = \mathrm{End}W$ acts on the set $\mathrm{Hom}(W,D)$. By the bijection this action is transferred to $\tilde{D}$. The system $\underline{M}_D^o$ of all subsets of $\tilde{D}$ aturally forms a pure Halmos algebra, and on the basis of the transiion above it can be provided with a structure of Halmos algebra, speialized in the variety Θ. This algebra turns out to be isomorphic to ne initial $\underline{M}_D$.

The permutation group of the system $D = (D_i, i \in \Gamma)$ we denote by G^o. irst of all, for the pure case it is proved that there is an isomorhism $G^o \to \mathrm{Aut}\underline{M}_D^o$. Further it is proved, that an element $g \in G^o$ commutes ith the action of the semigroup $S = \mathrm{End}W$, if and only if it belongs the group of automorphisms $G \subset G^o$. This leads to Theorem 1.

Theorem 2 is initially proved in the pure case, and then the desribed transitions are used. The fact that if a subalgebra R in V_D conains all the diagonal elements $d(w,w')$, then the subgroup R' in G^o is tually contained in $G = \mathrm{Aut}D$, plays the decisive role.

In Theorem 3 similar considerations are used. Two sets of elements R_1 and R_2 in V_D are called equivalent if they generate the same subalebra R. It is clear, that under the conditions in force the sets R_1 nd R_2 are equivalent if the groups R_1' and R_2' coincide. It may also be roved, that each set R is equivalent to one element, that is each subgebra in V_D is generated by one element. These circumstances may be aken into account in databases.

6. Galois theory of databases

It follows from the general remarks at the beginning of the previous section that the group G acts also on the boolean algebra $\underline{F}_D$. It is not difficult to understand that if $f \in \underline{F}_D$, $g \in G$ and $gf = f$, i.e. f is invariant under g, then g is an automorphism of the model (D, Φ, f).

The Galois correspondence for this action of the group G on $\underline{F}_D$ may be considered. If H is a subset of G, then $F = {}'H$ denotes the set of all $f \in \underline{F}_D$, which are invariant under each $g \in H$. F is a subalgebra of the boolean algebra $\underline{F}_D$. If, furthermore, F is a subset of $\underline{F}_D$, then $F' = H$ is the subgroup of G, consisting of all $g \in G$, which preserve each state from F (automorphisms of models from F).

Let us start from the conditions of the previous section. Let $D = (D_i, i \in \Gamma)$ be a finite algebra in Θ, and $D = \cup D_i$, $i \in \Gamma$. We shall take the set of variables-attributes $x_{\alpha_1}, \ldots, x_{\alpha_n}$, which is in one-to-one correspondence with the elements of the set D. This set is assumed to be ordered, and we shall take the corresponding type $\tau = (i_1, \ldots, i_n)$. Let us take the symbol of relations φ of type τ; we shall assume it belongs to the base set Φ. Then we have a formula $\varphi(x_{\alpha_1}, \ldots, x_{\alpha_n})$, and in the algebra U we get an element, which we call the supporting element, corresponding to it.

<u>Theorem 4</u>. Each subgroup $H \subset G$ is closed in the Galois correspondence under consideration, i.e. $H = ({}'H)'$.

In the proof of this theorem the previous results are used, and moreover we take into account the following. A subgroup $H \subset G$ can be considered as a subset-relation in the Cartesian product $D_1^{D_1} \times \ldots \times D_k^{D_k}$. The latter may be identified with $D_{i_1} \times \ldots \times D_{i_n}$, where $\tau = (i_1, \ldots, i_n)$ is the type of the supporting element. If, further, φ is the corresponding supporting symbol of relations, then we can take states $f \in \underline{F}_D$ with the condition $f(\varphi) = H$.

We cannot assert that each subalgebra of the boolean algebra $\underline{F}_D$ is closed in the Galois correspondence under consideration. However, the closed subalgebras can be characterized, using the structure of the automaton AtmD.

We shall start from the representation of G as a group of automorphisms of the automaton AtmD. It follows from the notes of the previous section, that this representation is naturally defined; that it is connected with the actions on V_D and $\underline{F}_D$; and that on the algebra of equests U the group G acts trivially. For this representation we hall also consider the Galois correspondence.

Let H be a subgroup of G. Taking F = 'H and R = H' we have $\underline{A}$ = 'H' = (F,U,R). It follows from the definitions that $\underline{A}$ is a subautomaton in he automaton AtmD. This subautomaton is complete in the sense that if $\in \underline{F}_D$ and $f*u \in R$ for each $u \in U$, then $f \in F$.

Let, further, $\underline{A}$ = (F,U,R) be a subautomaton of AtmD. By $\underline{A}'$ we enote the subgroup of G, which coincides with the intersection $F' \cap R'$. t can be proved that $\underline{A}' = R'$ always holds.

Theorem 5. The Galois correspondence considered above between the ubgroups of the group G and the complete automata in AtmD is one-to-ne.

This theorem easily follows from the previous one. It gives the lassification of databases by means of symmetries which was referred o in the Introduction. Indeed, all the subautomata of AtmD can easily e obtained from a description of the complete automata: we have to ake all complete automata (F,U,R) and then take all subsets in F.

Let us describe the characterization of the closed subsets of $\underline{F}_D$. or each $F \subset \underline{F}_D$, denote by F^* the subalgebra of V_D, generated by all $f*u$ $f \in F$, $u \in U$). If, further, R is a subalgebra of V_D, then $^*R$ denotes the et of all $f \in \underline{F}_D$, for which $f*u \in R$ for all $u \in U$.

Theorem 6. $'(F') = {}^*(F^*)$.

This formula follows directly from the preceding results.

Thus, finally, a subalgebra F in $\underline{F}_D$ is Galois-closed if and only f $F = {}^*(F^*)$ holds.

We shall give below the description of the group of automorphisms f the universal database AtmD. As a preliminary, we consider in deail the definition of a database homomorphism.

. Database homomorphisms

First of all, we consider homomorphisms in the given scheme. Let be a universal algebra of requests.

We take two databases $\underline{A} = (F,Q,R;\ U,D,\rho)$ and $\underline{A}' = (F',Q',R';U,D',\rho')$, here $\rho = (\alpha,\beta,\gamma)$ and $\rho' = (\alpha',\beta',\gamma')$. A homomorphism $\mu = (\nu,\delta):\underline{A} \to \underline{A}'$ onsists of a homomorphism of *-automata $\nu = (\nu_1,\nu_2,\nu_3):(F,Q,R)\to(F',Q',R')$ nd a homomorphism of data algebras $\delta:D'\to D$. These homomorphisms point in opposite directions. By the definition of ν, $(f*q)^{\nu_3} = f^{\nu_1} * q^{\nu_2}$, $f \in F$, $q \in Q$.

Recall about δ, that we simultaneously have the maps $\delta_*:\underline{F}_D \to \underline{F}_{D'}$ and $\delta_*:V_D \to V_{D'}$. These maps always are homomorphisms of boolean algebras, and if δ is an epimorphism, then $\delta_*:V_D \to V_{D'}$ is a monomorphism of Halmos algebras. The homomorphisms ν and δ have to satisfy the following conditions.

1. There is a commutative diagram

$$\begin{array}{ccc} & U & \\ {\scriptstyle\beta}\swarrow & & \searrow{\scriptstyle\beta'} \\ Q & \xrightarrow{\nu_2} & Q'. \end{array}$$

2. In the diagram

$$\begin{array}{ccc} F & \xrightarrow{\alpha} & \underline{F}_D \\ {\scriptstyle\nu_1}\downarrow & & \downarrow{\scriptstyle\delta_*} \\ F' & \xrightarrow{\alpha'} & \underline{F}_{D'} \end{array}$$

the following weakened commutative law holds:

$$f^{\nu_1\alpha'} \le f^{\alpha\delta_*}, \quad f \in F.$$

It can be verified that the condition from the second diagram means that a map $f^{\nu_1\alpha'} \to f^{\alpha}$ together with $\delta:D' \to D$ always is a model homomorphism.

We shall consider separately the case when the homomorphism $\mu = (\nu,\delta):\underline{A} \to \underline{A}'$ is an isomorphism of databases. Then, obviously, ν is an isomorphism of $*$-automata and δ is an isomorphism of algebras. Also, we have the isomorphism $\mu^{-1} = (\nu^{-1},\delta^{-1}):\underline{A}' \to \underline{A}$. In this case, the second diagram satisfies the usual commutative condition. The transitions in states give the isomorphism of models.

Further, the scheme will be changed. Actually, we only change the set of symbols of relations Φ and preserve the stratification of variables $n:X \to \Gamma$ and the variety Θ. This leads to such modifications of the Halmos algebra U, which do not remove it from the corresponding variety of Halmos algebras, specialized in Θ. So, we have two databases $\underline{A} = (F,Q,R;\ U,D,\rho)$ and $\underline{A}' = (F',Q',R';\ U',D',\rho')$. The homomorphism $\mu:\underline{A} \to \underline{A}'$ is now a triplet $\mu = (\nu,\zeta,\delta)$, where ν is the homomorphism of $*$-automata, $\delta:D' \to D$ is a homomorphism of algebras, and $\zeta:U' \to U$ is a homomorphism of Halmos algebras. Then, the commutative triangle above is replaced by a commutative square:

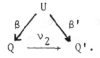

$$\begin{array}{ccc} U & \xleftarrow{\zeta} & U' \\ \beta\downarrow & & \downarrow\beta' \\ Q & \xrightarrow{\nu_2} & Q' \end{array}$$

We have to make some remarks concerning the second diagram. In the ase of changing scheme, we define a map $\delta_*:\underline{F}_D \to \underline{F}_{D'}$, taking into acount the homomorphism $\zeta:U' \to U$. As before, we start from $\delta_*:V_D \to V_{D'}$. If $\in \underline{F}_D$, then we have a homomorphism $\hat{f}:U \to V_D$. Applying $\delta_*:V_D \to V_{D'}$, we have $_*\hat{f}:U \to V_{D'}$. Finally, there is a homomorphism $\delta_*\hat{f}\zeta:U' \to V_{D'}$. So we deine $\widehat{\delta_* f} = \delta_*\hat{f}\zeta$.

Taking into account this definition of the map $\delta_*:\underline{F}_D \to \underline{F}_{D'}$, we preerve the second diagram in the definition of database homomorphisms.

In the case of modifying scheme the definition of model homomorhism is also changed. It is clear that if the scheme is fixed, then he homomorphism of algebras together with the correspondence of states $f' \to f$ is a homomorphism of models if and only if the inequality $'{*}u \le (f{*}u)^{\delta_*}$ holds for each $u \in U$ defined by an elementary formula. nder the new conditions the definition of model homomorphism is given y the inequality $f'{*}u \le (f{*}u^{\zeta})^{\delta_*}$, where u is an element in U' defined y an elementary formula.

Applying this to the states of databases $\underline{A}$ and $\underline{A}'$, one obtains $^{\nu_1\alpha'}{*}u \le (f^{\alpha}{*}u^{\zeta})^{\delta_*}$. This follows from the second diagram. Actually, he second diagram is equivalent to this condition.

Indeed, the condition $f^{\nu_1\alpha'} \le f^{\alpha\delta_*}$, $f \in F$, is an inequality in the oolean algebra $\underline{F}_{D'}$. It holds if and only if $f^{\nu_1\alpha'}{*}u \le f^{\alpha\delta_*}{*}u$ holds or each elementary $u \in U'$ in $V_{D'}$.

In this case we have

$$f^{\nu_1\alpha'}{*}u = (f^{\nu_1}{*}u^{\beta'})^{\gamma'} = (f^{\nu_1}{*}u^{\zeta\beta\nu_2})^{\gamma'} = (f{*}u^{\zeta\beta})^{\nu_3\gamma'} \le f^{\alpha\delta_*}{*}u =$$

$$= \widehat{\delta_* f^{\alpha}}\,\zeta(u) = (f^{\alpha}{*}u^{\zeta})^{\delta_*}.$$

It was mentioned above that each homomorphism of databases admits canonical expansion to a homomorphism, which does not modify the cheme, and a homomorphism connected only with the modification of he scheme.

Let us consider separately the case when $\mu = (\nu,\zeta,\delta):\underline{A} \to \underline{A}'$ is an somorphism. Here ν, δ, and ζ are isomorphisms, both diagrams are sual commutative diagrams, and the corresponding transitions in the tates are connected with the isomorphisms of models. In addition, we ay start from $\zeta^{-1}:U \to U'$ instead of $\zeta:U' \to U$.

8. The group of automorphisms of a universal automaton

First of all, note the fact that speaking about a modification of the scheme, we have in mind that there is a nontrivial homomorphism $\zeta: U' \to U$. The algebras U and U' may coincide. In this case ζ is an automorphism of the Halmos algebra U.

We shall consider further an automaton AtmD = $(\underline{F}_D, U, V_D)$. It was mentioned that the group G = AutD admits a representation as the group of automorphisms of this automaton. The group AutU of automorphisms of the Halmos algebra U is to be considered. Take a representation of this group as a group of automorphisms of the automaton AtmD. We shall use the reasons mentioned above.

The group AutU acts in each F_D in accordance with the rule $\widehat{f\zeta} = \hat{f}\zeta$, $\zeta \in \mathrm{AutU}$, $f \in \underline{F}_D$.

If we change this to a left action: $\zeta \circ f = f.\zeta^{-1}$, we have $f\zeta * u = f * \zeta u$ for each $u \in U$, and if $f_1 = f\zeta$, then $f_1 * u = f_1\zeta^{-1} * \zeta u = \zeta o f_1 * \zeta u$. Here f_1 is an arbitrary element in $\underline{F}_D$ and thus the actions of AutU in $\underline{F}_D$ and in U, together with the identical action in V_D, define a representation

$$\mathrm{AutU} \to \mathrm{Aut(AtmD)}.$$

This representation commutes on each pair of elements with the representation

$$\mathrm{AutD} \to \mathrm{Aut(AtmD)}.$$

It follows that there is a canonical homomorphism

$$\mathrm{AutU} \times \mathrm{AutD} \to \mathrm{Aut(AtmD)}.$$

Theorem 7. For each algebra D this homomorphism is actually an isomorphism.

This theorem is proved by direct verification and it is taken into account that for any algebra D each automorphism of the algebra V_D is induced by some $g \in \mathrm{AutD}$. Starting from this theorem, we can investigate the groups of automorphisms of arbitrary concrete databases.

References

1. Codd E.F., A relational model of data for large shared data banks, Commun. ACM 13 (1970), 377-387.

2. Codd E.F., Extending the database relational model to capture more meaning, ACM Trans. Database Syst. 3 ( ), 397-434.

3. Daigneault A., On automorphisms of polyadic algebras, Trans. Am. Math. Soc. 112 (1964), 84-130.

4. Halmos P.R. Algebraic logic, New York, 1962.

5. Henkin L., Monk I.D., Tarski A., Cylindric Algebras. North Holland, Amsterdam, London, 1971.

6. Krasner M., Une généralisation de la notion de corps. J. Math. Pures Appl. 17 (1938), 367-385.

7. Krasner M., Généralisation et analogues de la théorie de Galois, Congrès de la Victoire de l'Ass. Franc. Avancem. Sci., 1945, pp. 54-58.

8. Lawvere F.W., An elementary theory of the category of sets, Proc. Natl. Acad. Sci. USA 51 (1964), 1506-1510.

9. Lawvere F.W., Adjointness in foundations, Dialectica 23 (1969), 281-296.

0. Beniaminov E.M., Galois theory of complete relational subalgebras of relations; logical structures and symmetry, NTI, Ser. 2, Information processes and systems, VINITI, Moscow, 1980 (in Russian).

1. Boyko S.N., Galois theory of databases, in DBMS and program environments: problems of development and applications, Abstracts of talks presented at the republican conference, Riga, 1985; pp. 43-45 (in Russian).

2. Plotkin B.I., Algebraic models of databases, Latv. Mat. Ezheg. 27 (1983), 216-232 (in Russian).

ON THE CODIMENSIONS OF MATRIX ALGEBRAS

Amitai Regev

Department of Theoretical Mathematics
The Weizmann Institute of Science
Rehovot 76100, ISRAEL
and
Department of Mathematics
Pennsylvania State University
University Park, PA 16802, U.S.A.

§0. INTRODUCTION

Let F be a field of characteristic zero. We consider F-algebras which satisfy polynomial identities (P.I. algebras). The quantitative study of the identities of a P.I. algebra can be done by studying certain invariants which are associated to such algebras: these are the codimensions and the cocharacters - or equivalently, the Poincaré series.

In §1 and §4 we describe these sequences and the role played by the group algebra of the symmetric group in their studies.

Matrices have a trace, and the Procesi-Razmyslov theory [11], [12] of trace identities is briefly reviewed in §2, §3. We also review there some results of Formanek [6], [7], which relate the cocharacters with the trace cocharacters in a remarkable way. This allows us to calculate the codimensions of matrices asymptotically.

In §5 we turn to subalgebras of matrices over the Grassmann algebra. Results of Kemer [9] emphasize the importance of these algebras. Some of the results of the previous sections are extended here.

§1. THE GENERAL APPROACH

Let F be a field of characteristic zero. The representation theory of the symmetric group S_n (and sometimes of $GL(k,F)$) is our language for describing the identities of a P.I. algebra. We briefly summarize that theory.

Let $\mathrm{Par}(n)$ denote the partitions of n. Given $\lambda \in \mathrm{Par}(n)$ (or $\lambda \vdash n$), there is an explicit construction - via Young diagrams - of a minimal two-sided ideal I_λ in the group algebra FS_n, and of an irreducible character χ_λ, so that

$$FS_n = \oplus_{\lambda \vdash n} I_\lambda$$

and $\{\chi_\lambda | \lambda \vdash n\}$ are all the irreducible characters of S_n. The degree $d_\lambda = \deg(\chi_\lambda)$ can be calculated, for example, from the "hook formula", and it satisfies $\dim I_\lambda = d_\lambda^2$.

We turn now to P.I. algebras. Fix noncommutative variables

$$\{x\} = \{x_1, x_2, \ldots\}$$

and denote by

$$V_n = \left\{ \sum_{\sigma \in S_n} \alpha_\sigma x_{\sigma(1)} \cdots x_{\sigma(n)} \mid \alpha_\sigma \in F \right\}$$

the multilinear polynomials in $x_1, \ldots, x_n$. Identify $FS_n \equiv V_n$.

$$\sigma \equiv M_\sigma(x) = M_\sigma(x_1, \ldots, x_n) \overset{\text{def}}{=} x_{\sigma(1)} \cdots x_{\sigma(n)};$$

so V_n has an algebra as well as an FS_n module structure:

$$M_\sigma(x) M_\tau(x) \equiv \sigma M_\tau(x) \equiv M_\sigma(x)\tau \equiv \sigma\tau \equiv M_{\sigma\tau}(x).$$

It is easy to see that

$$\sigma M_\tau(x_1, \ldots, x_n) = M_\tau(x_{\sigma(1)}, \ldots, x_{\sigma(n)})$$

hence, if $f(x_1, \ldots, x_n) \in V_n$, then

$$\sigma\, f(x_1, \ldots, x_n) = f(x_{\sigma(1)}, \ldots, x_{\sigma(n)}) \qquad \ldots . \qquad (1)$$

Let $F\langle x\rangle = F\langle x_1, x_2, \ldots\rangle$ be the free (associative) algebra. The identities

$$id(A) = Q \subseteq F\langle x\rangle$$

of a P.I. algebra is a T-ideal in $F\langle x\rangle$. Construct

$$Q_n = Q \cap V_n.$$

Since $char(F) = 0$, the sequence $\{Q_n\}_{n=1}^{\infty}$ determines Q.

By (1), Q_n is an FS_n left (sub) module and such is also V_n/Q_n. The S_n character of that module, $\chi_{S_n}(V_n/Q_n)$, is the n-th cocharacter of A ($Q = id(A)$), and is denoted by $\chi_n(A)$. Also,

$$\deg \chi_n(A) = \dim(V_n/Q_n) = c_n(A)$$

are the codimensions of A.

Any character χ of S_n can be written as

$$\chi = \sum_{\lambda \vdash n} m_\lambda \cdot \chi_\lambda$$

where the m_λ's are the multiplicities of the χ_λ's, $0 \le m_\lambda \in \mathbf{Z}$. In particular,

$$\chi_n(A) = \sum_{\lambda \vdash n} m_\lambda(A) \cdot \chi_\lambda$$

and therefore

$$c_n(A) = \sum_{\lambda \vdash n} m_\lambda(A) \cdot d_\lambda, \quad d_\lambda = \deg \chi_\lambda .$$

In general, it is very difficult to obtain precise information about the m_λ's.

Some results which were obtained using the above approach can be found, for example, in [1], [5], [13], [14].

§2. TRACE IDENTITIES

The theory of trace identities was discovered independently by Procesi [11] and Razmyslov [12], and it can be based on a formula of Kostant [10] which we now describe.

Given a space V, define

$$T^n(V) = \underbrace{V \otimes \ldots \otimes V}_{n},$$

so

$$\mathrm{End}(T^n(V)) \cong T^n(\mathrm{End}(V)).$$

Define $\psi: FS_n \longrightarrow \mathrm{End}(T^n(V))$ via

$$\psi(\sigma)(v_1 \otimes \ldots \otimes v_n) = v_{\sigma^{-1}(1)} \otimes \ldots \otimes v_{\sigma^{-1}(n)},$$

where $\sigma \in S_n$, $v_1, \ldots, v_n \in V$ (extend ψ by linearity).

<u>Formula [10,p.256]:</u> Let $\sigma \in S_n$ admit the following cycle decomposition:

$$\sigma^{-1} = (i_1, \ldots, i_r)(j_1, \ldots, j_s)\ldots$$

(cycles of length one are included!), and let $A_1,\dots,A_n \in \mathrm{End}\ V$, then

$$\mathrm{tr}(\psi(\sigma)\cdot(A_1\otimes\dots\otimes A_n)) = \mathrm{tr}(A_{i_1}\cdots A_{i_r})\cdot\mathrm{tr}(A_{j_1}\cdots A_{j_s})\cdots .$$

Thus, let $a \in FS_n$, then

$$p_a(A_1,\dots,A_n) \overset{\mathrm{def}}{=} \mathrm{tr}(\psi(a)\cdot(A_1\otimes\dots\otimes A_n))$$

is a polynomial function of $A_1,\dots,A_n$ which involves the trace "tr".

The notion of a (general) trace polynomial is now clear, and $p_a(x_1,\dots,x_n)$ is a multilinear trace polynomial. For a formal construction of the free trace algebra $F\langle T(X),X\rangle$, see for example [8,§11]. A trace polynomial can be evaluated on any algebra with a trace function; it is a trace identity for that algebra if it vanishes in all such evaluations. As with the ordinary polynomial identities, we now restrict ourselves to the multilinear trace polynomials:

Definition: Denote

$$\mathrm{Tr}_n = \{p_a(x_1,\dots,x_n) \mid a \in FS_n\}.$$

This is the space of the multilinear trace polynomials of degree n in $x_1,\dots,x_n$.

Note that the map $a \longleftrightarrow p_a(x_1,\dots,x_n)$ is a vector space isomorphism: $FS_n \cong \mathrm{Tr}_n$. It is, in fact, as FS_n module isomorphism where FS_n acts on itself by conjugation and on Tr_n by substitutions: $\tau \in S_n$,

$$\tau p_a(x_1,\dots,x_n) = p_a(x_{\tau(1)},\dots,x_{\tau(n)}) = p_{\tau a\tau^{-1}}(x_1,\dots,x_n).$$

The trace identities of an algebra with a trace thus form an FS_n submodule of Tr_n and one can define, in an analogous way, the trace cocharacters and trace codimensions of such an algebra.

§3. THE ALGEBRA $M_k(F)$

We first consider the trace identities of $M_k(F)$; they form a two-sided ideal in $FS_n = \mathrm{Tr}_n$.

Notation: Denote

$$\Lambda_k(n) = \{(\lambda_1,\lambda_2,\dots) \vdash n \mid \lambda_{k+1} = 0\}.$$

A fundamental theorem of I. Schur and H. Weyl says

<u>Theorem:</u> Let $\psi: FS_n \longrightarrow End(T^n(V))$ as in §2, and let $\dim V = k$, then

$$\operatorname{Ker}\psi = \bigoplus_{\lambda\in Par(n)\setminus\Lambda_k(n)} I_\lambda .$$

<u>Corollary:</u> Let $a \in FS_n$, then $a \longleftrightarrow P_a(x_1,\dots,x_n)$ is a trace identity of $M_k(F)$ if and only if

$$a \in \bigoplus_{\lambda\in Par(n)\setminus\Lambda_k(n)} I_\lambda .$$

This follows from Kostant's formula (§2), from the non-degeneracy of the trace, and from the above theorem.

<u>Denote</u> the trace identities of $M_k(F)$ in Tr_n by $TI_n \subseteq Tr_n$. Since $TI_n = \ker\psi$, it follows that

$$Tr_n/TI_n \cong \bigoplus_{\lambda\in\Lambda_k(n)} I_\lambda .$$

Each I_λ is an FS_n submodule under conjugation, with character $\chi_\lambda \otimes \chi_\lambda$. It follows that the trace cocharacters of $M_k(F)$ are

$$T\chi_n(M_k(F)) = \sum_{\lambda\in\Lambda_k(n)} \chi_\lambda \otimes \chi_\lambda ,$$

and the trace codimensions are

$$t_n(M_k(F)) = \sum_{\lambda\in\Lambda_k(n)} d_\lambda^2 .$$

Using asymptotic analysis, it is possible to determine the asymptotic behavior of the sums $\Sigma_{\lambda\in\Lambda_k(n)} d_\lambda^2$, [17]. It follows that

$$t_n(M_k(F)) \simeq c \cdot \left(\frac{1}{n}\right)^g \cdot k^{2n}$$

where $c = (1/\sqrt{2\pi})^{k-1}\left[\frac{1}{2}^{(k^2-1)/2} \cdot 1!\dots(k-1)! \cdot k^{(k^2+4)/2}\right]$, $g = (k^2-1)/2$

and where $a_n \simeq b_n$ means that $\lim_{n\to\infty} \frac{a_n}{b_n} = 1$ [18].

The computation of the trace cocharacters of $M_k(F)$ is equivalent to the computation of the sum of Kronecker products

$$\sum_{\lambda\in\Lambda_k(n)} \chi_\lambda \otimes \chi_\lambda .$$

It can be shown [16] that

$$T\chi_n(M_k(F)) = \sum_{\lambda\in\Lambda_k(n)} \chi_\lambda \otimes \chi_\lambda = \sum_{\mu\in\Lambda_{k^2}(n)} \bar{m}_\mu \chi_\mu ,$$

for some coefficients $\bar{m}_\mu = \bar{m}_\mu(M_k(F))$. A formula for the $\bar{m}_\mu$'s is known, at present, only for $k = 2$, and is due to Procesi (see [19]). The general case seems difficult at present since a formula for the Kronecker (inner) product $\chi_\lambda \otimes \chi_\mu$ $(\lambda,\mu \vdash n)$ is unknown.

We now turn back to the ordinary polynomial identities of $M_k(F)$. The embedding $V_{n-1} \longrightarrow Tr_n$ via $f(x_1,\ldots,x_{n-1}) \longrightarrow tr(f(x_1,\ldots,x_{n-1})\cdot x_n)$, together with the nondegeneracy of the trace, imply

<u>Lemma</u>: $c_{n-1}(M_k(F)) \le t_n(M_k(F))$.

In fact, a stronger result holds here, as we shall soon see.

From the existence of Capelli identities for $M_k(F)$, it follows that the cocharacter of $M_k(F)$ satisfies a condition similar to that of the trace cocharacter:

$$\chi_n(M_k(F)) = \sum_{\mu\in\Lambda_{k^2}(n)} m_\mu \cdot \chi_\mu .$$

for some coefficients $m_\mu = m_\mu(M_k(F))$ [15].

The following important result was proved by Formanek [6], [7]:

<u>Theorem</u>: Let $\mu \in \Lambda_{k^2}(n)$ $\mu = (\mu_1,\mu_2,\ldots)$ such that $\mu_{k^2} \ge 2$, then $m_\mu(M_k(F)) = \bar{m}_\mu(M_k(F))$.

As an application of that theorem, one can deduce now

<u>Theorem [18]</u>: For all k, $c_{n-1}(M_k(F)) \simeq t_n(M_k(F))$. In particular

$$c_{n-1}(M_k(F)) \simeq c \cdot \left(\frac{1}{n}\right)^g \cdot k^{2n}$$

where $c = (1/\sqrt{2\pi})^{k-1}\left(\frac{1}{2}\right)^{(k^2-1)/2} \cdot 1!\ldots(k-1)! \cdot k^{(k^2+4)/2}$ and $g = (k^2-1)/2$.

Notice that no such information is known, at present, about the coefficients $m_\mu(M_k(F))$, if $k \geq 3$.

The above asymptotic results imply the following intriguing

Corollary [2]: Given a P.I. algebra A, let

$$g(t) = \sum_{n \geq 0} c_n(A) \cdot t^n$$

denote the generating function of its codimensions. Let $g_k(t) = \Sigma_{n\geq 0} c_n(M_k(F)) \cdot t^n$, then

a) $g_2(t)$ is algebraic

b) $g_k(t)$ is non-algebraic if $k \geq 3$ is odd (the cases $k \geq 4$ even are not known).

§4. COCHARACTERS IN GENERAL

Only partial information is available about general cocharacters, and we now briefly summarize it below.

Theorem [13], [14], [18]: Given any P.I. algebra A, there exists $a > 0$ such that for all n,

$$c_n(A) \leq a^n.$$

In fact, if A satisfies an identity of degree d, then $a \leq (d - 1)^2$.

Notation: Denote

$$H(k,\ell;n) = \{(\lambda_1,\lambda_2,\ldots) \vdash n \mid \lambda_{k+1} \leq \ell\}:$$

$H(k,\ell;n)$ is the "$k \times \ell$" hook of Young diagrams.

Theorem [1]: Given any P.I. algebra A, there exist k,ℓ such that

$$\chi_n(A) = \sum_{\lambda \in H(k,\ell;n)} m_\lambda(A) \cdot \chi_\lambda.$$

In fact, if A satisfies an identity of degree d, we can choose any $k = \ell > e \cdot (d - 1)^4$, where $e = 2.7\ldots$.

The homogeneous polynomial identities can be studied in a similar way, by applying the (polynomial) representation theory of the general linear Lie group (or algebra) $GL(k,\mathbb{C})$ ($g\ell(k,\mathbb{C})$) and of the general linear Lie super-algebra $p\ell(k,\ell)$ [4], [5]. The relation between these identities and the multilinear identities implies

<u>Theorem [5]:</u> Let A be a P.I. algebra, $\chi_n(A) = \sum_{\lambda\vdash n} m_\lambda(A)\chi_\lambda$ its (multilinear) cocharacter. Then there exists a fixed power h such that for all n and for all $\lambda \vdash n$,

$$m_\lambda(A) \le n^h.$$

Let E denote the Grassmann algebra. We remark that algebras of the form $A \otimes E$ play an important role in the above theorem.

§5. MATRICES OVER THE GRASSMANN ALGEBRA

Let $E = E(V)$ be the Grassmann (Exterior) algebra of a countable dimension vector space V over F. By considering the length of the basis elements of E we have that

$$E = E_0 \oplus E_1,$$

where E_0 (resp. E_1) is spanned by the elements of even (resp. odd) length. Given $k,\ell \ge 0$, we denote by $E_{k,\ell} = M_{k,\ell}(E)$ the following subalgebra of $M_k(E)$:

$$E_{k,\ell} = \left\{ \left(\begin{array}{c|c} A & B \\ \hline C & D \end{array}\right) \;\middle|\; A \in M_k(E_0),\ D \in M_\ell(E_0),\ B \text{ is } k \times \ell \text{ and } C \text{ is } \ell \times k, \text{ both with entries in } E_1. \right\}$$

We summarize now some important results of Kemer [9].

K-ideals (Kemer calls them T-ideals) are obtained from T-ideals by taking all possible evaluations.

The relatively free algebra in a given variety is called K-semiprime if it does not contain nilpotent K-ideals, and in that case, the variety itself is called K-semiprime. K-primeness is defined in an analogous way and a K-semiprime algebra is a finite direct sum of K-prime algebras. Equivalently,

<u>K-primeness:</u> The algebra A is K-prime if it satisfies the following property: Let $f(x_1,\dots,x_r), g(x_1,\dots,x_s)$ be polynomials such that $f(x_0,\dots,x_{r-1})x_r g(x_{r+1},\dots,x_{\bar{x}+s})$ is an identity for A, then either f or g is an identity for A. Equivalently, an algebra A is K-prime if no product of non-zero K ideals of A is zero.

<u>Theorem [9]:</u> Any K-prime variety is generated by one of the following algebras: $F_k = M_k(F)$; $E_k = M_k(E) = F_k \otimes E$; $E_{k,\ell}(E)$ where $\ell \le k$.

Let A,B be two P.I. algebras. Denote $A \sim B$ if they satisfy the same set of identities.

Theorem [9]: The next equivalences hold:

$$E_{1,1} \sim E \otimes E$$
$$E_{k,\ell} \otimes E \sim E_{k+\ell}$$
$$E_{k,\ell} \otimes E_{p,q} \sim E_{kq+\ell p, kp+\ell q}.$$

The importance of these algebras lies in the following:

Theorem [9]: Every relatively free algebra A has a maximal nilpotent K-ideal I such that A/I is k semiprime.

Kemer's results strongly motivate the study of these P.I. algebras.

As for hooks of Young diagrams, we have:

Theorem: (a) [20,§0]

$$\chi_n(M_k(E)) = \sum_{\lambda \in H(k^2,k^2;n)} m_\lambda(M_k(E)) \cdot \chi_\lambda .$$

(b) [3]

$$\chi_n(E_{k,\ell}) = \sum_{\lambda \in H(k^2+\ell^2, 2k\ell;n)} \tilde{m}_\lambda(E_{k,\ell})\chi_\lambda .$$

At present, we know very little about the coefficients $m_\lambda(M_k(E))$.

Theorem [20,04]: Let $\lambda \in H(k^2,k^2,n)$ have the Young diagram

λ :

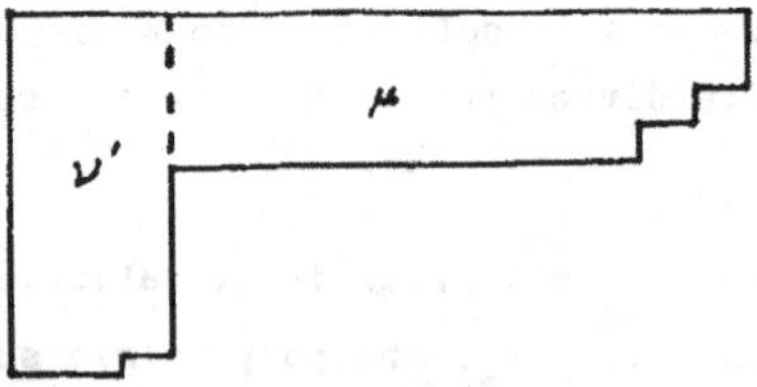

$\lambda = (\lambda_1, \lambda_2, \ldots)$ and $\lambda_{k^2} \geq k^2$. If both μ and ν are large enough, then $m_\lambda(M_k(E)) \geq 1$.

We can now deduce

Theorem [20,6.2]: There are (explicit) constants c_1, c_2, g_1 and g_2 such that

$$c_1 \cdot \left(\frac{1}{n}\right)^{g_1} \cdot (2 \cdot k^2)^n \le c_n(M_k(E)) \le c_2 \cdot \left(\frac{1}{n}\right)^{g_2} \cdot (2 \cdot k^2)^n.$$

For the algebras $E_{k,\ell} \subseteq M_{k+\ell}(E)$ we now deduce

Theorem [20,7.5]: There are (explict) constants c_1, c_2, g_1 and g_2 such that

$$c_1 \left(\frac{1}{n}\right)^{g_1} \cdot (k + \ell)^{2n} \le c_n(E_{k,\ell}) \le c_2 \cdot \left(\frac{1}{n}\right)^{g_2} \cdot (k + \ell)^{2n}.$$

We summarize: Define $\ell(A) = \lim_{n\to\infty} c_n(A)^{1/n}$ if the limit exists. Then

a) $\ell(M_k(F)) = k^2$

b) $\ell(M_k(E)) = 2 \cdot k^2$

c) $\ell(E_{k,\ell}) = (k + \ell)^2$

Moreover, if A and B are among these algebras, then

$$\ell(A \otimes B) = \ell(A) \cdot \ell(B).$$

REFERENCES

[1] Amitsur, S.A., Regev, A.: PI-algebras and their cocharacters. J. of Algebra, Vol. 78, No. 1, (1982) 248-254.

[2] Beckner, W., Regev, A.: Asymptotics and algebraicity of some generating functions. Advances in Math. Vol. 65 No. 1 (1987), 1-15.

[3] Berele, A.: Cocharacters of $\mathbb{Z}/2\mathbb{Z}$ graded algebras. Preprint.

[4] Berele, A., Regev, A.: Hook Young diagrams with applications to Combinatorics and to representations of Lie superalgebras. Advances in Math., Vol. 64, No. 2, (1987), 118-175.

[5] Berele, A., Regev, A.: Applications of hook Young diagrams to P.I. algebras. J. of Algebra, 82(1983) 559-567.

[6] Formanek, E.: Invariants of the ring of generic matrices. J. of Algebra 89(1984), 178-223.

[7] Formanek, E.: A conjecture of Regev on the Capelli polynomial. To appear in J. of Algebra.

[8] Formanek, E.: The invariants of $n \times n$ matrices. To appear in the Proceedings of a Conference on Invariant Theory, West Chester University, 1985, Springer Lecture Notes.

[9] Kemer, A.R.: Varieties and Z_2 graded algebras, Izv. Akad. Nauk SSSR Ser. Mat. 48(1984), 1042-1059 (Russian). Translation: Math. USSR Izv. 25(1985) 359-374.

[10] Kostant, B.: A theorem of Frobenius, a theorem of Amitsur-Levitski and cohomology theory. J. of Math. and Mech. 7(1958), 237-264.

[11] Procesi, C.: The invariant theory of n×n matrics. Advances in Math. 19(1976), 306-381.

[12] Razmyslov, Yu.P.: Trace identities of full matrix algebras over a field of characteristic zero. Izv. Akad. Nauk SSSR Ser. Mat. 38(1974), 723-756 (Russian). Translation: Math. USSR Izv. 8(1974), 727-760.

[13] Regev, A.: Existence of identities in $A \otimes B$. Israel J. of Math. 11(1972), 131-152.

[14] Regev, A.: The representation of S_n and explicit identities for P.I. algebras. J. of Algebra 51(1978), 25-40.

[15] Regev, A.: Algebras satisfying a Capelli identity. Israel J. of Math. 33(1979), 149-154.

[16] Regev, A.: The Kronecker product of S_n-characters and an $A \otimes B$ theorem for Capelli identities. J. of Algebra 66(1980), 505-510.

[17] Regev, A.: Asymptotic values for degrees associated with strips of Young diagrams. Advances in Math. 41(1981), 115-136.

[18] Regev, A.: Codimensions and trace-codimensions are asymptotically equal. Israel J. of Math. 47(1984), 246-250.

[19] Regev, A.: A combinatorial proof of a character formula of Procesi. J. "Linear and Multilinear Algebra", Vol. 21, No. 1, (1987), 29-39.

[20] Regev, A.: On the identities of subalgebras of matrices over the Grassmann algebra. To appear in Israel J. of Math.

"This paper is in final form and no version of it will be submitted for publication elsewhere"

HOMOLOGY OF FREE LOOP SPACES, CYCLIC HOMOLOGY AND NON-RATIONAL POINCARE-BETTI SERIES IN COMMUTATIVE ALGEBRA

JAN-ERIK ROOS
Department of Mathematics
University of Stockholm
Box 6701
S-113 85 STOCKHOLM, SWEDEN

Dedicated to the memory of Christer LECH (30.3.1926-9.2.1987)

0. INTRODUCTION.

his paper is a pure algebraic spin-off of mathematical work that we have started, nspired by modern string theory[1)]. In certain areas of string theory one needs an ndex theorem for e.g. the Dirac operator on the space of free loops $\mathcal{L}X$, where X is .g. a spin manifold and $\mathcal{L}X$ is the set $\mathrm{Hom}_{cont}(S^1,X)$ of all continuous maps from the ircle S^1 to X (no base points!), and where $\mathcal{L}X$ is equipped with the compact-open opology. For more details about this we refer the reader to the beautiful and nspiring papers [58],[59] by Edward Witten, and the literature cited there. In [53] .H. Taubes managed to put most of Witten's ideas in a rigorous form, by reducing verything to a study of operators in vector bundles over X. However, we felt that it ight also be interesting to work directly (in the spirit of Witten) on the infinite-imensional spaces $\mathcal{L}X$. We therefore started trying to develop a theory of characteris-ic classes for bundles over $\mathcal{L}Y$, when Y is (say) a manifold, in complete analogy with he way they are developed in [8] (the idea comes from Grothendieck [22]) for vector undles over ordinary finite-dimensional spaces. This led us to the study of the coho-ology of $\mathcal{L}Y$ and also of spaces like $\mathcal{L}P(V_Y)$, where $P(V_Y)$ is the projectivization of a ector bundle V_Y over Y. We then found that for (say) finite-dimensional, simply-onnected CW-complexes X, very few cases were known when the cohomology $H^*(\mathcal{L}X,k)$ or he homology $H_*(\mathcal{L}X,k)$ had been explicitly calculated, even if rational coefficients $k = \underline{\underline{Q}}$) were used, although much work had been done in connection with the celebrated romoll-Meyer theorem [21]. Recall that this theorem says that a closed Riemannian anifold M has infinitely many closed geodesics if, for some fixed field k, the umbers $\dim_k H^i(\mathcal{L}M,k)$ are unbounded for $i \geq 0$. Using a nice rational homotopy model for X, M. Vigué-Poirrier and D. Sullivan [54] proved that the numbers $\dim_{\underline{\underline{Q}}} H^i(\mathcal{L}X,\underline{\underline{Q}}), i \geq 0$, re unbounded if and only if the cohomology ring $H^*(X,\underline{\underline{Q}})$ needs more than one generator. he few explicit calculations of the power series

$$H^*(\mathcal{L}X,k)(Z) = \sum_{i\geq 0} \dim_k(H^i(\mathcal{L}X,k))\cdot Z^i \qquad (1)$$

---

1) Those readers that are mainly interested in algebra should read rapidly through the ext 54 lines, which are inserted with the aim of giving inspiration (?) and background.

that are known [54],[52],[28],[60],[48],[49],[39],[56] are mostly restricted to the case when $H^*(X,k)$ needs only one generator or when $H^*(X,k)$ is a graded complete intersection and when furthermore, the spectral sequence relating the Hochschild homology of $H^*(X,k)$ to $H^*(\mathcal{L}X,k)$ degenerates[48],[3] (when $k = \underline{\underline{Q}}$ or $\underline{\underline{Z}}/2\underline{\underline{Z}}$ [49])(more details about this are given in § 2 below). In the papers by A.S. Švarts [52], P. Klein [28], L. Smith [28], W. Ziller [60], J. McCleary [39] a few other special cases and methods are deployed. More details about this will be given in § 2 and we just remark here that it is well-known that if X is an H-space, then $\mathcal{L}X$ is homotopy equivalent to $X \times \Omega X$, where the space ΩX is the ordinary loop space of X (all maps $S^1 \longrightarrow X$ that respect fixed base points). In general (cf. § 2) $\mathcal{L}X$ sits in the middle of a non-trivial fibration with basis X and fibre ΩX. The spaces ΩX and their (co)homology have been much studied and it is natural to try to develop a similar theory for the more complicated spaces $\mathcal{L}X$. Recall that Lemaire [39] gave a nice theory of $H_*(\Omega X,k)$ when X is the mapping cone of a map between suspensions of connected spaces. Recall also that essentially any finite simply-connected 4-dimensional CW-complex can be obtained as such a mapping cone for a map from a finite wedge of 3-spheres to a finite wedge of 2-spheres. As a warm-up for the study of $H_*(\mathcal{L}X,\underline{\underline{Q}})$ for these 4-dimensional X, we therefore started with a series as "simple" as

$$(2) \qquad H_*(\mathcal{L}(S^2 \vee S^2),\underline{\underline{Q}})(Z) = \sum_{i\geq 0} \dim_{\underline{\underline{Q}}} H_i(\mathcal{L}(S^2 \vee S^2),\underline{\underline{Q}}).Z^i$$

In the literature there are many nice results about $H_*(\mathcal{L}SX,k)(Z)$ for any suspension SX of a connected space X [10],[18],[11],[14],[15],[12],[19],[20], but more work was needed even to make a series like (2) explicit. When we did that work (more details below) we found to our genuine surprise that (2) did not represent a rational function of Z Similar results are also true for more general series than (2) (cf. § 2, where explicit formulae are given, and indeed we have a general problem 2 in § 3 that would clarify this completely). It was now clear that one should use a well-known recipe [40] and tran form all this into examples in commutative algebra. Our amazement was great when we found e.g. the following (readers interested only in algebra can start reading more attentively here):

For any field k, there exists a local commutative noetherian ring $(S,\underline{m})$, with maximal ideal $\underline{m}$ and residue field k, such that $\underline{m}^3 = 0$, $\dim_k(\underline{m}/\underline{m}^2) = 4$ (the embedding dimension) and an S-module M, whose length over S ($\ell_S(M)$) is 3, such that the Tor-series

$$(3) \qquad \mathrm{Tor}_*^S(M,M)(Z) = \sum_{i\geq 0} \ell_S(\mathrm{Tor}_i^S(M,M))\cdot Z^i$$

does *not* represent a rational function of Z.

We can e.g. take

$$(4) \qquad S = \frac{k[x_1,x_2,x_3,x_4]}{(x_1^2,x_2^2,x_3^2,x_4^2,x_1x_2,x_3x_4)} \quad \text{and} \quad M = S/_{(x_1-x_3,x_2-x_4)} .$$

An explicit formula for the corresponding series (3) is given in § 1 (Corollary 1 to Theorem 1), at least when the characteristic of k is 0.

However, the corresponding Ext-series (same S, same M):

$$\mathrm{Ext}_S^*(M,M)(Z) = \sum_{i\geq 0} \ell_S(\mathrm{Ext}_S^i(M,M))\cdot Z^i$$

is rational, and so are also the series $\mathrm{Tor}_*^S(k,M)$ and $\mathrm{Tor}_*^S(k,\tilde{M})$, where $\tilde{M}$ is the Matlis dual of M ($\tilde{M} = \mathrm{Hom}_S(M,I(k))$, where I(k) is the injective envelope of k). However, since the Matlis dual of $\mathrm{Tor}_i^S(M,N)$ is isomorphic to $\mathrm{Ext}_S^i(M,\tilde{N})$, we also have that $\mathrm{Ext}_S^*(M,\tilde{M})(Z)$ is a non-rational function.

We have a short exact sequence (soc(M) = the socle of M):

$$(5) \qquad 0 \longrightarrow \mathrm{soc}(M) \longrightarrow M \longrightarrow M/\mathrm{soc}(M) \longrightarrow 0$$

In our case $\mathrm{soc}(M) = k \oplus k$ and $M/\mathrm{soc}(M) = k$. Applying the functor $\mathrm{Tor}_*^S(\cdot,M)$ to the exact sequence (5) we get a long exact sequence of S-modules:

$$(6) \quad \ldots \longrightarrow \mathrm{Tor}_{*+1}(k,M) \xrightarrow{\partial_{*+1}} \mathrm{Tor}_*^S(k \oplus k,M) \longrightarrow \mathrm{Tor}_*^S(M,M) \longrightarrow \mathrm{Tor}_*^S(k,M) \longrightarrow \ldots$$

In view of the preceding remarks the boundary operator ∂_{*+1} in (6) must be highly non-trivial and indeed we will see in § 1 that the cokernel of ∂_{*+1} can be identified with the reduced cyclic homology (in the sense of J.-L. Loday and D. Quillen [32] of the trivial extension $k \times V$ ($\dim_k V = 2$) - more details and generalizations in § 1. Thus this cyclic homology has a non-rational series (Coker ∂_{*+1})(Z).

All this is of interest for several reasons:

a) The examples are very simple and everything is (as we will see) in principle easy to calculate from first principles. Thus the examples can be presented with proofs in an introductory course in homological and/or commutative algebra.

b) In [23] T. Gulliksen proved that if $(R,\underline{m})$ is a local commutative noetherian ring, which is a local complete intersection and if M and N are any noetherian R-modules such that $\ell_R(M \otimes_R N) < \infty$, then the series $\mathrm{Tor}_*^R(M,N)(Z)$ is a rational function of the form(polynomial in Z)$/(1-Z^2)^t$ (t = embedding dimension of R - Krull dimension of R) and he showed later [24] that $\lim_{Z\to -1} \mathrm{Tor}_*^R(M,N)(Z)\cdot(1-Z^2)^t$ could be taken as the natural generalization of the Serre multiplicity [44] when R is not regular. One might ask what happens for local rings that are not complete intersections. In [6] L. Avramov, A.R. Kustin and M. Miller proved (they were inspired by earlier work by J. Weyman [57]), that if $(R,\underline{m})$ is any local commutative (complete) noetherian ring of embedding dimension 3, there exists a local complete intersection R′ and a local ring map onto R: R′ $\longrightarrow$ R that is a so-called Golod map [31]. It should be possible to deduce from this and Gulliksen′s complete intersection result that $\mathrm{Tor}_*^R(M,N)(Z)$ is always rational in the embedding dimension 3 case for all noetherian modules M and N such that $\ell_R(M \otimes_R N) < \infty$ [and in any case this is true if M or N is k, according to a more general result of Levin ([7] Corollary 4, p. 112, combined with Theorem 1, loc.cit. gives Levin′s result)]. Our example shows that there is no hope of going beyond the embedding dimension 3 case with this general rationality assertion. Thus we have here the first examples that show that something goes wrong with rationality of $\mathrm{Tor}_*^R(M,N)(Z)$ already in the embedding dimension 4 case. Recall that if the embedding dimension of R is 5 (or higher) it is known that already the series $\mathrm{Tor}_*^R(k,k)(Z)$ can be non-rational [1], [33].

3) Gulliksen has also proved ([25], to appear) that if $(R,\underline{m})$ is any noetherian commutative, local ring, M, N R-modules of length ≤ 2, then $\mathrm{Tor}_*^R(M,N)(Z)$ is an explicit rational function of $\mathrm{Tor}_*^H(k,k)(Z)$ except perhaps - this case has not yet been decided - when M and N are isomorphic and of length 2. In our example S in (4) we have $\mathrm{Tor}_*^S(k,k)(Z) = (1-2Z)^{-2}$ and $\ell_S(M) = 3$ ...

4) We have now a denumerable set of "canonical", "minimal" non-rational series $\mathrm{Tor}_*^R(M,M)(Z)$ (explicitly given in § 1). Maybe they "rationally generate" [42] a substantial part of the set of possible series $\mathrm{Tor}_*^R(L,N)(Z)$?

5) Our examples show that $\{R \mid \mathrm{Tor}_*^R(M,N)(Z)$ is rational for all M, N with $\ell_R(M \otimes_R N)<\infty\}$ is a very restricted class. Is it reasonable to hope for a classification of these R:s?

After this lengthy introduction, here is a brief summary of the contents of this paper: In § 1 we calculate the Hochschild (co)homology of a "trivial" ring extension. This calculation is essentially in Loday-Quillen [32], but we have preferred to make the paper self-contained. The more explicit form for the corresponding (co)homology series, the proof that it is irrational in general and the applications to local algebra are new. We also make some remarks about cyclic (co)homology and its applications in local algebra. In § 2 we discuss the (co)homology of free loop spaces and we also make some calculations. We also give a corrected version of McCleary's general form of the "Švarts trick" (more details below) and we make some remarks about the 2-torsion part of $H_*(\mathcal{L}X,\underline{Z})$. Finally in § 3 we discuss the relations between the following problems (X is a finite, simply-connected CW-complex, $(R,\underline{m})$ is a local commutative noetherian ring which is (say) equicharacteristic):

i) Calculate the homology of the free loop space $\mathcal{L}X$ (at least with coefficients in $\underline{Q}$)

ii) Study in detail how the graded Lie algebra $\pi_*(\Omega X) \otimes \underline{Q}$ (Whitehead product) operates on its enveloping algebra $H_*(\Omega X,\underline{Q})$ by means of the adjoint representation.

iii) Study the Hochschild (co)homology of $(R,\underline{m})$.

iv) Study in detail how the homotopy Lie algebra of $(R,\underline{m})$ operates on its enveloping algebra $\mathrm{Ext}_R^*(k,k)$ by means of the adjoint representation.

v) Problems similar to iii) and iv) for $(R,\underline{m})$, with $\underline{m}^3 = 0$.

vi) Problems similar to i),ii) for finite, simply-connected 4-dimensional X:s.

vii) Problems similar to iv) and ii) for Hopf algebras that have generators in degree 1 and relations in degree 2.

viii) Similar problems for the cyclic (co)homology of X and $(R,\underline{m})$.

etc.... We also discuss some other open problems.

A few words about the style: The § 1 is essentially self-contained and in order not to bscure the simplicity of the examples, we have not tried to carry through the theory here in greatest possible generality.

wish to thank Luchezar Avramov, Jörgen Backelin, Lars Brink, Ralf Fröberg, Tor H. ulliksen and Bengt E.W. Nilsson for stimulating discussions.

1. THE HOCHSCHILD AND CYCLIC (CO)HOMOLOGY OF TRIVIAL RING EXTENSIONS. APPLICATIONS O LOCAL ALGEBRA.

et K be a commutative field, V a finite-dimensional vector space over K and $\Lambda = K \ltimes V$ he trivial extension of K by V { i.e. the set of pairs (k,v), $k\in K$, $v\in V$ with pairwise ddition and multiplication $(k_1,v_1)\cdot(k_2,v_2) = (k_1k_2,\ k_1v_2+k_2v_1)$}. Let $\Lambda^e = \Lambda \otimes_K \Lambda^o$, here Λ^o is the opposite ring ($\Lambda^o = \Lambda$ in our case), and let Λ be considered as a Λ^e-odule in the natural way. We wish to calculate explicitly the Hochschild homology $\mathrm{Tor}_*^{\Lambda^e}(\Lambda,\Lambda)$ and the Hochschild cohomology $\mathrm{Ext}^*_{\Lambda^e}(\Lambda,\Lambda)$, and in particular the series

$$\mathrm{Tor}_*^{\Lambda^e}(\Lambda,\Lambda)(Z) = \sum_{i\geq 0} \ell(\mathrm{Tor}_i^{\Lambda^e}(\Lambda,\Lambda))\cdot Z^i \tag{7}$$

nd

$$\mathrm{Ext}^*_{\Lambda^e}(\Lambda,\Lambda)(Z) = \sum_{i\geq 0} \ell(\mathrm{Ext}^i_{\Lambda^e}(\Lambda,\Lambda))\cdot Z^i \tag{8}$$

here ℓ denotes length of Λ^e-modules. Note further that here Λ^e is a local commutative rtinian ring $(S,\underline{m}_S)$ with $m_S^3 = 0$. Taking length over S in (7) and (8) is the same as alculating the dimension over K, but the S-modules in (7) and (8) are not in general nnihilated by $\underline{m}_S$. Here are our results:

HEOREM 1.- Let Λ as above be the trivial extension of a field K with a finite-dimenional vector space V. Then:

(i) If the characteristic of K is 0, we have the following explicit formula for the Hochschild homology series:

$$\mathrm{Tor}_*^{\Lambda\otimes_K\Lambda^o}(\Lambda,\Lambda)(Z) = 1 + (1+Z^{-1})\cdot\sum_{n=1}^{\infty}\frac{1}{n}\left(\sum_{s|n}\varphi(\tfrac{n}{s})\cdot(-|V|)^s\right)\cdot(-Z)^n \tag{9}$$

where φ is the Euler φ-function [$\varphi(n)$ = # of invertible elements in the ring $\underline{\underline{Z}}/n\underline{\underline{Z}}$] and where $|V| = \dim_K V$. The series (9) (which converges for $|Z| < |V|^{-1}$) represents a non-rational function if $|V| \geq 2$, and is equal to $\frac{2-Z}{1-Z}$ if $|V|=1$.

(ii) If the characteristic of K is arbitrary (characteristic 0 is also included!) we have at least the following formula (cf.[32],[10]...):

$$\mathrm{Tor}_*^{\Lambda\otimes_K\Lambda^o}(\Lambda,\Lambda)(Z) = 1 + (1+Z^{-1})\sum_{n\geq 1}|(V^{\otimes n})^{\underline{\underline{Z}}/n\underline{\underline{Z}}}|\cdot Z^n \tag{10}$$

where the generator $t_n = \overline{1}$ of $\underline{\underline{Z}}/n\underline{\underline{Z}}$ operates on $V^{\otimes n} = V\otimes_K\ldots\otimes_K V$ (V n times) by $t_n(v_1\otimes\ldots\otimes v_n) = (-1)^{n-1}v_n\otimes v_1\otimes\ldots\otimes v_{n-1}$ and where the vector space of vectors in $V^{\otimes n}$ that are invariant under this action is denoted by $(V^{\otimes n})^{\underline{\underline{Z}}/n\underline{\underline{Z}}}$.
The series (10) also converges for $|Z| < |V|^{-1}$, represents a non-rational function

if $|V| \geq 2$ and is equal to $\frac{2-Z}{1-Z}$ if char $K \neq 2$ and $2/(1-Z)$ if char $K = 2$. If char$K=p>0$ and if $\Phi_V(Z) = \sum_{n\geq 1} |(V^{\otimes n})^{\mathbb{Z}/n\mathbb{Z}}| \cdot Z^n$ is the series in (10), then if $\zeta = e^{2\pi i/p}$ we have

(11) that $\Phi_V(Z) - \frac{1}{p}\left(\Phi_V(Z) + \Phi_V(\zeta Z) + \dots + \Phi_V(\zeta^{p-1}Z)\right) = \sum_{n\not\equiv 0(p)} \frac{1}{n}\left(\sum_{s|n} \varphi(\tfrac{n}{s})(-|V|)^s\right)(-Z)^n$

which can be treated in the same way as (9) when non-rationality is studied.

(iii) For the Hochschild cohomology series we always have the following explicit formulae:

$$\mathrm{Ext}^*_{\Lambda\otimes_K\Lambda^o}(\Lambda,\Lambda)(Z) = 1 + |V| + |V|Z + \frac{|V|Z^2(|V|^2-1)}{1-|V|Z} \quad \text{if } |V| \geq 2;$$

$$(12) \qquad = \frac{2-Z}{1-Z} \quad \text{if } |V| = 1 \text{ and } \mathrm{char}(K) \neq 2;$$

$$= \frac{2}{1-Z} \quad \text{if } |V| = 1 \text{ and } \mathrm{char}(K) = 2.$$

PROOF: Recall (cf. Cartan-Eilenberg [13], Chapter IX, § 6, p. 174-176) that if Λ is a general associative K-algebra with unit, then the normalized standard free resolution

$$(13) \quad \dots \longrightarrow N_n(\Lambda) \xrightarrow{d_n} N_{n-1}(\Lambda) \longrightarrow \dots \longrightarrow N_1(\Lambda) \longrightarrow N_o(\Lambda) \xrightarrow{\varepsilon} \Lambda \longrightarrow 0$$

of the left $\Lambda^e = \Lambda \otimes_K \Lambda^o$ -module Λ can be constructed as follows: Put $\overline{\Lambda} = \mathrm{Coker}(K \to \Lambda)$, where $K \longrightarrow \Lambda$ is defined by the unit element in Λ, and introduce for $n \geq 0$

$$(14) \qquad N_n(\Lambda) = \Lambda \otimes_K \underbrace{\overline{\Lambda} \otimes_K \dots \otimes_K \overline{\Lambda}}_{n} \otimes_K \Lambda = \Lambda^e \otimes_K \overline{\Lambda}^{\otimes n}$$

where a typical element will be denoted by $\lambda[\lambda_1,\dots,\lambda_n]\mu$ (which is zero if λ_i is in the image of K) if $n \geq 1$ and by $\lambda[\ ]\mu$ if $n = 0$. The left Λ^e-module (i.e. Λ-bimodule) map $N_n(\Lambda) \xrightarrow{d_n} N_{n-1}(\Lambda)$ $(n \geq 1)$ is defined by:

$$(15) \quad d_n[\lambda_1,\dots,\lambda_n] = \lambda_1[\lambda_2,\dots,\lambda_n] + \sum_{0<i<n} (-1)^i[\lambda_1,\dots,\lambda_i\cdot\lambda_{i+1},\dots,\lambda_n] + (-1)^n[\lambda_1,\dots,\lambda_{n-1}]\lambda$$

and $N_o(\Lambda) \xrightarrow{\varepsilon} \Lambda$ by $\lambda[\ ]\mu \longmapsto \lambda\cdot\mu$. Therefore, if M is any Λ-bimodule, considered as a right Λ^e-module, we can use the resolution (13) to calculate the Hochschild homolog $H_n(\Lambda,M) = \mathrm{Tor}_*^{\Lambda^e}(M,\Lambda)$ as the homology of the complex:

$$(16) \qquad \begin{array}{ccccc} & & M \otimes_K \overline{\Lambda}^{\otimes n} & \dashrightarrow & M \otimes_K \overline{\Lambda}^{\otimes(n-1)} \\ & & | & & | \\ \dots \longrightarrow M \otimes_{\Lambda^e}\Lambda^e \otimes_K \overline{\Lambda}^{\otimes(n+1)} & \xrightarrow{id_M \otimes d_{n+1}} & M \otimes_{\Lambda^e}\Lambda^e\otimes_K\overline{\Lambda}^{\otimes n} & \xrightarrow{id_M \otimes d_n} & M \otimes_{\Lambda^e}\Lambda^e\otimes_K\overline{\Lambda}^{\otimes(n-1)} \longrightarrow \dots \end{array}$$

and here $id_M \otimes d_n$ can be identified with the map

$$(17) \quad \begin{aligned} m \otimes_K[\lambda_1,\dots,\lambda_n] \longmapsto m\lambda_1 \otimes_K[\lambda_2,\dots,\lambda_n] &+ \sum_{0<i<n}(-1)^i m \otimes_K[\lambda_1,\dots,\lambda_i\cdot\lambda_{i+1},\dots,\lambda_n] + \\ &+ (-1)^n\lambda_n m \otimes_K[\lambda_1,\dots,\lambda_{n-1}] \qquad (m \in M). \end{aligned}$$

Returning now to the special case when $\Lambda = K \ltimes V$, we have $\overline{\Lambda} = V$ and $\lambda_i\cdot\lambda_{i+1} = 0$ in (17). Therefore, if $M = K \ltimes V$ we easily obtain from (16) and (17) that:

18) $$\mathrm{Tor}_n^{\Lambda^e}(\Lambda,\Lambda) = \frac{V^{\otimes(n+1)}}{\mathrm{Im}(V^{\otimes(n+1)} \xrightarrow{\tau_{n+1}} V^{\otimes(n+1)})} \coprod \mathrm{Ker}(V^{\otimes n} \xrightarrow{\tau_n} V^{\otimes n})$$

or $n \geq 1$, where $\tau_n : V^{\otimes n} \longrightarrow V^{\otimes n}$ is defined by:

19) $$v_1 \otimes \ldots \otimes v_n \longmapsto v_1 \otimes \ldots \otimes v_n + (-1)^n v_n \otimes v_1 \otimes \ldots \otimes v_{n-1}$$

note that $\tau_1 = 0$). Clearly $\mathrm{Tor}_0^{\Lambda^e}(\Lambda,\Lambda) = \Lambda = V \oplus K$, which is (18) for n = 0, with the nterpretation $\tau_0 = 0$, $V^{\otimes 0} = K$.

ow Ker τ_n in (18) is exactly the vector space $(V^{\otimes n})^{\underline{\underline{Z}}/n\underline{\underline{Z}}}$ of invariants in $V^{\otimes n}$ for the roup $\underline{\underline{Z}}/n\underline{\underline{Z}}$, acting through its generator $t_n = \overline{1}$ by

20) $$t_n(v_1 \otimes \ldots \otimes v_n) = (-1)^{n-1} v_n \otimes v_1 \otimes \ldots \otimes v_{n-1}$$

nd,in a similar way, Coker τ_{n+1} in (18) is the space of coinvariants for the nalogous action of $\underline{\underline{Z}}/(n+1)\underline{\underline{Z}}$ in $V^{\otimes(n+1)}$. But the exact sequence

21) $$0 \longrightarrow \mathrm{Ker}\,\tau_n \longrightarrow V^{\otimes n} \xrightarrow{\tau_n} V^{\otimes n} \longrightarrow \mathrm{Coker}\,\tau_n \longrightarrow 0$$

hows that Ker τ_n and Coker τ_n always have the same dimension over K. Therefore it ollows from (18) and the preceding discussion that

$$\ell(\mathrm{Tor}_n^{\Lambda \otimes_K \Lambda^o}(\Lambda,\Lambda)) = |(V^{\otimes n})^{\underline{\underline{Z}}/n\underline{\underline{Z}}}| + |(V^{\otimes(n+1)})^{\underline{\underline{Z}}/(n+1)\underline{\underline{Z}}}| \quad (\text{for } n \geq 1),$$

ich gives the formula (10) for the Hochschild homology series in Theorem 1, (ii). ne more explicit formulae in (i) and (ii), are trivial when $|V| = 1$. Assume therefore rom now on that $|V| > 1$. Assume also furthermore that the characteristic of K is 0. en the endomorphism $\overline{t}_n = \frac{1}{n}(\mathrm{Id} + t_n + (t_n)^2 + \ldots + (t_n)^{n-1})$ of $V^{\otimes n}$ (t_n is defined (20)) is a projection of $V^{\otimes n}$ onto $(V^{\otimes n})^{\underline{\underline{Z}}/n\underline{\underline{Z}}}$ (a general fact - Maschke...- about nite groups operating linearly on vector spaces, when the characteristic of the ound field does not divide the order of the group). It follows that:

22) $$|(V^{\otimes n})^{\underline{\underline{Z}}/n\underline{\underline{Z}}}| = \mathrm{trace}\,\overline{t}_n = \frac{1}{n}\sum_{i=0}^{n-1} \mathrm{trace}(t_n)^i$$

ere $(t_n)^0 = \mathrm{Id}_{V^{\otimes n}}$ clearly has trace $|V|^n$. In order to find the other traces in (22) have to work a little harder. Fix for the moment n, and write for simplicity $t = t_n$, $= \overline{t}_n$. Let $e_1,\ldots,e_m$ be a K-basis for V ($m = |V|$). Then the

3) $$e_{i_1} \otimes \ldots \otimes e_{i_s} \otimes \ldots \otimes e_{i_n}, \quad 1 \leq i_s \leq m,\ 1 \leq s \leq n$$

rm a K-basis for $V^{\otimes n}$. Now t has a simple form with respect to the basis (23): $e_{i_1} \otimes \ldots \otimes e_{i_n}) = (-1)^{n-1} e_{i_n} \otimes e_{i_1} \otimes \ldots \otimes e_{i_{n-1}}$,and therefore

4) $$\mathrm{trace}(t) = (-1)^{n-1} \cdot \mathrm{Card}\{e_{i_1} \otimes \ldots \otimes e_{i_n} \mid e_{i_n} = e_{i_1},\ e_{i_1} = e_{i_2}, \ldots, e_{i_{n-1}} = e_{i_n}\}$$

us only the elements $e_i \otimes e_i \otimes \ldots \otimes e_i$, $1 \leq i \leq m$ can occur in (24) and

$\mathrm{trace}(t) = (-1)^{n-1}|V|$. In a similar way one sees that

(25) $$\mathrm{trace}(t^i) = (-1)^{(n-1)i}|V|^{(i,n)} \qquad (1 \leq i \leq n-1)$$

where (i,n) = the greatest common divisor of i and n and (25) is also valid for $i = n$, when $t^n = t^0 = \mathrm{Id}_{V^{\otimes n}}$. Thus $|(V^{\otimes n})^{\underline{\underline{Z}}/n\underline{\underline{Z}}}| = \frac{1}{n}\sum_{i=1}^{n}(-1)^{(n-1)i}|V|^{(i,n)}$ and therefore

(26) $$\sum_{n\geq 1}|(V^{\otimes n})^{\underline{\underline{Z}}/n\underline{\underline{Z}}}|Z^n = \sum_{n=1}^{\infty}\left(\frac{1}{n}\sum_{i=1}^{n}(-|V|)^{(i,n)}\right)\cdot(-Z)^n$$

We rewrite (26) by noting that for fixed s, $1 \leq s \leq n$ there is an i with $1 \leq i \leq n$ such that $(i,n) = s$ if and only if $s|n$ and then there are exactly $\varphi(n/s)$ such i:s where φ is the Euler φ-function. Thus the series in (26) is equal to

(27) $$f_V(Z) \overset{\mathrm{def}}{=} \sum_{n=1}^{\infty}\frac{1}{n}\left(\sum_{s|n}\varphi(\tfrac{n}{s})(-|V|)^s\right)\cdot(-Z)^n$$

and the formula (9) in Theorem 1 (i) is proved. Let us prove that $f_V(Z)$ does not represent a rational function of Z if $|V| > 1$. If $f_V(Z)$ were rational, then

$$H_V(Z) \overset{\mathrm{def}}{=} \frac{Z}{2}\frac{d}{dZ}\Big(f_V(Z)-f_V(-Z)\Big) = \sum_{\substack{n\geq 1\\ n \text{ odd}}}\left(\sum_{s|n}\varphi(\tfrac{n}{s})|V|^s\right)\cdot Z^n$$

would also be so. But the nth coefficient in $H_V(Z)$ is $\geq |V|^n + (n-1)|V|$ (n odd) with equality if and only if n is an odd prime. Furthermore $\sum_{n\geq 1, n \text{ odd}}(|V|^n + (n-1)|V|)Z^n =$

$= \dfrac{|V|Z}{1-|V|^2Z^2} + \dfrac{2|V|Z^3}{(1-Z^2)^2}$ which is a rational function which we will denote by $R_V(Z)$.

It follows that $H_V(Z)$ and

(28) $$H_V(Z) - R_V(Z) = \sum_{n\geq 1, n \text{ odd}}\Big[\sum_{s|n}\varphi(\tfrac{n}{s})|V|^s - (|V|^n + (n-1)|V|)\Big]\cdot Z^n$$

are rational and irrational at the same time. But the coefficients a_n in (28) are ≥ 0 and are zero if and only if n is even or an odd prime. But the set of these n does not form a periodic sequence for big n:s and therefore, by the Skolem-Mahler-Lech theorem [46],[37],[29],[45], used here almost as in [45], we have that $H_V(Z)$ and (therefore) $f_V(Z)$ do not represent rational functions of Z. Thus Theorem 1 (i) is completely proved (the assertion about the convergence is easy).

Assume now that the characteristic of K is $p \neq 0$. In this case we know already that the Hochschild homology series is $1 + (1+Z^{-1})\cdot\sum_{n\geq 1}|(V^{\otimes n})^{\underline{\underline{Z}}/n\underline{\underline{Z}}}|\cdot Z^n$. Put $\Phi_V(Z) =$

$= \sum_{n\geq 1}|(V^{\otimes n})^{\underline{\underline{Z}}/n\underline{\underline{Z}}}|\cdot Z^n$ and let ζ be any non-trivial p^{th} root of unity. If $\Phi_V(Z)$ were

rational, then $\Phi_V(Z) - \frac{1}{p}\Big(\Phi_V(Z) + \Phi_V(\zeta Z) + \ldots + \Phi_V(\zeta^{p-1}Z)\Big) = \sum_{\substack{n\neq 0(p)\\ n\geq 1}}|(V^{\otimes n})^{\underline{\underline{Z}}/n\underline{\underline{Z}}}|\cdot Z^n =$

$= \sum_{\substack{n\neq 0(p)\\ n\geq 1}}\frac{1}{n}\left(\sum_{s|n}\varphi(\tfrac{n}{s})(-|V|)^s\right)\cdot(-Z)^n$ would also be so, where the last equality comes from

the fact that if $n\neq 0(p)$ we can use the Maschke trick (22) as before, since the order or the group $\underline{\underline{Z}}/n\underline{\underline{Z}}$ is not divisible by $p = \mathrm{Char}(K)$. We can now repeat the previous easoning and it follows that $\Phi_V(Z)$ is not rational. Thus the assertion (ii) of Theorem is proved (the assertion about convergence is easy).

inally the proof of part (iii) of Theorem 1 is very easy: just take $\mathrm{Hom}_{\Lambda\otimes_K\Lambda^\circ}(\ ,\Lambda)$ of e resolution (13) and calculate (this is easier than for homology). This ends the oof of Theorem 1.

ROLLARY 1.- Let k be any field, m any integer ≥ 2 and let $(S,\underline{m})$ be the commutative cal ring $k[X_1,\ldots,X_m,Y_1,\ldots,Y_m]/(\ldots X_iX_k\ldots,\ldots Y_jY_s\ldots)$ (we divide by all squares the X_i:s and all squares in the Y_j:s) of embedding dimension 2m, having $\underline{m}^3 = 0$. t M be the S-module $S/(X_1-Y_1,X_2-Y_2,\ldots,X_m-Y_m)$ of length m+1 (it is annihilated by $\underline{m}^2$). en with the notations of the introduction (§ 0; in particular $\tilde{M}$ is the Matlis dual M):

) If $\mathrm{Char}(k)=0$, then $\mathrm{Tor}_*^S(M,M)(Z) = 1 + (1+Z^{-1})\sum_{n=1}^{\infty}\log(1+m(-Z)^n)^{-\frac{\varphi(n)}{n}}$, which is non-rational (φ is the Euler φ-function).

If $\mathrm{Char}(k)\neq 0$, then $\mathrm{Tor}_*^S(M,M)(Z)$ is also non-rational.

i) The series $\mathrm{Tor}_*^S(k,M)(Z)$, $\mathrm{Tor}_*^S(k,\tilde{M})(Z)$ and $\mathrm{Tor}_*^S(M,\tilde{M})(Z)$ are all rational.

ii) If $\mathrm{Tor}_{*+1}^S(M/\mathrm{soc}(M),M) \xrightarrow{\partial_{*+1}} \mathrm{Tor}_*^S(\mathrm{soc}(M),M)$ is the boundary operator associated to the short exact sequence:

$$0 \longrightarrow \mathrm{soc}(M) \longrightarrow M \longrightarrow M/\mathrm{soc}(M) \longrightarrow 0 \tag{29}$$

then if $\mathrm{char}(k) = 0$, we have that Coker ∂_{*+1} is the cyclic homology of $k[t_1,\ldots,t_m]/(t_1,\ldots,t_m)^2$ in the sense of Loday and Quillen [32].

v) Since S and M are graded, Tor:s are bigraded and the corresponding double series is:

$$\mathrm{Tor}_{*,*}^S(M,M)(X,Y) = \sum_{p,q\geq 0}|\mathrm{Tor}_{p,q}^S(M,M)|\cdot X^pY^q = 1 + (1+X^{-1})\sum_{n\geq 1}|(V^{\otimes n})^{\underline{Z}/n\underline{Z}}|(XY)^n .$$

OOF: Most of the assertions in Corollary 1 can be easily deduced from Theorem 1 and s proof. We have rewritten $\mathrm{Tor}_*^S(M,M)(Z)$ in (i) in a form slightly different from that Theorem 1, (i). This will be used in Remark 2 below. The cyclic homology assertion llows from Example 4.3 of [32]: Tensor the exact sequence (29) with the complex $N_*(\Lambda)$ 3) for the trivial extension and make the boundary explicit. The cyclic homology rns out to be $V^{\otimes(n+1)}/\mathrm{Im}\,\tau_{n+1}$ and this graded vector space has a non-rational Hilbert-ries.

MARK 1.- As we have said in the introduction, our first approach to § 1 came through e study of $H^*(\mathcal{L}X,\underline{Q})$, and we first met the combinatorial problems of § 1 in the dual tting for graded Lie algebras. When we consulted Jörgen Backelin he soon came up with rely combinatorial solutions. At the same time we had found the reasoning related group representations which is presented here. We later found out that similar mbers as those in the series in (9) had already been encountered in combinatorics by chard Stanley and others in connection with e.g. the enumeration of necklaces of fferent types. Cf. Exercise 27, p. 48 and its solution on p. 59-60 of [51] and the terature cited there.

MARK 2.- If we take the derivative of the series $\sum_{n=1}^{\infty}\log(1+m(-Z)^n)^{-\frac{\varphi(n)}{n}}$ in Corollary

1 (i),we obtain $\sum\limits_{n\geq 1} \frac{m\varphi(n)(-Z)^{n-1}}{1+m(-Z)^n}$ $(m \geq 2)$ and it is easy to see that this series represents a function, which can be continued to a meromorphic function for $|Z| < 1$. It can not represent a rational function because it has infinitely many poles for $|Z| < 1$. This gives an alternative proof of non-rationality.

REMARK 3.- There are many attempts to make explicit calculations of cyclic (co)homology and its connections with Hochschild (co)homology in the literature. Several papers are devoted to just the case $k[X]/(X^n)$ $(n \geq 2)$[56],[38],[50],[26].It should be interesting for commutative algebraists to start a more systematic study, using older homological results in local algebra.

§ 2. THE (CO)HOMOLOGY OF FREE LOOP SPACES.

Recall (cf.e.g.[48],[3]) that if X is (say) a finite, simply-connected CW-complex, and $\mathcal{L}X$ the free loop space of X, i.e. the space of all continuous maps $S^1 \longrightarrow X$ with the compact-open topology, then the "fiber homotopy pull-back diagram"

$$(30)\qquad \begin{array}{ccc} \mathcal{L}X & \longrightarrow & X \\ \downarrow & & \downarrow \Delta \\ X & \xrightarrow{\ \Delta\ } & X \times X \end{array}$$

where Δ is the diagonal map, gives rise to an Eilenberg-Moore spectral sequence in the second quadrant:

$$E_2^{-p,q} = \mathrm{Tor}_p^{H^*(X\times X,\underline{\underline{Q}})}(H^*(X,\underline{\underline{Q}}),H^*(X,\underline{\underline{Q}}))^q \Rightarrow H^n(\mathcal{L}X,\underline{\underline{Q}})$$

where $H^*(X\times X,\underline{\underline{Q}}) = H^*(X,\underline{\underline{Q}}) \otimes H^*(X,\underline{\underline{Q}})$ operates on the $H^*(X,\underline{\underline{Q}})$:s in the natural way.
We also have a dual spectral sequence:

$$E^2_{-p,q} = \mathrm{Ext}^p_{H^*(X\times X,\underline{\underline{Q}})}(H^*(X,\underline{\underline{Q}}),\ \mathrm{Hom}_{\underline{\underline{Q}}}(H^*(X,\underline{\underline{Q}}),\ \underline{\underline{Q}}))_q \Rightarrow H_n(\mathcal{L}X,\underline{\underline{Q}})$$

where $H^*(X\times X,\underline{\underline{Q}})$ operates on $\mathrm{Hom}_{\underline{\underline{Q}}}(H^*(X,\underline{\underline{Q}}),\underline{\underline{Q}})$ in the natural way. Note also that $\mathrm{Hom}_{\underline{\underline{Q}}}(H^*(X,\underline{\underline{Q}}),\underline{\underline{Q}}) \simeq H_*(X,\underline{\underline{Q}})$.

If X is formal, e.g. if X = SY (the suspension of a connected CW-complex Y; example: $X = \bigvee\limits_{i=1}^{m} S^{n_i}$ - a wedge of spheres identified on one point, where the $n_i \geq 2$) or if X is 4-dimensional, or if X is a compact Kähler manifold or a compact Riemannian symmetric space, both these spectral sequences degenerate and we therefore get (cf.[3], p.489) e.g.

$$(31)\qquad \coprod_{t\geq 0} \mathrm{Tor}_t^{H^*(X,\underline{\underline{Q}}) \otimes H^*(X,\underline{\underline{Q}})}(H^*(X,\underline{\underline{Q}}),H^*(X,\underline{\underline{Q}}))^{t+n} \simeq H^n(\mathcal{L}X,\underline{\underline{Q}})$$

where on the left hand side we have the graded Hochschild homology. This Hochschild homology can be calculated by means of a normalized standard free resolution as in § 1, but with extra signs that take into account the extra grading that we have [9],[27].

[I]n order not to complicate the exposition, we here just study the case when $H^*(X,\underline{\underline{Q}})$ [i]s concentrated in even degrees, so that the theory of § 1 applies without change. [T]hus we obtain as before that (cf. (18))

$$(32)\quad \operatorname{Tor}_t^{H^*(X\times X,\underline{\underline{Q}})}(H^*(X,\underline{\underline{Q}}),H^*(X,\underline{\underline{Q}})) = \frac{(H^+(X,\underline{\underline{Q}}))^{\otimes(t+1)}}{\operatorname{Im}\tau_{t+1}} \coprod \operatorname{Ker}\left((H^+)^{\otimes t} \xrightarrow{\tau_t} (H^+)^{\otimes t}\right)$$

[w]here $H^+ = H^+(X,\underline{\underline{Q}}) = \bigoplus_{i>0} H^i(X,\underline{\underline{Q}})$ and where τ_n is the map $h_1\otimes\ldots\otimes h_n \longmapsto h_1\otimes\ldots\otimes h_n + (-1)^n h_n\otimes h_1\otimes\ldots\otimes h_{n-1}$ as before. Furthermore (32) is now compatible with the extra [g]rading coming from $H^*(X,\underline{\underline{Q}})$. We are therefore back to our old problem of calculating [t]he invariants of $\underline{\underline{Z}}/n\underline{\underline{Z}}$ operating on a tensor power $V^{\otimes n}$ of a vector space V, through [i]ts generator $t_n = \bar{1} \in \underline{\underline{Z}}/n\underline{\underline{Z}}$ as $t_n(v_1\otimes\ldots\otimes v_n) = (-1)^{n-1}v_n\otimes v_1\otimes\ldots\otimes v_{n-1}$ but now V is also [g]raded. Clearly t_n respects this extra grading, and if we go through the theory of § 1 [c]arefully we can obtain explicit formulae, but in order to simplify the exposition we [n]ow assume that V is concentrated in one degree d (think of the case when X is the [w]edge of m spheres S^d!). Now it follows as before that $(V^{\otimes n})^{\underline{\underline{Z}}/n\underline{\underline{Z}}}$ is concentrated in [d]egree nd, and that its dimension there is $\frac{1}{n}\sum_{i=1}^{n}(-1)^{(n-1)i}|V|^{(i,n)}$. Summing up, [c]ombining (32) and (31) with the theory of § 1, we therefore obtain e.g. the following [t]heorem, which clearly can be generalized:

[T]HEOREM 2.- Let $X = S^d\vee\ldots\vee S^d$ be the wedge of m d-spheres $(d \geq 2$, and - in order to [s]implify - d even), and let $\mathcal{L}X$ be the free loop space of X. Then we have the following [e]xplicit formula for the generating series of the (co)homology of $\mathcal{L}X$:

$$(33)\qquad H_*(\mathcal{L}X,\underline{\underline{Q}})(Z) = 1 + (1+Z)\cdot\sum_{n=1}^{\infty}\log(1+m(-Z^{d-1})^n)^{-\frac{\varphi(n)}{n}}$$

[w]here φ is the Euler φ-function. This series represents a non-rational function of Z [i]f $m \geq 2$ and is equal to $(1+Z^{d-1}+Z^d-Z^{2(d-1)})/(1-Z^{2(d-1)})$ if n = 1.

[P]ROOF: For given n, the Tor_t in (31) can, according to (32) , only occur if t satisfies [e]ither $t+n = d\cdot t$ or $t+n = d\cdot(t+1)$, i.e. only if $t = n/(d-1)$ $(t\geq 1)$ or $t = (n-d)/(d-1)$ [(]$t\geq 0$) which requires either n or n-1 to be divisible by d-1. In the first case the [c]ontribution to $|H^n(\mathcal{L}X,Q)|$ is $|(V^{\otimes t})^{\underline{\underline{Z}}/t\underline{\underline{Z}}}|$ (where $V = H^+(X,\underline{\underline{Q}})$), which has been deter[m]ined above. In the second case the contribution to $|H^n(\mathcal{L}X,\underline{\underline{Q}})|$ is $|(V^{\otimes(t+1)})^{\underline{\underline{Z}}/(t+1)\underline{\underline{Z}}}|$ [(]recall that $|\operatorname{Coker}\tau_{t+1}| = |\operatorname{Ker}\tau_{t+1}|$), which also has been determined. We therefore [g]et the formula (33). The non-rationality assertion for $m \geq 2$ follows from § 1, and [t]he explicit formula for m = 1 easily follows from (33) and the Theorem 2 is completely [p]roved.

[R]EMARK 1.- It follows in particular that $H_*(\mathcal{L}(S^2\vee S^2),\underline{\underline{Q}})(Z)$ is irrational. In [2] [D]. Anick shows that the smallest finite, simply-connected CW-complex X such that ΩX [(]the space of based loops in X) has an irrational series $H_*(\Omega X,\underline{\underline{Q}})(Z)$ must be 6-dimen[s]ional with at least 4 cells. Thus, using the terminology of [2], we can say that the smallest $\mathcal{L}$-irrational CW-complex is $S^2 \vee S^2$.

REMARK 2.- The case m = 1 of Theorem 2 was of course known before [52],[28],[60].

REMARK 3.- In [55] M. Vigué-Poirrier makes estimates for the series $H_*(\mathcal{L}X,\underline{\underline{Q}})(Z)$. These estimates also follow from our explicit formulae.

We now turn to some remarks about the study of $H^*(\mathcal{L}X,k)$ with other coefficients, mostly $k=\underline{\underline{Z}}$ or $k=\underline{\underline{Z}}/2\underline{\underline{Z}}$. Recall that the "fiber homotopy pull-back diagram" (30) is obtained by replacing the right vertical map Δ by a homotopy equivalent fibration $X^I \xrightarrow{p} X \times X$, where $X^I = \mathrm{Hom}_{cont}([0,1],X)$ and where $p(\psi) = (\psi(0),\psi(1))$ for $\psi \in X^I$, and then taking the ordinary pull-back. The fiber of p is ΩX (the based loops) and therefore the left vertical fibration in (30) is a well-known fibration (X is now for a while quite general)

(34) $$\Omega X \longrightarrow \mathcal{L}X \longrightarrow X$$

and the corresponding Serre spectral sequence [43] has been studied with different coefficients in e.g. [52],[28],[48]. Originally we had planned to apply Lemma 4 of [39] (cf. also Lemma 2 of p. 773 of the McCleary-Ziller paper in Amer.J.Math.109, 1987 - I thank L. Avramov for this last reference) to e.g. $S^2 \vee S^2$. This would have given

(35) $$H^*(\mathcal{L}X,\underline{\underline{Z}}/2\underline{\underline{Z}}) = H^*(X,\underline{\underline{Z}}/2\underline{\underline{Z}}) \otimes H^*(\Omega X,\underline{\underline{Z}}/2\underline{\underline{Z}})$$

for e.g. $X = S^2 \vee S^2$, and thereby the generators of the 2-torsion $T_*(X)$ of $H_*(\mathcal{L}X,\underline{\underline{Z}})$ in view of the well-known formula that follows from the universal coefficients theorem:

(36) $$(1+Z)\cdot(T_*(Y)/2T_*(Y))(Z) = H^*(Y,\underline{\underline{Z}}/2\underline{\underline{Z}})(Z) - H^*(Y,\underline{\underline{Q}})(Z)$$

(Y "any" space). However, D. Anick told me that a formula like (35) could never hold for $S^2 \vee S^2$ and he then found the following corrected version of Lemma 4 in [39]:
"THE TRICK OF ŠVARTS": Assume that in the Serre fibration (34) X is (r-1)-connected ($r \geq 2$). Assume furthermore that:

(37) $H_*(\Omega X,\underline{\underline{Z}}/2\underline{\underline{Z}})$ is commutative and that in $H^+(\Omega X,\underline{\underline{Z}}/2\underline{\underline{Z}})$ we have that $x^2 = 0$ for all x. Then in the Serre spectral sequence of (34): $E_2^{p,q}=H^p(X,H^q(\Omega X,\underline{\underline{Z}}/2\underline{\underline{Z}})) \Rightarrow H^n(\mathcal{L}X,\underline{\underline{Z}}/2\underline{\underline{Z}})$ we have that the differential d_r is 0.
This modified trick can be applied to S^d and to all the cases studied by McCleary and Ziller, but the 2-torsion of $H_*(\mathcal{L}(S^2 \vee S^2),\underline{\underline{Z}})$ is unknown. We later learned that McCleary also had found that (37) is needed to make his $T^* = \mathrm{Id}$. (If $T^* \neq \mathrm{Id}$ his arguments do not work.)

§ 3. RELATIONS BETWEEN THE (CO)HOMOLOGY OF FREE LOOP SPACES AND HOCHSCHILD HOMOLOGY OF LOCAL RINGS AND RELATED PROBLEMS.

Recall that if $(R,\underline{m})$ is a local ring, which is an algebra over its residue field $k = R/\underline{m}$, we denote its Hochschild homology by $H_n(R,R) = \mathrm{Tor}_n^{R \otimes_k R}(R,R)$ and the series (defined if R is artinian) $\sum \ell_R(H_n(R,R))\cdot Z^n$ by $H_*(R,R)(Z)$. For any R-R-bimodule M, in particular for any R-module M, considered as a symmetric bimodule, we also have a corresponding series $H_*(R,M)(Z)$ if the lengths involved are defined.

Here is an evident consequence of the preceding theory: Let X be a finite wedge of

-spheres and let $(R,\underline{m})$ be the local ring $H^*(X,\underline{\underline{Q}})$ (forget the grading!). Then

$$38) \qquad H_*(\mathcal{L}X,\underline{\underline{Q}})(Z) - Z\cdot H_*(R,R)(Z) = 1 - Z$$

combine formula (33) with Corollary 1,(i) in § 1 above). Recall [40], that if X is a eneral finite, simply-connected CW-complex of dimension four and if $R = H^*(X,\underline{\underline{Q}})$ forget the grading!) as before (this time the maximal ideal satisfies $\underline{m}^3 = 0$) then:

$$39) \left(H_*(\Omega X,\underline{\underline{Q}})(Z)\right)^{-1} - Z\cdot\left(\mathrm{Tor}_*^R(\underline{\underline{Q}},\underline{\underline{Q}})(Z)\right)^{-1} = (1 - Z)\cdot(1 - |H^2(X,\underline{\underline{Q}})|Z + |H^4(X,\underline{\underline{Q}})|Z^2)$$

here are clear analogies between (38) and (39) and it is not unreasonable to believe hat for general X (4-dimensional...) some variant of (38) (like (39)) should hold rue. In some cases, where we have made the calculations, the right-hand side of (38) urns out to be $1 - Z - |H^4(X,\underline{\underline{Q}}|Z\cdot(1-Z)^{-1}$. (In our examples we had $(H^2(X,\underline{\underline{Q}})^2=H^4(X,\underline{\underline{Q}})$!) ll this suggests a number of problems, motivated by [40],[34],[4] , and mentioned in he introduction. We recall furthermore that the "holonomy spectral sequence" of Felix-homas ([16], Remarque following Proposition 8.1) applied to the fibration

$$40) \qquad \Omega X \longrightarrow \mathcal{L}X \longrightarrow X$$

ives rise to a spectral sequence:

$$41) \qquad E^2_{p,q} = \mathrm{Tor}_{p,q}^{H_*(\Omega X,\underline{\underline{Q}})}(\underline{\underline{Q}},H_*(\Omega X,\underline{\underline{Q}})) \Rightarrow H_*(\mathcal{L}X,\underline{\underline{Q}})$$

here $H_*(\Omega X,\underline{\underline{Q}})$ operates on itself by the adjoint map). If X is a suspension, $H_*(\Omega X,\underline{\underline{Q}})$ s a free algebra, thus of global dimension 1, and this spectral sequence degenerates, nd we get an explicit relation between $H_*(\mathcal{L}X,\underline{\underline{Q}})$ and $H_*(\Omega X,\underline{\underline{Q}})/[H_*(\Omega X,\underline{\underline{Q}}),H_*(\Omega X,\underline{\underline{Q}})]$, nich is a part of Proposition C in [10]. (Added in proof: D. Anick told the author nat M. Vigué-Poirrier and he several years ago also made calculation with the Serre pectral sequence coming from (40), when X was a wedge of spheres, leading to the tudy of the commutator quotient above. They did not study the rationality of the orresponding series.). Now several cases are known when the homological dimension f the algebra $H_*(\Omega X,\underline{\underline{Q}})$ is small (2,3,..) and in these cases it is reasonable to use 1) for calculations. Of course there is also a similar spectral sequence relating he dual of) Hochschild homology of local rings (equicharacteristic, artinian)and the $\mathrm{or}_{p,q}^{\mathrm{Ext}_R^*(k,k)}(k, \mathrm{Ext}_R^*(k,k))$, where $\mathrm{Ext}_R^*(k,k)$ operates on itself by the adjoint represen-tion. But now there are many cases studied when the algebra $\mathrm{Ext}_R^*(k,k)$ has simple roperties (the case when R is a Golod ring, when the Ext-algebra is an extension- the sense of Hopf algebras - of an exterior algebra with a free algebra - [34], 5],[41], or e.g. the case when R is a Fröberg ring [17],[34]), and it is quite easonable to hope for a calculation of Hochschild homology by means of this spectral equence in these cases.

end with three open problems which are not stated here in their greatest possible nerality:

OBLEM 1.- Let $(R,\underline{m})$ be a local commutative equicharacteristic artinian (say) ring. Is it true that the Hochschild homology series of (R,m) is rational if and only if (R,m) is a local complete intersection ?

PROBLEM 2.- Let X be a finite, simply-connected CW-complex, $\mathcal{L}X$ the free loop space on X. Is it true that the series

$$H^*(\mathcal{L}X,\underline{Q})(Z) = \sum_{i\geq 0} |H^i(\mathcal{L}X,\underline{Q})|\cdot Z^i$$

is a rational function of Z, if and only if X is $\underline{Q}$- elliptic (i.e. $\pi_*(X)\otimes\underline{Q}$ is finite-dimensional).

PROBLEM 3.- If the local rings are not equicharacteristic, it is more natural to study Mac Lane-Shukla homology instead of Hochschild homology[35-6,47] Do the same calculations! Can the corresponding groups be used in commutative algebra, when non-equicharacteristic questions are studied (Hochster's questions etc.) ?

BIBLIOGRAPHY

[1] D. ANICK, Construction d'espaces de lacets et anneaux locaux à séries de Poincaré-Betti non rationnelles, Comptes Rendus Acad. Sc. Paris, 290, série A, p. 729-732. 1980 (Cf. also D. ANICK, A counterexample to a conjecture of Serre, Ann.Math., 115, 1982, p. 1-33. Correction: Ann.Math., 116, 1983, 661.)

[2] D. ANICK, The smallest Ω-irrational CW-complex, J. Pure Appl. Algebra, 28, 1983, p. 213-222.

[3] D. ANICK, A model of Adams-Hilton type for fiber squares, Ill. J. Math., 29, 1985, p. 463-502.

[4] D. ANICK - T.H. GULLIKSEN, Rational dependence among Hilbert and Poincaré series, J. Pure Appl. Algebra, 38, 1985, p. 135-157.

[5] L.L. AVRAMOV, Golod homomorphisms, Lecture Notes in Mathematics, 1183, 1986, p. 59-78, Springer-Verlag, Berlin, Heidelberg, New York, Tokyo.

[6] L.L. AVRAMOV, A.R. KUSTIN, M. MILLER, Poincaré series of modules over local rings of small embedding codepth or small linking number, preprint 1986, to appear in J. Algebra.

[7] J. BACKELIN - J.-E. ROOS, When is the double Yoneda Ext-algebra of a local noetherian ring again noetherian ?, Lecture Notes in Mathematics, 1183, 1986, p. 101-119, Springer-Verlag, Berlin, Heidelberg, New York, Tokyo.

[8] R. BOTT - L.W. TU, Differential forms in algebraic topology, Graduate Texts in Mathematics, n° 82, Springer-Verlag, Berlin, Heidelberg, New York, 1982.

[9] D. BURGHELEA, Cyclic homology and the algebraic K-theory of spaces I, Contemporary Mathematics, 55, 1986, p. 89-115. American Mathematical Society.

[10] D. BURGHELEA - Z. FIEDOROWICZ, Cyclic homology and the algebraic K-theory of spaces II, Topology, 25, 1986, p. 303-317.

[11] G.E. CARLSSON - R.L. COHEN, The cyclic groups and the free loop space, Comment. Math. Helv., 62, 1987, p. 423-449.

[12] G.E. CARLSSON, R.L. COHEN, T. GOODWILLIE, W.C. HSIANG, The free loop space and the algebraic K-theory of spaces, K-theory, 1, 1987, 53-82.

[13] H. CARTAN - S. EILENBERG, Homological algebra, Princeton University Press, Princeton, 1956.

[14] R. L. COHEN, A model for the free loop space of a suspension, Lecture Notes in Mathematics, 1286, 1987, p. 193-207, Springer-Verlag, Berlin, Heidelberg, New York, Tokyo.

15] R. L. COHEN, Pseudo-isotopies, K-theory and homotopy theory, London Mathematical Society Lecture Notes Series, 117, 1987, p. 35-71, Cambridge University Press, Cambridge.

16] Y. FELIX - J.C. THOMAS, Sur l'opération d'holonomie rationnelle, Lecture Notes in Mathematics, 1183, 1986, p. 136-169, Springer-Verlag, Berlin, Heidelberg, New York, Tokyo.

17] R. FRÖBERG, Determination of a class of Poincaré series, Math. Scand., 37, 1975, p. 29-39.

18] T.G. GOODWILLIE, Cyclic homology, derivations and the free loop space, Topology, 24, 1985, p. 187-215.

19] T.G. GOODWILLIE, On the general linear group and Hochschild homology, Ann. Math., 121, 1985, p. 383-407. Corrections: Ann. Math., 124, 1986, p. 627-628.

20] T.G. GOODWILLIE, Relative algebraic K-theory and cyclic homology, Ann. Math., 124, 1986, p. 247-402.

21] D. GROMOLL - W. MEYER, Periodic geodesics on compact Riemannian manifolds, J. Differential Geometry, 3, 1969, p. 493-510.

22] A. GROTHENDIECK, Sur la théorie des classes de Chern, Bull. Soc. Math. France, 86, 1958, p. 137-154.

23] T.H. GULLIKSEN, A change of ring theorem with applications to Poincaré series and intersection multiplicity, Math. Scand., 34, 1974, p. 167-183.

24] T.H. GULLIKSEN, A note on intersection multiplicities, Lecture Notes in Mathematics, 1183, 1986, p. 192-194, Springer-Verlag, Berlin, Heidelberg, New York, Tokyo.

25] T.H. GULLIKSEN, Homology of modules of length 2 (to appear).

26] P. HANLON, Cyclic homology and the Macdonald conjectures, Invent. Math., 86, 1986, p. 131-159.

27] C. KASSEL, A Künneth formula for the cyclic cohomology of $\mathbb{Z}/2$-graded algebras, Math. Ann., 275, 1986, p. 683-699.

28] P. KLEIN, Über die Kohomologie des freien Schleifenraums, Bonner Mathematische Schriften, Nr 55, 1972, Bonn.

29] C. LECH, A note on recurring series, Arkiv f. Matematik, 2, 1953, p. 417-421.

30] J.-M. LEMAIRE, Algèbres connexes et homologie des espaces de lacets, Lecture Notes in Mathematics, 422, 1974, Springer-Verlag, Berlin, Heidelberg, New York.

31] G. LEVIN, Local rings and Golod homomorphisms, J. Algebra, 37, 1975, p. 266-289.

32] J.-L. LODAY - D. QUILLEN, Cyclic homology and the Lie algebra homology of matrices, Comment. Math. Helv., 59, 1984, p. 565-591.

33] C. LÖFWALL - J.-E. ROOS, Cohomologie des algèbres de Lie graduées et séries de Poincaré-Betti non rationnelles, Comptes rendus Acad. Sc. Paris, 290, série A, 1980, p. 733-736.

34] C. LÖFWALL, On the subalgebra generated by the one-dimensional elements in the Yoneda Ext-algebra , Thesis, Stockholm University, 1976, and Lecture Notes in Mathematics, 1183, 1986, p. 291-338, Springer-Verlag, Berlin, Heidelberg, New York, Tokyo.

35] S. Mac LANE, Homologie des anneaux et des modules, Colloque de Topologie algébrique, tenu à Louvain les 11,12 et 13 juin 1956, p. 55-80, Centre Belge de Recherches Mathématiques, Thone, Liège and Masson & Cie, Paris, 1957.

36] S. Mac LANE, Homology, Die Grundlehren der Math. Wiss., Band 114, Springer-Verlag, Berlin, Göttingen, Heidelberg, 1963.

37] K. MAHLER, On the Taylor coefficients of rational functions, Proc. Cambridge Phil. Soc., 52, 1956, p. 39-48. Addendum: Proc. Cambr.Ph.Soc. 53, 1957, p. 544. (Cf. also K. MAHLER, Proc.Nederl. Akad. Sci., 38, 1935, p. 50-60).

[38] T. MASUDA - T. NATSUME, Cyclic cohomology of certain affine schemes, Publ. RIMS, Kyoto University, 21, 1985, p. 1261-1279.

[39] J. McCLEARY, Closed geodesics on Stiefel manifolds, Lecture Notes in Mathematics, 1172, 1985, p. 157-162, Springer-Verlag, Berlin, Heidelberg, New York, Tokyo.

[40] J.-E. ROOS, Relations between the Poincaré-Betti series of loop spaces and of local rings, Lecture Notes in Mathematics, 740, 1979, p. 285-322, Springer-Verlag, Berlin, Heidelberg, New York.

[41] J.-E. ROOS, On the use of graded Lie algebras in the theory of local rings, London Mathematical Society Lecture Notes Series, 72, 1982, p. 204-230, Cambridge University Press, Cambridge.

[42] J.-E. ROOS, A mathematical introduction, Lecture Notes in Mathematics, 1183, 1986, p. III-VIII, Springer-Verlag, Berlin, Heidelberg, New York, Tokyo.

[43] J.-P. SERRE, Homologie singulière des espaces fibrés. Applications, Ann. Math., 54, 1951, p. 425-505.

[44] J.-P. SERRE, Algèbre locale. Multiplicités (rédigé avec la collaboration de P. GABRIEL), Lecture Notes in Mathematics, 11, 3e édition, 1975, Springer-Verlag, Berlin,Heidelberg, New York.

[45] J.-P. SERRE, Un exemple de série de Poincaré non rationnelle, Proc. Nederland. Acad. Sci., 82, 1979, p. 469-471 [= Indag. Math. 41, 1979, p. 469-471].

[46] Th. SKOLEM, Ein Verfahren zur Behandlung gewisser exponentialer Gleichungen und diophantischer Gleichungen, Comptes rendus 8ème Congrès Scandinave à Stockholm 1934, Lund, 1935, p. 163-188.

[47] U. SHUKLA, Cohomologie des algèbres associatives, Ann. Sci. Ecole Norm. Sup., 78, 1961, p. 163-209.

[48] L. SMITH, On the characteristic zero cohomology of the free loop space, Amer. J. Math., 103, 1981, p. 887-910.

[49] L. SMITH, The Eilenberg-Moore spectral sequence and the mod 2 cohomology of certain free loop spaces, Ill. J. Math., 28, 1984, 516-522.

[50] R.E. STAFFELDT, Rational algebraic K-theory of certain truncated polynomial rings, Proc. Amer. Math. Soc., 95, 1985, p. 191-198.

[51] R.P. STANLEY, Enumerative Combinatorics, volume I, Wadsworth & Brooks/Cole, Advanced Books & Software, Monterey, Calif., 1986.

[52] A.S. ŠVARTS, Gomologii prostranstv zamknutich krivich, Trudy Moskovsk. Mat. Obščestva, 9, 1960, p. 3-44.

[53] C.H. TAUBES, S^1-actions and elliptic genera, preprint, Harvard University 1986-87.

[54] M. VIGUÉ-POIRRIER and D. SULLIVAN, The homology theory of the closed geodesic problem, J. Differential Geometry, 11, 1976, p. 633-644.

[55] M. VIGUÉ-POIRRIER, Homotopie rationnelle et croissance du nombre de géodésiques fermées, Ann. Scient. Ec. Normale Sup., 4e série, 17, 1984, p. 413-431.

[56] M. VIGUÉ-POIRRIER and D. BURGHELEA, A model for cyclic homology and algebraic K-theory of 1-connected topological spaces, J. Differential Geometry, 22, 1985, p. 243-253.

[57] J. WEYMAN, On the structure of free resolutions of length 3. Preprint 1985 (to appear).

[58] E. WITTEN, The index of the Dirac operator in loop space, to appear in the Proceedings of the Conference on Elliptic Curves and Modular Forms in algebraic topology (Princeton 1986) in Springer Lecture Notes. Cf. also Colloquium Lectures given by E. WITTEN at the 836th meeting of the American Mathematical Society in Salt Lake City, Utah, August 5-8, 1987.

[59] E. WITTEN, Elliptic genera and quantum field theory, Communications in Math. Physics, 109, 1987, p. 525-

60] W. ZILLER, The free loop space of globally symmetric spaces, Inv. Math., 41, 1977, p. 1-22.

"This paper is in final form and no version of it will be submitted for publication elsewhere"

INCOMPLETE SUMS AND TWO APPLICATIONS OF DELIGNE'S RESULT

A. Tietäväinen
Department of Mathematics
University of Turku
20500 Turku, Finland

In this paper I consider some problems which relate to polynomials, equations and character sums over finite fields and which can be studied by appealing to a classical method of Vinogradov [29] and its modifications. The proofs depend, in an essential manner, on estimates of Weil [30], Carlitz and Uchiyama [4], and Deligne [11]. It is interesting to observe that rather similar methods can be used also in a covering radius problem of BCH codes propounded by Helleseth [13].

1. The Vinogradov inequality

First I must define what we mean by an incomplete sum. Let n be a positive integer and let F be $\mathbf{Z}_n$, the set of residues (mod n). Let ϕ be a mapping from F to $\mathbf{C}$, the field of complex numbers. Then

$$\sum_{x\in F} \phi(x)$$

is called a complete sum, and of course

$$\sum_{x\in B} \phi(x),$$

where B is a proper subset of F, is called an incomplete sum. Throughout the paper we shall abbreviate complete sums $\sum_{x\in F}$ by just $\sum_x$. If $|B|$, the cardinality of B, is equal to b and if the values of ϕ are in a sense well distributed then

$$\sum_{x\in B} \phi(x) = \frac{b}{n}\sum_x \phi(x) + \text{something}$$

where this something, say $A(\phi,B)$, depends on ϕ and B and is small.

In order to estimate $A(\phi,B)$ we may use finite Fourier transforms. Put, for $z\in F$,

$$\psi(z) = \sum_x \phi(x)e_n(xz) \tag{1.1}$$

where $e_n(\alpha) = e^{2\pi i\alpha/n}$. By the well known result

$$\sum_z e_n((x-t)z) = \begin{cases} n \text{ if } x = t \text{ in } F, \\ 0 \text{ otherwise,} \end{cases} \tag{1.2}$$

we have

$$\sum_z \psi(z)e_n(-tz) = \sum_x \phi(x)\sum_z e_n((x - t)z) = n\phi(t).$$

Thus

$$\begin{aligned}\sum_{t\in B} \phi(t) &= \frac{1}{n}\sum_z \psi(z) \sum_{t\in B} e_n(-tz)\\ &= \frac{b}{n}\psi(0) + \frac{1}{n}\sum_{z\neq 0}\sum_x \phi(x)e_n(xz) \sum_{t\in B} e_n(-tz) \qquad (1.3)\\ &= \frac{b}{n}\sum_x \phi(x) + \frac{\theta}{n}\Delta(\phi)E(B).\end{aligned}$$

Here and hereafter, θ is complex number with $|\theta| \leq 1$,

$$\Delta(\phi) = \max_{z\neq 0}\left|\sum_x \phi(x)e_n(xz)\right|,$$

and

$$E(B) = \sum_{z\neq 0} \left| \sum_{t\in B} e_n(tz)\right|.$$

Often we use the result (1.3) in the form

$$\left| \sum_{t\in B} \phi(t) - \frac{b}{n}\sum_x \phi(x)\right| \leq \frac{1}{n}\Delta(\phi)E(B) \qquad (1.4)$$

and call it the Vinogradov inequality [29].

Above F was the set of residues (mod n) but we could prove the inequality (1.4) - or, in fact, the appropriate modifications of (1.4) - also in the cases where F is a finite prime field $\mathbf{F}_p$, or a general finite field $\mathbf{F}_q$, $q = p^m$, or the linear space $\mathbf{F}_q^s$, or $\mathbf{Z}_n^m$, the set of m-tuples over $\mathbf{Z}_n$. Of course, in the latter cases the product xz must be replaced by the corresponding dot product.

Though the inequality (1.4) is easy to understand and easy to prove, it has several interesting applications which will be considered in the sequel. Then in order to use (1.4) we must be able to calculate or to estimate the functions $\Delta(\phi)$ and $E(B)$.

2. Estimation of E(B)

In general it is very difficult to find good estimates of the sum

$$\sum_{t\in B} e_n(tz) \qquad (2.1)$$

where $z \neq 0$. However, if B is a set of consecutive integers, say

$$B = \{0,1,\dots,b - 1\},$$

or, more generally, if

$$B = \{h + 1, h + 2,\dots, h + b\} \qquad (2.2)$$

for some integer h, then the estimating is very easy because the sum

(2.1) is geometric and therefore

$$\left|\sum_{t\in B} e_n(tz)\right| = \left|\frac{1 - e_n(bz)}{1 - e_n(z)}\right| \le \frac{n}{2\|z\|}$$

where $\|z\|$ is the modulus of the absolutely smallest residue of z (mod n). Hence

$$\sum_{z\neq 0}\left|\sum_{t\in B} e_n(tz)\right| \le \frac{n}{2}\cdot 2\cdot\sum_{a=1}^{\frac{n-1}{2}}\frac{1}{a} < n\cdot\sum_{a=1}^{\frac{n-1}{2}}\log\frac{2a+1}{2a-1} = n\log n$$

if n is odd. If n is even, the problem can be treated on the same lines, with the same result, and we obtain the following old result of Vinogradov [29].

If B is defined by (2.2), we have

$$E(B) < n \log n. \tag{2.3}$$

Of course, more careful calculations would yield slightly better results ([14, p. 447], [21]).

It is not very difficult to extend (2.3) to the case where $F = \mathbf{Z}_n^m$ (or $F = \mathbf{F}_q^s$). If

$$B = \{\underline{x} = (x_1,\dots,x_m) \mid \forall i:\ h_i + 1 \le x_i \le h_i + b\}$$

then [6]

$$E(B)\begin{cases} = O((n\log n)^m) \\ < (n\log n)^m \quad \text{if } n \text{ is a large prime.}\end{cases} \tag{2.4}$$

3. Multiplicative characters

Let F be a prime field $\mathbf{F}_p$ and let ϕ be χ, a nontrivial multiplicative character of F. Then

$$\sum_x \chi(x)e_p(xz) \tag{3.1}$$

is a Gaussian sum. So it is well known (see, e.g., [14, p. 199]) that for $z \neq 0$ the modulus of (3.1) is equal to $\sqrt{p}$. Hence

$$\Delta(\phi) = \max_{z\neq 0}\left|\sum_x \chi(x)e_p(xz)\right| = \sqrt{p}\ .$$

Since

$$\sum_x \chi(x) = 0,$$

we thus have, by (1.4), the following classical inequality of Vinogradov [28] and Pólya [20]:

$$\left| \sum_{t=h+1}^{h+b} \chi(t) \right| < \sqrt{p} \log p. \tag{3.2}$$

The fact that the left-hand side of (3.2) can be $\Omega(\sqrt{p} \log\log p)$ was shown by Bateman, Chowla and Erdös [1]. On the other hand, an upper bound of this order of magnitude was proved by Montgomery and Vaughan [15] under the assumption of the Riemann Hypothesis for Dirichlet functions. Further, Burgess (see [3] for the quadratic case) showed that there are nontrivial upper bounds for $b > p^{1/4+\varepsilon}$ by establishing the following result:

$$\forall\varepsilon>0\ \exists\delta>0:\ [p > p_0(\varepsilon),\ b > p^{1/4+\varepsilon}] \ \Rightarrow \left| \sum_{t=h+1}^{h+b} \chi(t) \right| < \frac{b}{p^{\delta}}.$$

A consequence of the inequality (3.2) is the following corollary:

If d is a factor of $p - 1$ and greater than 1 then the least positive d th power non-residue (mod p) is less than $\sqrt{p}\log p$.

This is obvious, using (3.2), since the sum $\sum\chi(t)$ over d th power residues $1, 2, \ldots, \lceil\sqrt{p}\log p\rceil$ would be greater than $\sqrt{p}\log p$.

A natural generalization is the following. Let F be $\mathbf{F}_q$ where $q = p^m$, and let $\{\alpha_1, \alpha_2, \ldots, \alpha_m\}$ be a basis of F_q over $\mathbf{F}_p$. If

$$B = \{x_1\alpha_1 + \ldots + x_m\alpha_m \mid \forall i:\ h_i + 1 \le x_i \le h_i + b_i\}$$

then [10]

$$\left| \sum_{t\in B} \chi(t) \right| < \sqrt{q}(1 + \log p)^m.$$

A consequence of this result is that large boxes B necessarily contain d th power non-residues.

4. The least nonnegative residue of a polynomial

In section 3 we considered residues and non-residues of a d th power (mod p). In this section we replace x^d by a general polynomial f of degree d, $2 \le d < p$. Let $\phi(a)$ denote the number of solutions of the congruence $f(x) \equiv a \pmod{p}$ and let F be $\mathbf{F}_p$. Then

$$\sum_a \phi(a) e_p(az) = \sum_x e_p(zf(x)).$$

Further, by a well known result of Weil [30],

$$\Big|\sum_x e_p(zf(x))\Big| \leq (d-1)\sqrt{p},$$

provided $z \not\equiv 0 \pmod p$. Thus

$$\Delta(\phi) \leq (d-1)\sqrt{p}.$$

Since

$$\sum_a \phi(a) = p,$$

the inequalities (1.4) and (2.3) imply

$$\sum_{a=0}^{b-1} \phi(a) = b + \Theta(d-1)\sqrt{p}\log p. \tag{4.1}$$

An immediate consequence of (4.1) is the following theorem of Mordell [17]:

κ, the least nonnegative residue of $f \pmod p$, satisfies the inequality

$$\kappa < (d-1)\sqrt{p}\log p. \tag{4.2}$$

Using similar but deeper and more complicated arguments Bombieri and Davenport [2] proved the following.

Let f be a polynomial over $\mathbb{F}_p$ of degree d (≥ 2), and suppose that at least one of the irreducible factors of $\frac{f(x) - f(y)}{x - y}$ over $\mathbb{F}_p$ is absolutely irreducible. Then there exists a number $C(d)$ depending only on d such that, for every large p, there is some λ with

$$0 < \lambda < C(d)\sqrt{p}\log p \tag{4.3}$$

for which the congruence $f(x) \equiv \lambda \pmod p$ is insoluble.

Mordell's theorem extends to general finite fields without any significant difficulties. Let q be equal to p^m, let $\phi(a)$ be the number of solutions $x \in \mathbb{F}_q$ of the congruence $tr(f(x)) \equiv a \pmod p$ when the trace mapping $tr: \mathbb{F}_q \to \mathbb{F}_p$ is defined by

$$\forall \xi \in \mathbb{F}_q: \; tr(\xi) = \xi + \xi^p + \xi^{p^2} + \ldots + \xi^{p^{m-1}},$$

and let $e(\alpha)$ be equal to $e^{2\pi i\, tr(\alpha)/p}$. Then it follows from a well known result of Carlitz and Uchiyama [4] that

$$\Big|\sum_a \phi(a) e_p(az)\Big| = \Big|\sum_{x \in \mathbb{F}_q} e(zf(x))\Big| \leq (d-1)\sqrt{q}$$

if $z \not\equiv 0 \pmod p$ and $f \neq g^p - g + \beta$ for all g in $\mathbb{F}_q[x]$ and for all β in $\mathbb{F}_q$. Hence

$$\Delta(\phi) \leq (d - 1)\sqrt{q}.$$

ince

$$\sum_a \phi(a) = q,$$

he inequalities (1.4) and (2.3) imply

$$\sum_{a=0}^{b-1} \phi(a) = bp^{m-1} + \Theta(d - 1)p^{\frac{m}{2}} \log p.$$

hus we have the following rasult of Cavior [5].

If $q = p^m$, $f \in \mathbb{F}_q[x]$ and f is not of the form $g^p - g + \beta$ with $g \in F[x]$, $\in \mathbb{F}_q$ then μ, the least nonnegative trace of f, satisfies the inequality

$$\mu < (d - 1)p^{1-\frac{m}{2}} \log p. \tag{4.4}$$

5. The distribution of zeros of a polynomial

Assume that $f(\underline{x}) = f(x_1,\ldots,x_s)$ is a polynomial of s (≥ 3) variables nd of degree d (≥ 2) over $\mathbb{F}_p$. Let N(B) be the number of solutions of he congruence $f(x) \equiv 0 \pmod p$ in the set

$$B = \{(x_1,\ldots,x_s) \mid \forall i: h_i + 1 \leq x_i \leq h_i + b\}.$$

hen it is natural to denote $\mathbb{F}_p^s$ by F and to write

$$\phi(\underline{x}) = \left\{\begin{matrix} 1 \text{ if } f(\underline{x}) = 0 \\ 0 \text{ otherwise} \end{matrix}\right\} = \frac{1}{p}\sum_{k\in\mathbb{F}_p} e_p(kf(\underline{x})).$$

ence

$$N(B) = \sum_{\underline{x}\in B} \phi(x), \qquad N(F) = \sum_{\underline{x}} \phi(x),$$

nd,by (1.4) and (2.4),

$$N(B) = \frac{|B|}{p^s}N(F) + \frac{\Theta}{p^s}\Delta(\phi)(p \log p)^s \tag{5.1}$$

f p is large. In this application

$$\Delta(\phi) = \max_{\underline{z}\neq\underline{0}} \left|\sum_{\underline{x}} \phi(\underline{x})e_p(\underline{x}\cdot\underline{z})\right|$$

$$= \max_{\underline{z}\neq\underline{0}} \left|\frac{1}{p} \sum_{k\in\mathbb{F}_p} \sum_{\underline{x}} e_p(kf(\underline{x}) + \underline{x}\cdot\underline{z})\right|.$$

In the special case where the variables can be separated, e.g. in ne case

$$f(\underline{x}) = x_1^d + \ldots + x_s^d, \quad d \mid (p - 1),$$

we can estimate $\Delta(\phi)$ by the result of Weil in the following way:

$$\Delta(\phi) \leq \max_{\substack{\underline{z} \neq \underline{0} \\ k \neq 0}} \left| \prod_{j=1}^{s} \sum_{x_j \in \mathbb{F}_p} e_p(kx_j^d + z_j x_j) \right|$$

$$\leq (d - 1)^s p^{\frac{s}{2}}.$$

Further,

$$N(F) = \frac{1}{p} \sum_{\underline{x}} \sum_{k \in \mathbb{F}_p} e_p(kf(\underline{x}))$$

$$= p^{s-1} + \frac{\Theta_1}{p} \sum_{k \neq 0} \prod_{j=1}^{s} \left| \sum_{x_j \in \mathbb{F}_p} e_p(kx_j^d) \right|$$

$$= p^{s-1} + \Theta_2 (d - 1)^s p^{\frac{s}{2}},$$

and hence, by (5.1), for any $\varepsilon > 0$ and any $p > p_0(\varepsilon)$,

$$N(B) = \frac{|B|}{p} + \Theta_3 (1 + \varepsilon)(d - 1)^s p^{\frac{s}{2}} (\log p)^s$$

where $|\Theta_i| \leq 1$ $(i = 1, 2, 3)$. Thus the polynomial $x_1^d + \ldots + x_s^d$ has a zero in B if p is large and

$$|B| > (1 + \varepsilon)(d - 1)^s p^{\frac{s+2}{2}} (\log p)^s,$$

i.e., if

$$b > (1 + \varepsilon)^{\frac{1}{s}} (d - 1) p^{\frac{s+2}{2s}} \log p. \tag{5.2}$$

These investigations are due to Mordell ([16], [18]) and to Chalk and Williams ([6], [8]). For later results, see [26], [22], [7], [23], [19], and [9].

6. A modification

We have seen that the Vinogradov inequality (1.4) has several applications. In some applications we ask:

How many important elements are there in a box B? (6.1)

In some other applications we ask:

How large is the smallest important element? (6.2)

n the sections 3, 4, and 5 the "important elements" were residues or on-residues of a power or of a polynomial or zeros of a polynomial but lso distribution properties of linear recurring sequences could be estiated by the inequality (1.4) ([14], [21]).

In order to calculate the number of important elements in a box B it s natural to define

$$\phi(x) = \begin{cases} 1 & \text{if } x \text{ is an important element,} \\ 0 & \text{otherwise,} \end{cases}$$

nd to calculate the incomplete sum

$$\sum_{x \in B} \phi(x)$$

y appealing to the Vinogradov inequality (1.4). Further, in order to use hat inequality one has to calculate or to estimate the functions $\Delta(\phi)$ nd $E(B)$. The former of these functions was estimated by using deep reults of Weil [30] and Carlitz and Uchiyama [4]. Certainly it is very ifficult to improve those results. $E(B)$ was estimated by using the inquality (2.3) and analogous results, and the logarithm factor in these stimates looks natural and even unavoidable.

However, in applications of the form (6.2) the situation is different. et $\alpha(t)$ be a function on F such that $\alpha(t) > 0$ if t is important and $(t) \le 0$ otherwise,. Then in order to show that a box A contains an imortant element it suffices to show that $\sum_{t \in A} \alpha(t) > 0$. For instance, if we an write the set A in the form $B + B = \{x + y \mid x \in B,\ y \in B\}$, we may define (t) to be the number of ways of expressing t as the sum of two (not ecessarily different) elements of B. Then we have, by the argument givn in Section 1,

$$\begin{aligned} \sum_{t \in A} \alpha(t) &= \sum_{x \in B} \sum_{y \in B} \phi(x + y) \\ &= \frac{1}{n} \sum_{z} \psi(z) \sum_{x \in B} \sum_{y \in B} e_n(-(x + y)z) \qquad (6.3) \\ &= \frac{b^2}{n} \sum_{x} \phi(x) + \frac{\Theta}{n} \Delta(\phi) E_2(B) \end{aligned}$$

here

$$E_2(B) = \sum_{z \neq 0} \Big| \sum_{t \in B} e_n(tz) \Big|^2 .$$

urther, $E_2(B)$ can be calculated very easily:

$$E_2(B) = \sum_{z} \sum_{t \in B} e_n(tz) \sum_{u \in B} e_n(-uz) - b^2$$

$$= \sum_{t \in B} \sum_{u \in B} \sum_{z} e_n((t - u)z) - b^2$$

nd so, by (1.2),

$$E_2(B) = bn - b^2 < bn.$$

Thus

$$\Big| \sum_{t \in A} \alpha(t) - \frac{b^2}{n} \sum_x \phi(x) \Big| < b\Delta(\phi).$$

Using this modification of the Vinogradov inequality we observe that in the solution of the problem (6.2) the logarithm factor of E(B) can be avoided.

For example, the inequality (4.1) can be replaced by

$$\left| \sum_{t=0}^{2b-2} \alpha(t) - b^2 \right| < b(d - 1)\sqrt{p}.$$

Thus in the set $\{0,1,2,...,2b-2\}$ there is a t such that $\alpha(t) > 0$ provided $b \geq (d - 1)\sqrt{p}$, and we may replace the inequality (4.2) by

$$\kappa < 2(d - 1)\sqrt{p}.$$

In the same way ([24], [25], and [26]) in the inequalities (4.3) and (4.4) $\log p$ and in the inequality (5.2) $(1 + \varepsilon)^{1/s} \log p$ can be replaced by 2.

This method was extended by Chalk [7] and Cochrane[9].

7. The distribution of solutions of a system of equations

Consider the problem of Section 5 in a more general form [9]. Let $\mathbf{F}_q$ be the finite field of $q = p^m$ elements and let V be set of solutions of

$$\begin{cases} f_1(x_1,...,x_s) = 0 \\ f_2(x_1,...,x_s) = 0 \\ \quad . \quad . \quad . \quad . \quad . \\ f_t(x_1,...,x_s) = 0; \end{cases} \tag{7.1}$$

In other words,

$$V = \{\underline{x} \in \mathbf{F}_q^s \mid \underline{f}(\underline{x}) = \underline{0}\}$$

where $\underline{x} = (x_1,...,x_s)$ and $\underline{f}(\underline{x}) = (f_1(\underline{x}),...,f_t(\underline{x}))$. The problem is whethe we can assert the existence of some b, relatively small compared with q^s, for which $|B| \geq b$ necessarily implies $(B + B) \cap V \neq \emptyset$.

We extend the method of Section 5 in two respects: The original Vinogradov method will be replaced by its modification and the character sum estimate of Weil by a result of Deligne [11].

A polynomial $g(\underline{x}) = g(x_1,...,x_s)$ is said to be nonsingular at infinity over $\mathbf{F}_q$ if its maximal degree homogeneous part is nonsingular as a form over the algebraic closure of $\mathbf{F}_q$. Then Deligne's estimate of exponential sums can be written in the following form.

If $g(\underline{x}) = g(x_1,\dots,x_s)$ is a polynomial over $\mathbb{F}_q$ of degree d not divisible by p and is nonsingular at infinity then

$$\left|\sum_{\underline{x}} e(g(\underline{x}))\right| \le (d-1)^s q^{\frac{s}{2}}.$$

Let

$$\phi(\underline{x}) = \left\{\begin{array}{l} 1 \text{ if } f(\underline{x}) = \underline{0} \\ 0 \text{ otherwise} \end{array}\right\} = \frac{1}{q^t} \sum_{\underline{k}\in\mathbb{F}_q^t} e(\underline{k}\cdot\underline{f}(\underline{x})),$$

$$S(\underline{z}) = \sum_{\underline{x}\in B} e(-\underline{x}\cdot\underline{z}),$$

and

$$\psi(\underline{z}) = \sum_{\underline{x}\in\mathbb{F}_q^s} \phi(\underline{x}) e(\underline{x}\cdot\underline{z}).$$

Then it follows from the equation (6.3) and from the observation

$$\sum_{\underline{z}\in\mathbb{F}_q^s} |S(\underline{z})|^2 = |B| q^s$$

that

$$\begin{aligned}\sum_{\underline{x}\in B}\sum_{\underline{y}\in B} \phi(\underline{x}+\underline{y}) &= \frac{1}{q^s}\sum_{\underline{z}\in\mathbb{F}_q^s} \psi(\underline{z})(S(\underline{z}))^2 \\ &= \frac{1}{q^{s+t}}\sum_{\underline{k}\in\mathbb{F}_q^t}\sum_{\underline{z}\in\mathbb{F}_q^s}(S(\underline{z}))^2 \sum_{\underline{x}\in\mathbb{F}_q^s} e(\underline{k}\cdot\underline{f}(\underline{x}) + \underline{x}\cdot\underline{z}) \\ &= \frac{|B|^2}{q^t} + \theta|B| \max_{\underline{z}\in\mathbb{F}_q^s}\max_{\substack{\underline{k}\in\mathbb{F}_q^t \\ \underline{k}\ne\underline{0}}}\left|\sum_{\underline{x}\in\mathbb{F}_q^s} e(\underline{k}\cdot\underline{f}(\underline{x}) + \underline{x}\cdot\underline{z})\right|.\end{aligned}$$

By [9], a k-tuple $\underline{f}(\underline{x}) = (f_1(\underline{x}),\dots,f_t(\underline{x}))$ is "nonsingular" at infinity over $\mathbb{F}_q$ if every polynomial in the pencil

$$\{\underline{k}\cdot\underline{f}(\underline{x}) = \sum_{i=1}^{t} k_i f_i(\underline{x}) \mid \underline{k}\in\mathbb{F}_q^t,\ \underline{k}\ne\underline{0}\}$$

is of degree greater than 1 and not divisible by p and is nonsingular at infinity. If $\underline{f}(\underline{x})$ is "nonsingular" at infinity then it follows from the theorem of Deligne that

$$\max_{\underline{z}\in\mathbb{F}_q^s}\max_{\substack{\underline{k}\in\mathbb{F}_q^t \\ \underline{k}\ne\underline{0}}}\left|\sum_{\underline{x}\in\mathbb{F}_q^s} e(\underline{k}\cdot\underline{f}(\underline{x}) + \underline{x}\cdot\underline{z})\right| \le (d-1)^s q^{\frac{s}{2}}$$

where d is the maximum of degrees of polynomials $f_1(\underline{x}),\dots,f_t(\underline{x})$. Thus in this case

$$\left|\sum_{\underline{x}\in B}\sum_{\underline{y}\in B}\phi(\underline{x}+\underline{y}) - \frac{|B|^2}{q^t}\right| \le |B|(d-1)^s q^{\frac{s}{2}}$$

and we have the following theorem.

Assume that every polynomial in the set $\{\underline{k}\cdot\underline{f}(\underline{x}) \mid \underline{k}\in\mathbf{F}_q^t,\ \underline{k}\neq\underline{0}\}$ is of degree less than or equal to d, greater than 1 and not divisible by p and is nonsingular at infinity. Then the system (7.1) has a solution in the set $B+B$ provided that

$$|B| > (d-1)^s q^{\frac{s+2t}{2}}.$$

8. The covering radius of a binary BCH code

Let t and M be fixed positive integers and let m be a positive integer such that $M|(2^m-1)$. Denote 2^m by q. Let α be a primitive element of $\mathbf{F}_q$ and let $m_i(x)$ be the minimal polynomial for α^i over $\mathbf{F}_2$. Assume that $C = C(t,M,m)$ is the binary code of length $n = \frac{2^m-1}{M}$ and with generator $m_M(x)m_{3M}(x)\cdots m_{(2t-1)M}(x)$; i.e., C is the binary BCH code of length n and with designed distance $2t+1$. Denote the covering radius of C by $R = R(t,M,m)$. Thus R is the smallest integer s such that the balls of radius s and with centres at code words cover the space $\mathbf{F}_2^n$. The problem is whether $R(t,M,m)$ is equal to $2t-1$ when m is large.

This problem relates to equations over finite fields because [13] $R(t,M,m)$ is also the smallest s such that, for each $(b_1,\ldots,b_t)\in\mathbf{F}_q^t$, the system

$$\begin{cases} x_1^M + \cdots + x_s^M = b_1 \\ x_1^{3M} + \cdots + x_s^{3M} = b_2 \\ \cdots\cdots\cdots\cdots \\ x_1^{(2t-1)M} + \cdots + x_s^{(2t-1)M} = b_t \end{cases}$$

is soluble in $\mathbf{F}_q$. Considering the case where $(b_1,\ldots,b_t) = (0,\ldots,0,1)$ we can see [13] that

$$R(t,M,m) \geq 2t-1. \tag{8.1}$$

On the other hand, Helleseth [13] showed that

$$R(t,M,m) \leq 2t+1 \quad \text{if} \quad 2^m \geq ((2t-1)M-1)^{4t+2}. \tag{8.2}$$

Further [12], $R(t,1,m) = 2t-1$ when $t = 1,2,3$.

In fact, $R(t,M,m)$ is also the smallest integer s such that, for each $(b_1,\ldots,b_t)\in\mathbf{F}_q^t$, the system

$$\begin{cases} x_1^M + \cdots + x_s^M = b_1 z^M \\ x_1^{3M} + \cdots + x_s^{3M} = b_2 z^{3M} \\ \cdots\cdots\cdots \\ x_1^{(2t-1)M} + \cdots + x_s^{(2t-1)M} = b_t z^{(2t-1)M} \end{cases} \tag{8.3}$$

has a solution $(x_1,\ldots,x_s,z)$ with $z \neq 0$. If we put

$$\underline{f}(x) = (x^M, x^{3M}, \ldots, x^{(2t-1)M})$$

and

$$\underline{b}(z) = (b_1 z^M, b_2 z^{3M}, \ldots, b_t z^{(2t-1)M}),$$

we may write (8.3) in the form

$$\sum_{j=1}^{s} \underline{f}(x_j) = \underline{b}(z). \tag{8.4}$$

Denote $(2t-1)M$ by D and $\mathbb{F}_q\setminus\{0\}$ by $\mathbb{F}_q^*$. Assume that $s = 2t$ and $2^m \geq D^{4t+2}$. Let $\mathbb{N}_0$ be the number of solutions $(x_1,\ldots,x_{2t},z) \in \mathbb{F}_q^{2t} \times \mathbb{F}_q^*$ of (8.4). Then, by the argument given in earlier sections, $q^t\mathbb{N}_0$ can be expressed in the form

$$\sum_{x_1\in\mathbb{F}_q} \cdots \sum_{x_{2t}\in\mathbb{F}_q} \sum_{z\in\mathbb{F}_q^*} \sum_{\underline{k}\in\mathbb{F}_q^t} e(\underline{k}\cdot(\sum_{j=1}^{2t} \underline{f}(x_j) + \underline{b}(z)))$$

$$= \sum_{\underline{k}\in\mathbb{F}_q^t} \prod_{j=1}^{2t} (\sum_{x_j\in\mathbb{F}_q} e(\underline{k}\cdot\underline{f}(x_j))) \sum_{z\in\mathbb{F}_q^*} e(\underline{k}\cdot\underline{b}(z))$$

$$= \sum_{\underline{k}\in\mathbb{F}_q^t} (\sum_{x\in\mathbb{F}_q} e(\underline{k}\cdot\underline{f}(x)))^{2t} \sum_{z\in\mathbb{F}_q^*} e(\underline{k}\cdot\underline{b}(z)).$$

Since, by the result of Carlitz and Uchiyama [4],

$$(\sum_{x\in\mathbb{F}_q} e(\underline{k}\cdot\underline{f}(x)))^{2t} \begin{cases} = q^{2t} \text{ for } \underline{k} = \underline{0}, \\ \geq 0 \text{ for all } \underline{k}, \\ \leq (D-1)^{2t}q^t \text{ for } \underline{k} \neq \underline{0} \end{cases}$$

and

$$\sum_{z\in\mathbb{F}_q^*} e(\underline{k}\cdot\underline{b}(z)) \begin{cases} = q - 1 \text{ for } \underline{k} = \underline{0} \\ \geq -(D-1)\sqrt{q} - 1 \text{ for all } \underline{k}, \end{cases}$$

we thus have

$$\begin{aligned} q^tN_0 &\geq q^{2t}(q-1) - (q^t-1)(D-1)^{2t}q^t((D-1)\sqrt{q}+1) \\ &> q^{2t+1} - (D-1)^{2t}Dq^{2t+\frac{1}{2}} - q^{2t} \\ &> q^{2t+\frac{1}{2}}(\sqrt{q} - D^{2t+1}) \geq 0. \end{aligned}$$

Hence the system (8.3) has a solution $(x_1,\dots x_{2t},z)$ with $z \neq 0$. Together with (8.1) this implies [27]

$$2t - 1 \leq R(t,M,m) \leq 2t \quad \text{if } 2^m \geq ((2t-1)M)^{4t+2}.$$

So we have closed the gap between (8.1) and (8.2) halfway, and it is natural to ask:

For which values of t and M

$$R(t,M,m) = 2t - 1$$

when m is large?

Thus assume that $s = 2t - 1$. Consider first the case $M = 1$. Then one can show that in (8.3) it is sufficient to consider only the case where $b_1 = 0$. Solving x_{2t-1} from the first equation (8.3) and substituting it to the others, we get

$$\begin{cases} x_1^3 + \dots + x_{2t-2}^3 + (x_1 + \dots + x_{2t-2})^3 = b_2 z^3 \\ \dots\dots\dots\dots\dots\dots\dots\dots \\ x_1^{2t-1} + \dots + x_{2t-2}^{2t-1} + (x_1 + \dots + x_{2t-2})^{2t-1} = b_t z^{2t-1}. \end{cases} \tag{8.5}$$

When we use the character sum method described earlier, we must consider the sum

$$\sum_{\underline{x} \in \mathbf{F}_q^{2t-2}} e(f(\underline{x}))$$

where $f(\underline{x}) = \sum_{i=2}^{t} k_i f_i(\underline{x})$ and $f_i(\underline{x})$ is the left-hand side of the $(i-1)$th equation (8.5). If $k_t \neq 0$ then the maximal degree homogeneous part of $f(\underline{x})$ is $k_t f_t(\underline{x})$. Further, if t is of the form $2^u + 1$ then the equations

$$\frac{\partial f_t(\underline{x})}{\partial x_j} = 0 \quad (j = 1,2,\dots,2t-2) \tag{8.6}$$

imply

$$x_j^{2t-2} = (x_1 + \dots + x_{2t-2})^{2t-2} \quad (j = 1,\dots,2t-2),$$

i.e.,

$$x_j^{2^{u+1}} = (x_1 + \dots + x_{2t-2})^{2^{u+1}} \quad (j = 1,\dots,2t-2),$$

and so the only solution of (8.6) in the algebraic closure of $\mathbf{F}_q$ is

$$x_1 = \dots = x_{2t-2} = 0.$$

Thus in case $k_t \neq 0$

$$\left|\sum_{\underline{x}\in F_q^{2t-2}} e(f(\underline{x}))\right| \le (2t-2)^{2t-2} q^{t-1}$$

y Deligne's thorem, and after complicated calculations we get the following result.

$R(t,M,m) = 2t - 1$ if (i) m is large, (ii) $M = 1$, and (iii) t is of he form $2^u + 1$.

Certainly some hypothesis of the form (i) is necessary. On the other and, perhaps hypothesis (ii) and very probably hypothesis (iii) can be eakened without affecting the conclusion but they were needed at that tage of the proof above where we applied Deligne's result.

REFERENCES

[1] Bateman, P.T., Chowla, S., and Erdös, P.: Remarks on the size of $L(1,\chi)$, Publ. Math. Debrecen 1, 165-182 (1950).

[2] Bombieri, E., and Davenport, H.: On two problems of Mordell, Amer. J. Math. 88, 61-70 (1966).

[3] Burgess, D.A.: The distribution of quadratic residues and non-residues, Mathematika 4, 106-112 (1957).

[4] Carlitz, L., and Uchiyama, S.: Bounds for exponential sums, Duke Math.J. 24, 37-41 (1957).

[5] Cavior, S.R.: On the least non-negative trace of a polynomial over a finite field Boll. Un. Mat. Ital. (3)20, 120-121 (1965).

[6] Chalk, J.H.H.: The number of solutions of congruences in incomplete residue systems, Canad.J. Math. 15, 291-296 (1963).

[7] Chalk, J.H.H.: The Vinogradov-Mordell-Tietäväinen inequalities, Indag. Math. 42, 367-374 (1980).

[8] Chalk, J.H.H., and Williams, K.S.: The distribution of solutions of congruences, Mathematika 12, 176-192 (1965);

[9] Cochrane, T.: The distribution of solutions to equations over finite fields, Trans. Amer. Math. Soc. 293, 819-825 (1986).

10] Davenport, H., and Lewis, D. J.: Character sums and primitive roots in finite fields, Rend. Circ. Mat. Palermo (2)12, 129-136 (1963).

[11] Deligne, P.:
La conjecture de Weil. I,
Inst. Hautes Etudes Sci. Publ. Math. 43, 273-307 (1974).

[12] Helleseth, T.:
All binary 3-error correcting BCH codes of length $2^m - 1$ have covering radius 5,
IEEE Trans. Inform. Theory 24, 257-258 (1978).

[13] Helleseth, T.
On the covering radius of cyclic linear codes and arithmetic codes,
Discrete Appl. Math. 11, 157-173 (1985).

[14] Lidl, R., and Niederreiter, H.:
Finite Fields,
Addison-Wesley, 1983.

[15] Montgomery, H. L., and Vaughan, R.C.,:
Exponential sums with multiplicative coefficients,
Invent. Math. 43, 69-82 (1977).

[16] Mordell, L. J.:
On the number of solutions in incomplete residue sets of quadratic congruences,
Arch. Math. 8, 153-157 (1957).

[17] Mordell, L. J.:
On the least residue and non-residue of a polynomial,
J. London Math. Soc. 38, 451-453 (1963).

[18] Mordell, L. J.:
Incomplete exponential sums and incomplete residue systems for congruences,
Czechoslovak Math. J. 14, 235-242 (1964).

[19] Myerson, G.:
The distribution of rational points on varieties defined over a finite field,
Mathematika 28, 153-159 (1981).

[20] Pólya, G.:
Über die Verteilung der quadratischen Reste und Nichtreste,
Göttinger Nachr. 1918, 21-29.

[21] Sarwate, D. V.:
An upper bound on the aperiodic autocorrelation function for a maximal-length sequence,
IEEE Trans. Inform. Theory 30, 685-687 (1984).

[22] Smith, R. A.:
The distribution of rational points on hypersurfaces defined over a finite field,
Mathematika 17, 328-332 (1970).

[23] Spackman, K. W.:
On the number and distribution of simultaneous solutions to diagonal congruences,
Canad. J. Math. 33, 421-436 (1981).

[24] Tietäväinen, A.:
On the trace of a polynomial over a finite field,
Ann. Univ. Turku Ser. AI 87 (1966), 7 pp.

[25] Tietäväinen, A.:
On non-residues of a polynomial,
Ann. Univ. Turku Ser. AI 94 (1966), 6 pp.

[26] Tietäväinen, A.:
On the solvability of equations in incomplete finite fields,
Ann. Univ. Turku Ser. AI 102 (1967), 13 pp.

[27] Tietäväinen, A.:
On the covering radius of long binary BCH codes,
Discrete Appl. Math. 16, 75-77 (1987).

[28] Vinogradov, I. M.:
Sur la distribution des résidus et des non-résidus des puissances,
Ž. Fiz.-Mat. Obŝĉ. Permsk. Gos. Univ. 1918, no 1, 94-98.

[29] Vinogradov, I. M.:
Elements of number theory,
Dover, 1954.

[30] Weil, A.:
Sur les courbes algébriques et les variétés qui s'en déduisent,
Actualités Sci. Ind., no. 1041, Hermann, Paris, 1948.

"This paper is in final form and no version of it will be submitted for publication elsewhere"

EIGENVALUES OF MATRICES OF COMPLEX REPRESENTATIONS OF FINITE GROUPS OF LIE TYPE

A.E. Zalesskii
Institute of Mathematics of the Academy
of Sciences of Byelorussian SSR
Minsk, 220604, USSR

Let $G = G(p^{\alpha})$ be a finite group of Lie type. Let φ be an irreducible representation of G over $\mathbb{C}$ (the field of complex numbers) or over an algebraically closed field F of characteristic $f \neq p$.

Let $Z = Z(G)$ be the centre of G. For $g \in G$ let $m(g)$ be the order of g modulo Z. The symbol $\deg\varphi(g)$ denotes the degree of the minimal polynomial of the matrix $\varphi(g)$. It is clear that $\deg\varphi(g) \leq m(g)$.

In this paper I should like to discuss the following problem.

Problem 1. Describe the triples (G, φ, g) with $1 < \deg\varphi(g) < m(g)$.

One may ask whether this problem can have a reasonable solution. The results which will be presented below allow one to think that an explicit description of such triples is quite possible and that their list should not be long.

First I want to make some preliminary comments. Let Spec x be the set of eigenvalues of the matrix x. If $(m(g), f) = 1$, then $\deg\varphi(g) = |\mathrm{Spec}\ \varphi(g)|$, so Problem 1 is a problem on the eigenvalues of $\varphi(g)$. Of course if the character of φ is known (the Brauer one if $f > 0$) then one can easily determine the eigenvalues of $\varphi(g)$ and their multiplicities. However, explicit formulas are not known for all irreducible characters and many of the known formulas are too complicated for practical calculations of multiplicities of eigenvalues. Therefore it would be useful to develop alternative methods for estimating these multiplicities. Of course, Problem 1 concerns only the question of how many roots of the equation $x^{m(g)} = \lambda$ are really eigenvalues of $\varphi(g)$, where λ is determined by the equality $\varphi(g^{m(g)}) = \lambda.\mathrm{id}$. Nevertheless the regularities in this case seem to be a manifestation of some general ones.

There are many examples demonstrating the importance of knowing the eigenvalues and minimal polynomials of matrices of linear groups.

However, investigations of this topic have not been active for a long time after the classical results of Blichfeldt (see [2]). Much later, profound ideas were proposed by Ph. Hall and G. Higman [6] and by J. Thompson [16] and were developped further by E. Shult [12], T.R. Berger [1], C.Y. Ho [8], etc. C. Hering [7] considered a particular case of Problem 1 with dim $\varphi < |g|$ for g of prime order.

Note that the condition $(f,p) = 1$ is essential since for $f = p$ the problem is of a quite different nature.

Of course Problem 1 may be stated for faithful representations of arbitrary groups under some restrictions which would prevent a proliferation of isolated examples. Thus the general problem probably reduces to groups close to simple ones. It is clear that groups of Lie type present a central part of the latter problem.

The absence of eigenvalue 1 may be interpreted geometrically as the absence of fixed points in the underlying space. The latter is essential for a number of applications. Therefore the following particular case of Problem 1 is of independent interest:

Problem 1'. Determine the triples (G, φ, g) with $1 \notin \operatorname{Spec} \varphi(g)$.

Below we shall discuss some results concerning Problem 1 and Problem 1' and outline their proofs.

1. Eigenvalues of elements of order p in representations of the groups $G(p^\alpha)$.

Notation. $G = G(p^\alpha)$ is a finite group of Lie type, F is an algebraically closed field of characteristic $f \neq p$, $\operatorname{Irr}_F G$ is a set of representatives of the equivalence classes of irreducible representations of G over F.

Theorem 1 [20]. Let $\varphi \in \operatorname{Irr}_F G$ and let $g \in G$ be an element of order p. Suppose that $1 < \deg\varphi(g) < p$. Then $p > 2$, and one of the following holds:

(i) $G \cong {}^2A_2(p)$, g is a transvection;

(ii) $G \cong A_1(p^2)$;

(iii) $G \cong C_n(p)$ $(n \geq 1)$, g is a transvection;

(iv) $G \cong C_2(p)$, g is not a transvection.

Furthermore, in each case there exists a $\varphi \in \operatorname{Irr}_F G$ with $1 < \deg\varphi(g) < p$.

For cases (i), (ii), (iv) and $F = \mathbb{C}$ the list of all φ can easily be obtained from the character tables of the groups in question.

Problem 2. Give a complete description of the representations φ from Theorem 1.

I conjecture that $\dim\varphi = (p^n \mp 1)/2$ in case (iii). Such representations of the groups $C_n(p)$ were first constructed by Ward [18], see also Seitz [11]. I observed [19] that $\deg\varphi(g) = (p+1)/2$ for Ward's representations φ with $n > 1$.

Theorem 2 [21]. Let G, φ, g be as in Theorem 1. Suppose that $\deg\varphi(g) < p-1$. Then $p > 3$ and $G \cong C_n(p)$ $(n \geq 1)$. Furthermore, if $\deg\varphi(g) \leq (p-1)/2$, then $n = 1$.

The last assertion of Theorem 2 seems to hold for arbitrary groups in the following form:

Problem 3. Let H be a finite group, let $h \in H$ be an element of prime order p, and let N be the normal closure of h in H. Let $\varphi \in \mathrm{Irr}_{\mathbb{C}} H$. Suppose that $\deg\varphi(h) < (p-1)/2$. Is it true that $[N,N] \subset \mathrm{Ker}\varphi$, where $[N,N]$ is the commutator subgroup of N?

According to Feit and Thompson [5] this is true if $\dim\varphi < (p-1)/2$.

Theorem 3 [21]. Let G, φ, g, F be as in Theorem 1. Suppose that $G = [G,G]$ and that $\varphi(g)$ has no eigenvalue equal to 1. Then $p > 2$ and G is one of groups $A_1(p)$, $A_1(p^2)$, $C_2(p)$, ${}^2A_2(p)$.

If $F = \mathbb{C}$ then $\dim\varphi$ takes only the following values:

(i) $G \cong A_1(p)$, $\dim\varphi = p-1$ or $(p-1)/2$;

(ii) $G \cong A_1(p^2)$, $\dim\varphi = (p^2-1)/2$;

(iii) $G \cong {}^2A_2(p)$, $\dim\varphi = p(p-1)/2$;

(iv) $G \cong C_2(p)$, $\dim\varphi = p(p-1)^2/2$ or $(p^2-1)/2$.

2. Outlines of proofs.

Lemma 1. Let $H \subset GL_n(F)$ be a finite group and let $h \in H$ with $|h| = p$, where p is prime. Let A be an abelian group such that $(f, |A|) = 1$. Suppose that $h \in N_H(A)$ and $h \notin C_H(A)$. Then $\deg h = p$.

This assertion is certainly well known, but I can indicate no pre-ise reference. To prove it, diagonalize A and consider an irreducible omponent ψ of dimension $d > 1$ of the group $\langle A,h\rangle$. Then $d = p$ and h perates transitively on the weight spaces of A. It follows that $\psi|\langle h\rangle$ s a regular representation of the group $\langle h\rangle$.

Corollary 1. Let $G = G(p^{\alpha})$ and $g \in G$, $|g| = p$. If g normalizes, but oes not centralize an abelian p-group A, then $\deg\varphi(g) = p$ for any $\in \mathrm{Irr}_F G$ with $(f,p) = 1$ and $\dim\varphi > 1$.

It turns out that most elements of order p satisfy the assumptions f Corollary 1. How does one analyze the remaining ones? Fortunately hey normalize, but do not centralize a group of extraspecial type except for certain groups of rank 1 whose character tables are avai-able). Therefore we may try to find an analogue of Corollary 1 for xtraspecial groups. This will be done in Section 3. Now we explain how ne can clarify what elements $g \in G$ normalize nontrivially an abelian -group.

Lemma 2. Let $G = G(p^{\alpha})$ $(p > 2)$ be a group of type E_6, E_7, E_8, F_4, $_2$, ${}^2E_6$ or ${}^3D_4$. Then every element $g \in G$ of order p satisfies the assump-ion of Corollary 1.

The proof of this lemma is obtained in [20] by means of a careful nalysis of relations between root subgroups. Details do not seem to e interesting enough to reproduce them here. There is an explicit cha-acter table for ${}^2G_2(3^{2\alpha+1})$, so only classical groups have to be consi-ered in order to finish the proof of Theorem 1. Note that the asser-ion $p > 2$ is obvious since $1 < \deg\varphi(g) \le p-1$. Thus we assume from now n that $p > 2$.

Lemma 3. Let $p > 2$, $G \cong A_n(p^{\alpha})$ $(n > 1)$, $D_n(p^{\alpha})$ $(n > 3)$, or ${}^2D_n(p^{\alpha})$ $n > 3)$. If $g \in G$ with $|g| = p$ then there exists an abelian p-group $A \subset G$ uch that $gAg^{-1} = A$ and $ga \neq ag$ for some $a \in A$.

Proof. The case of $A_n(p^{\alpha}) \cong SL_{n+1}(p^{\alpha})$ is trivial. Let V be the pace over $\mathbb{F}_{p^{\alpha}}$ where G operates in the natural way and let f be a bi-inear form on V such that $G = \{g \in GL(V) \mid f(gx,gy) = f(x,y)$ for any $,y \in V\}$. Note that $\dim V = 2n$. Set $m=n$ for $G=D_n$, and $m = n-1$ for $G = {}^2D_n$. et $v_1,\dots,v_{2n}$ be a Witt basis of V, that is $f(v_i,v_{2n-m+j}) = \delta_{ij}$ for or $1 \le i,j \le m$, and $f(v_i,v_j)=0$ for $1 \le i,j \le m$ as well as for $2n-m+1 \le i,j \le 2n$

and for $m<j<2n-m$, $i\in\{1,\dots,m, 2n-m+1,\dots,2n\}$. In particular $W=\langle v_1,\dots,v_m\rangle$ is a maximal completely isotropic subspace of V. Let $P=\{g\in G\,|\,gW=W\}$, $R=O_p(P)$, $Z=Z(R)$. Then $P=H'R$ where H,R,Z consist of matrices of the form

$$H=\begin{pmatrix} S & 0 & 0 \\ 0 & T & 0 \\ 0 & 0 & {}^tS^{-1} \end{pmatrix},\quad R=\begin{pmatrix} 1_m & L & N \\ 0 & 1_{2n-2m} & M \\ 0 & 0 & 1_m \end{pmatrix},\quad Z=\begin{pmatrix} 1_m & 0 & Q \\ 0 & 1_{2n-2m} & 0 \\ 0 & 0 & 1_m \end{pmatrix}$$

if $G={}^2D_n$ (t denotes the transpose); and of the form

$$H=\begin{pmatrix} S & 0 \\ 0 & {}^tS^{-1} \end{pmatrix},\quad R=Z=\begin{pmatrix} 1_m & Q \\ 0 & 1_m \end{pmatrix}$$

if $G=D_n$. Here all matrix coefficients are again matrices whose sizes correspond to the decomposition $2n=m+(2n-2m)+m$. Moreover, S, T, L, M, N, Q are arbitrary matrices satisfying the conditions ${}^tT\Psi=\Psi T^{-1}$; $Q+{}^tQ=0$, ${}^tL=M\Psi$, $N+{}^tN=-{}^tM\Psi M$, where $\Psi=f(v_{m+k},v_{m+1})_{0<k,1<2n-2m}$. In particular S runs over $GL_m(\mathbb{F}_{p^\alpha})$ and Q runs over all the skew-symmetric matrices. Since W is maximal completely isotropic then the order of the group $\langle T\rangle$ is prime to p.

It is clear P contains a Sylow p-subgroup of G. Therefore we may assume that $g\in P$. Note that Q is transformed into $SQ{}^tS$ under the action of P on Z by conjugation. It follows that either g normalizes, but does not centralize Z, or $g\in R$. Obviously $[R,R]\subset Z$. Therefore either g normalizes, but does not centralize an abelian subgroup of R, or $g\in Z$.

Let Q be the submatrix defined by $g\in Z$. Taking xgx^{-1} instead of g for an appropriate $x\in H$ we can assume that the entry of Q in position (m,1) is not 0. Let $h=1_{2n}-e_{3,m}-e_{2n,2n-m+2}$, where $e_{i,j}$ are matrix units. Then $[x,g]=[x,h]=1$ where $x=[h,g]$. Therefore g normalizes, but does not centralize the abelian p-group $\langle x,h\rangle$.

Thus it remains to consider groups of type ${}^2A_n$ (n>1) and C_n. We shall do it in the next section.

3. Linear groups with a normal extraspecial subgroup

Certain parabolic subgroups of finite groups of Lie type contain normal subroups very similar to extraspecial ones (many details can be found in [3, §4]). They are called groups of extraspecial type by Landazuri and Seitz [9]. Furthermore, the study of irreducible representations of such parabolic subgroup is reduced to a large extent to groups with a normal extraspecial subgroup. Note also that linear groups with a normal extraspecial subgroup appear naturally when considering primitive groups.

A group A is called of extraspecial type if $Z(A)=[A,A]$, and if for any proper subgroup $X\subset Z(A)$ we have $Z(A/X)=Z(A)/X$. Observe that $Z(A)$ and $A/Z(A)$ are elementary p-groups. When $Z(A)$ is cyclic then A is called extraspecial.

If ψ is an irreducible representation of A with $\psi(Z(A))\neq 1$ then $\psi(A)$ is extraspecial.

Let E_p denote an extraspecial group. Let $Z=Z(E_p)$. We may consider E_p/Z as a space over $\mathbb{F}_p$. The map $(x,y)\to[x,y]=xyx^{-1}y^{-1}$ induces on $E_p/Z(E_p)$ an alternating bilinear form t with values in $\mathbb{F}_p$. It is known that t is nondegenerate, so $|E_p/Z|=p^{2k}$. Therefore $|E_p|=p^{2k+1}$. If we need to emphasize the order of E_p, we write E_p^k. Therefore if S is an automorphism group of E_p operating trivially on Z, we then have a natural homomorphism $S\to Sp_{2k}(p)$. Furthermore if E_p^k is a normal subgroup of a group H then we obtain a natural homomorphism of H/E_p^k in $Sp_{2k}(p)$.

<u>Lemma</u> 4 [15,4]. Let $E_p^k\subset GL_n(F)$ be irreducible. Then $n=p^k$ and every automorphism of E_p^k trivial on Z can be realized by some element of $GL_n(F)$.

<u>Proposition</u> 1. Let $H\subset GL_n(F)$ be an irreducible group with a normal extraspecial subgroup E_p^k and $Z(E_p^k)=Z(H)$. Then $n=p^km$ and H is contained in a Kronecker product $H'\otimes GL_m(F)$, where $H'\subset GL_{p^k}(F)$ and H' contains a normal subgroup E' isomorphic to E_p^k. Furthermore $H/E_p^k\cong H'/E'$ and for any $h\in H$ we have $h=h'\otimes\tau(h)$ where $\tau: H/E_p^k\to GL_m(F)$ is a projective irreducible representation. If E' has a complement in H' then τ can be chosen ordinary.

This proposition follows from Clifford's theorem, Lemma 4 and some simple facts of representation theory.

<u>Lemma</u> 5. If a,b are matrices, then Spec $a\otimes b$ = (Spec a)(Spec b).

The proof is obvious.

Proposition 1 and Lemma 5 allow one to reduce the description of the spectrum of h to that of h' and $\tau(h)$. To study Spec h' we have to consider linear groups with an irreducible extraspecial subgroup. This will be done in Theorem 4 below.

Lemma 6. Let $H \subset GL_n(F)$, $h \in H$, $|h| = p^k$ and $(p,f)=1$. Let X be the conjugacy class of h and $I = \{i, 0<i<|h| \mid h^i \in X\}$. Then $\deg h \geq |I|$.

Indeed there exists an $s \in \mathrm{Spec}\, h$ with $|s|=|h|$. Then $s^i \in \mathrm{Spec}\, h^i$; if $i \in I$ then $\mathrm{Spec}\, h^i = \mathrm{Spec}\, h$ so $s^i \in \mathrm{Spec}\, h$.

Theorem 4 [19]. Let $G \subset GL_n(F)$, $(f,p) = 1$, $g \in G$, $|g| = p > 2$. Suppose that G contains an irreducible extraspecial normal subgroup E_p^k. Then the following assertions are equivalent:

(i) $\deg g < p$;

(ii) g centralizes a maximal abelian subgroup of E_p^k, and the image $\bar{g}$ of g under the canonical map $G \to Sp_{2k}(p)$ is a transvection.

Moreover, (i) implies $\deg g = (p+1)/2$.

Proof. For k=1 the theorem can be proved by a direct calculation which we omit.

Recall some properties of representations of the extraspecial groups E_p^k. An irreducible representation is either 1-dimensional or of degree p^k. Let $G_1 = G \cdot S$ where S is the group of scalar matrices. Then (for p>2) G_1 contains a normal extraspecial subgroup of exponent p. The addition of scalars to G has no influence on the questions of interest to us, so we shall assume that E_p^k is of exponent p (we could replace in all subsequent considerations E_p^k by $E_p^k S$ (cf. [19]). Any two irreducible extraspecial subgroups are conjugate in $GL_n(F)$.

Thus, let E_p^k be of exponent p. Let N_k be its normalizer in $GL_n(F)$. It is well known [15, 4] that $N_k/SE_p^k \cong Sp_{2k}(p)$ and this extension splits: $N_k = SE_p^k \cdot H_k$ where $H_k \cong Sp_{2k}(p)$. The Kronecker product operation yields the embedding $N_k \otimes N_1 \subset N_{k+1}$. The natural projection $N_{k+1} \to Sp_{k+1}(p)$ induces the standard embedding $Sp_{2k}(p) \times Sp_{2l}(p) \to Sp_{2(k+1)}(p)$. Fix isomorphisms $\theta_r : Sp_{2r} \to H_r$ $(r = 1,2,\ldots)$ in such a way that $\theta_{k+1}(H_k \otimes H_1) = \theta_k(H_k) \otimes \theta_1(H_1)$. The embedding $N_1 \otimes 1_{p^{k-1}} \to N_k$ proves (ii) $\Rightarrow$ (i).

Let $Sp_{2k}(p)$ operate naturally on the space V over $\mathbb{F}_p$ and let $U \subset V$ be a maximal isotropic subspace. Let $P = \{s \in Sp_{2n}(p) \mid sU=U\}$. Then P=TR where the subgroups T,R can be written in an appropriate basis $v_1,\ldots,v_{2k}$ in the form

$$T = \begin{pmatrix} L & 0 \\ 0 & {}^tL^{-1} \end{pmatrix}, \quad R = \begin{pmatrix} 1_k & M \\ 0 & 1_k \end{pmatrix}.$$

Here L runs over $GL_k(p)$ and M runs over symmetric matrices. P contains a Sylow p-subgroup of $Sp_{2k}(p)$. Hence we may assume $\bar{g} \in P$. Let D be the nverse image of U under the natural projection $E_p^k \to E_p^k/Z$. Then D is a aximal abelian subgroup of E_p^k. If $\bar{g}\notin R$ then g operates nontrivially on ; therefore g normalizes, but does not centralize an abelian p-group . Then deg g = p by Corollary 1. It follows $\bar{g}\in R$.

The action of T on R by conjugation transforms M into LM^tL. It is nown that LM^tL is diagonal for an appropriate L. Hence M may be assu-ed to be diagonal. Then the hyperbolic planes $\langle v_i, v_{i+k}\rangle$ are stable nder $\bar{g}$. Therefore $\bar{g}\in Sp_2\times\ldots\times Sp_2$ (k times), so $g\in N_1\otimes\ldots\otimes N_1$. Write $= x_1\otimes x_2\otimes\ldots\otimes x_k$ where $x_i\in N_1$ and x_i is either a transvection or $_2$. By Lemma 5, deg $x_i < p$. If $\bar{x}_i = 1_2$ then $x_i\in E_p^1.S$ and $x_i\in S$ because eg $x_i < p$. By Theorem 4 for k = 1, deg $x_i = (p+1)/2$ provided $x_i\notin S$. herefore deg g = (p+1)/2 if $\bar{g}$ is a transvection.

Suppose now $\bar{g}$ is not a transvection. It suffices to prove that eg $(x_1\otimes x_2)$ = p when x_1,x_2 are not scalars. A careful analysis of N_1 hows that Spec $x_1 = \{\varepsilon^i\}$ where $\varepsilon\in F$, $\varepsilon^p = 1$, $\varepsilon\neq 1$, i runs over $(y)\pmod p$, and $f(y) = ay^2+by+c$ $(a,b,c,y\in\mathbb{Z}, a\neq 0)$. Therefore Spec $x_1)$(Spec $x_2) = \{\varepsilon^j\}$ where j runs over $f_1(y_1)+f_2(y_2)\pmod p$. It s known that for any such f_1,f_2 the equation $f_1(y_1)+f_2(y_2) = i$ is so-uble in a finite field. Therefore Spec$(x_1\otimes x_2) = \{1,\varepsilon,\varepsilon^2,\ldots,\varepsilon^{p-1}\}$. his finishes the proof that (i) implies (ii).

Corollary 2. Let R be a p-group of extraspecial type and let g be n automorphism of R of order p > 2, which is trivial on Z(R). Let $= \{l\in R\,|\,g(l).l^{-1}\in Z(R)\}$. Let $H=R\rtimes\langle g\rangle$ be the semidirect product and let $\in$ Irr H, with $Z(R)\not\subseteq Ker\varphi$. Suppose that $|R:S| > p$. Then deg $\varphi(g) = p$.

Proof. $\varphi(Z(R))$ is scalar so $|\varphi(Z(R))| = p$. Hence $\varphi(R)$ is extra-pecial. It is clear that $|R:S| = |\varphi(R):\varphi(S)| > p$. It follows from Pro-osition 1 that $\varphi(R)$ is irreducible. Apply Theorem 4 to $\varphi(H)$. Then eg$\varphi(g) = p$, unless $\varphi(g)$ operates on $\varphi(R)/\varphi(Z(R))$ as a transvection or, quivalently, $|\varphi(R):\varphi(S)| = p$.

Corollary 3. Let p > 2, G be $C_n(p^\alpha)$ (n > 1) or $^2A_n(p^\alpha)$ (n > 2). Let be the stabilizer of a non-zero vector, and $R = O_p(G)$. Let $\varphi\in$ IrrP nd $Z(P)\not\subseteq Ker\varphi$. Let $g\in P$, $g\notin Z(P)$, $g^p=1$. Then deg$\varphi(g)$ = p unless $G = Sp_{2n}(p)$.

Proof. It is known that R is of extraspecial type. It is easy to heck that $|R:S| > p$ where $S=C_R(g)$ unless $G \cong Sp_{2n}(p)$. It remains to se Corollary 2.

Proof of Theorem 1. According to Section 1, it remains to consider the cases $G = C_n(p^\alpha)$ $(n > 1)$ or $^2A_n(p^\alpha)$ $(n > 2)$. Let P,R be as in Corollary 3. Since P contains a Sylow p-subgroup of G, we may assume $g \in P$. It is clear that g is conjugate in G with a non-central element of P. By Corollary 3, $G = C_n(p)$. Moreover, by Corollary 2, g has to operate on $R/Z(P)$ as a transvection. It follows that $g=h.r$ where h is a transvection, $r \in R$. If $\deg g < p$ then $r \in Z(R) = Z(P)$. Therefore either h,r are commuting transvections, or $r = 1$. If $n > 2$ then g can be included in $C_2(p) \subset C_n(p)$ (the natural embedding). Moreover $C_2(p)$ is conjugate in G with a subgroup of P. Let $C_2(p) \subset P$. Applying again Corollary 2 we see that g is a transvection, so $r = 1$ if $n > 2$.

The groups $C_2(p)$, $A_1(p^\alpha)$, $C_1(p^\alpha)$, $^2A_n(p^\alpha)$ can be examined by means of their character tables.

Proof of Theorem 2. Use Theorem 1. If $G \cong {}^2A_2(p)$ or $A_1(p^2)$ then all g^i $(0 < i < p)$ are conjugate; this is true for $g \in C_2(p)$ as well, provided g is not a transvection. By Lemma 6, $\deg g = p-1$. Let $G = C_n(p)$, g is a transvection. Then $\{g^i\}$ splits into 2 conjugacy subsets. Each one is of cardinality $(p-1)/2$. Hence by Lemma 6, $\deg\varphi(g)$ can take only the values p, p-1, (p+1)/2, (p-1)/2. This implies the first part of Theorem 2.

Let $\deg\varphi(g) = (p-1)/2$. The case $n = 1$ is examined directly using the character table. Let $n > 1$ and let P,R be as in Corollary 3. Write $P = HR$ where $H \cong C_{n-1}(P)$. Replacing g by some conjugate we may assume that $g \in H$. According to Proposition 1 and Lemma 5, $\deg g' < (p-1)/2$. This contradicts Theorem 4.

Proof of Theorem 3. Use Theorem 1. Show that $C_n(p)$ $(n > 2)$ does not satisfy the assumptions of Theorem 3. The remaining groups in Theorem 1 can be examined with the aid of character tables. Let $R \subset P \subset C_n(p)$ be as in Corollary 3. Then $P = HR$ where $H \cong C_{n-1}(p)$. It is clear that $\varphi|P$ contains a representation $\psi \in \mathrm{Irr}P$ such that $Z(P) \not\subset \mathrm{Ker}\psi$. We may assume $g \in H$. Apply Proposition 1 to $h = \psi(g)$. Then $\deg h' < p$ and $\deg\tau(h) < p$. By Theorem 4, $\deg h' = (p+1)/2$. By Theorem 2 either $\tau = 1$ or $\deg\tau(h) >$ $> (p+1)/2$ since $n > 2$. In the latter case it follows from Lemma 6 that $\mathrm{Spec}\,\tau(h)$. and $\mathrm{Spec}\,h'$ contain $\{\varepsilon^i\}$ where i runs over squares or nonsquares of the field $\mathbb{F}_p$. If $\deg\tau(h) > (p+1)/2$ then $\deg\tau(h) = p-1$ and then $\deg h'\otimes\tau(h) = p$. If $\deg\tau(h) = (p+1)/2$ then $1 \in \mathrm{Spec}\,\tau(h)$ as well as $1 \in \mathrm{Spec}\,h'$. Therefore $\mathrm{Spec}\,h'\otimes\tau(h)$ is of the form $\{\varepsilon^j\}$ where j runs over values of a quadratic form $ax^2 + by^2$ in $\mathbb{F}_p$. It is known that these values contain 0. If $\tau = 1$, then $1 \in \mathrm{Spec}\,h' = \mathrm{Spec}\,h$.

My program in the near future is to extend the arguments of this section to elements of proper parabolic subgroups of groups of Lie type. Lemma 6 may be helpful for examining elements which are contained in a product of proper elementwise commuting noncentral subgroups. To consider other elements one needs some new ideas. One of them, to be discussed in the next section, is based on modular representation theory. Unfortunately this idea is effective only for examining eigenvalue 1 when $F = \mathbb{C}$. Moreover we cannot analyze effectively enough groups of type different from G_2, F_4, E_8.

4. Eigenvalue 1 of semisimple elements of groups of Lie type in complex representations.

Theorem 5 [22] Let G be a finite group of Lie type E_8, F_4, G_2, ${}^2F_4$, or ${}^2G_2$ (excluding $G_2(2^m)$). If $g \in G$ is semisimple, then $1 \in \mathrm{Spec}\,\varphi(g)$ for any complex representation φ of G.

Theorem 6 [22] Let $Z(G) = 1$ and let G be of one of the following types:

(i) $A_n(p)$, $p \equiv 1 \pmod{n+1}$;

(ii) ${}^2A_n(p)$, $p \equiv -1 \pmod{n+1}$;

(iii) $D_n(p)$, $p \equiv 1 \pmod 4$;

(iv) ${}^2D_n(p)$, $p \equiv -1 \pmod 4$;

(v) $E_6(p)$, $p \equiv 1 \pmod 3$;

(vi) ${}^2E_6(p)$, $p \equiv 2 \pmod 3$;

(vii) $B_n(p)$, $C_n(p)$, $E_7(p)$, $p \equiv 1 \pmod 2$.

If $g \in G$ is semisimple, then $1 \in \mathrm{Spec}\,\varphi(g)$ for any complex representation φ of G.

We do not see any way to extend Theorem 6 to groups $G(p^\alpha)$ with $\alpha > 1$. An attempt to argue similarly to the proof of Theorem 5 leads to the following question. Let $\mathrm{Comp}\,\varphi^{(p)}$ be the set of composition factors of a reduction of φ modulo p; what are the representations $\varphi \in \mathrm{Irr}_{\mathbb{C}} G(p^\alpha)$ ($\alpha > 1$) with $Z(G) \subset \mathrm{Ker}\,\varphi$ for which $\mathrm{Comp}\,\varphi^{(p)}$ contains no representation which is a Steinberg product of representations ψ parametrized by radical weights. (According to R. Steinberg every irreducible representation of $G(p^\alpha)$ over an algebraically closed field of characteristic p is a tensor product of representations obtained from infinitesimally irreducible ones by Frobenius automorphisms. These can be parametrized by highest weights of corresponding representations of algebraic groups). One may conjecture that such representations are very rare.

Proof of Theorem 5. Let P be an algebraically closed field of characteristic p. Let G^* be the algebraic group of the same type as G so that $G \subset G^*$.

Let $\varphi^{(p)}$ be a reduction of φ modulo p. Let ψ be an arbitrary composition factor of $\varphi^{(p)}$. It suffices to prove that $1 \in \mathrm{Spec}\,\psi(g)$.

It is known that ψ can be extended to G^* and that there exists a maximal torus $T \subset G^*$ such that $g \in T$. Therefore it suffices to prove that $\psi(G^*)$ possesses the weight 0. Since $\psi(G^*)$ is a Steinberg product of infinitesimally irreducible representations it suffices to examine infinitesimally irreducible ψ.

Now I need the following results of A. Premet [10]. Let π be an infinitesimally irreducible representation of a simple algebraic group X and let $w = a_1w_1+\ldots+a_nw_n$ be its highest weight. Let $X_{\mathbb{C}}$ be the algebraic group over $\mathbb{C}$ of the same type as X and let $\pi_{\mathbb{C}}$ be an irreducible representation of $X_{\mathbb{C}}$ with highest weight w. Suppose that $p > 2$ if X is of type F_4, B_n, C_n $(n > 1)$ and $p > 3$ if X is of type G_2. Then the lists of weights of π and $\pi_{\mathbb{C}}$ are the same (that is the coefficients of the decompositions of the weights in terms of the fundamental ones are the same). It is known that 0 is a weight of a representation $\pi_{\mathbb{C}}$ if and only if w is radical. Since every weight is radical for groups of type G_2, F_4, E_8, it follows π has weight 0 unless G^* is of type F_4 and $p = 2$ or G is of type G_2 and $p = 2,3$. For the latter cases the weights of infinitesimally irreducible representations are known completely (see [14, 17]). After inspection we see that 0 is a weight of ψ for $G^* = F_4$, $p = 2$, $G^* = G_2$, $p = 3$.

The proof of Theorem 6 is similar. Observe that for the groups of Theorem 6, $\psi(G^*)$ is infinitesimally irreducible [14]. The assumptions on p are chosen so that the highest weight of $\psi(G^*)$ would be radical.

To finish I mention some cases where a semisimple element does not have the eigenvalue 1.

Proposition 2 [22]. Let $g \in G$ be one of the following pairs:

(i) $G = PSL_n(q)$, $(n,q-1) = 1$, $|g| = (q^n-1)/(q-1)$;

(ii) $G = PSp_{2n}(q)$, $q^n \equiv 1 \pmod 4$, $|g| = (q^n+1)/2$;

(iii) $G = PSU_n(q)$, $(n,q+1) = 1$, $|g| = q+1$.

Then there exists a complex irreducible representation φ of G such hat $1 \notin \mathrm{Spec}\varphi(g)$, for some g of the order specified above.

REFERENCES

1. T.R. Berger, Hall-Higman type theorems, IV. Proc. Amer. Math. Soc. 37 (1973), 317-325.

2. H.F. Blichfeldt, Finite collineation groups, Univ. Chicago Press, Chicago, 1917.

3. C.W. Curtis, W.M. Kantor and G.M. Seitz, The 2-transitive permutation representations of the finite Chevalley groups, Trans. Amer. Math. Soc. 218 (1976), 1-59.

4. W. Feit and J. Tits, Projective representations of minimum degree of group extensions, Can. J. Math. 30 (1978), 1092-1102.

5. W. Feit and J. Thompson, Finite groups which have a faithful representation of degree less than (p-1)/2, Pacific J. Math. 11 (1961), 1257-1262.

6. Ph. Hall and G. Higman, On the p-length of p-soluble groups and reduction theorems for Burnside's problem, Proc. London Math. Soc. 6 (1956), 1-42.

7. Ch. Hering, Transitive linear groups and linear groups which contain irreducible subgroups of prime order, J. Algebra 93 (1985), 151-164.

8. C.Y. Ho, On quadratic pairs, J. Algebra 43 (1976), 338-358.

9. V. Landazuri and G. Seitz, On the minimal degree of projective representations of the finite Chevalley groups, J. Algebra 32 (1974), 418-443.

10. A.A. Premet, The weights of irreducible rational representations of semisimple algebraic groups over a field of prime characteristic (in Russian), Mat. Sb. 133 (175) (1987), 167-183. (English. translation: Math. USSR, Sb.).

11. G.M. Seitz, Some representations of classical groups, J. London Math. Soc. 10 (1975), 115-120.

12. E. Shult, On groups admitting fixed point free abelian operator groups, Illinois J. Math. 9 (1965), 701-720.

13. R. Steinberg, Lectures on Chevalley groups, mimeographed lecture notes, Yale Univ. Math. Dept. (1968).

14. R. Steinberg, Representations of algebraic groups, Nagoya J. Math. 22 (1963), 33-56.

15. D.A. Suprunenko, Soluble and nilpotent linear groups (in Russian), Minsk, 1958. (English translation: Amer. Math. Soc., Providence, R.I., 1963.)

16. J. Thompson, Quadratic pairs, Proc. Intern. Congr. Math. (Nice, 1970), Vol. 1, Gauthier-Villars, Paris (1971), 375-376.

17. F.D. Veldcamp, Representations of algebraic groups of type F_4 in characteristic 2, J. Algebra 16 (1970), 326-339.

18. H.N. Ward, Representations of symplectic groups, J. Algebra 20 (1972), 182-195.

19. A.E. Zalesskii, Normalizer of an extraspecial linear group (in Russian), Vestsi Akad. Navuk BSSR, Ser. fiz.-mat. navuk, 6 (1985), 11-16.

20. A.E. Zalesskii, Spectra of elements of order p in complex representations of finite Chevalley groups of characteristic p (in Russian) Vestsi Akad. Navuk BSSR, Ser. fiz.-mat. navuk, 6, (1986), 20-25.

21. A.E. Zalesskii, Fixed points of elements of order p in complex representations of finite Chevalley groups of characteristic p (in Russian), Dokl. Akad. Nauk BSSR, 31 (1987), 104-107.

22. A.E. Zalesskii, Eigenvalue 1 of matrices of complex representations of finite Chevalley groups, (in Russian), Trudy Mat. Inst. AN SSSR, 187 (1988) (to appear) (English translation: Proc. Steklov Inst. Math.).

Remark. After this paper was prepared I was informed that Problem 3 was solved by G.R. Robinson for the case $\deg\varphi(g) < (p+1)/4$ (see G.R. Robinson, Remarks on reduction (mod p) of finite complex linear groups. J. Algebra, 83 (1983), 477-483.)

"This paper is in final form and no version of it will be submitted for publication elsewhere"

NORMAL FITTING CLASSES OF GROUPS AND GENERALIZATIONS

Guido Zappa
Istituto Matematico "Ulisse Dini"
Università degli Studi
Viale Morgagni 67/A
50134 Firenze, Italia

The aim of this survey is to present some classical results concerning classes of finite groups (in particular formations, Schunck and Fitting classes), to expose briefly the theory of normal Fitting classes and to present some recent generalizations.

All groups considered hereafter are assumed to be finite.

. A historical approach to the theory of Fitting classes

In 1928, Philip Hall [7] proved a celebrated theorem which generalizes, for soluble groups, the Sylow theorem:

Hall theorem. Let G be a soluble group of order mn (with m and n relatively prime). Then G contains subgroups of order m; they form a conjugacy class; moreover, every subgroup, whose order divides m, is contained in some subgroup of order m.

The subgroups of order m are called Hall subgroups.

In 1961, R. Carter [3] found in soluble groups another class of subgroups enjoying properties similar to those of the Hall subgroups:

Carter theorem. Let G be a soluble group. Then

a) G contains some nilpotent subgroup H such that $N_G(H) = H$;

b) The subgroups H of G satisfying a) form a conjugacy class.

These subgroups are called Carter subgroups.

In 1963, W. Gaschütz developed a general theory which includes the Hall and Carter theorems as particular cases: the theory of formations of groups.

A formation is a class $\underline{F}$ of groups such that:

a) If $G \in \underline{F}$, then every epimorphic image of G is also in $\underline{F}$;

b) If N_1, N_2 are normal subgroups of G and $\frac{G}{N_1}, \frac{G}{N_2} \in \underline{F}$, then also $\frac{G}{N_1 \cap N_2} \in \underline{F}$.

Examples of formations: the class of abelian groups; the class of π-groups (π being a set of primes); the class of nilpotent groups; the class of supersoluble groups; the class of soluble groups.

Recall that the Frattini subgroup $\phi(G)$ of a group G is the intersection of all maximal subgroups of G.

A formation $\underline{F}$ of groups is said to be saturated if $\frac{G}{\phi(G)} \in \underline{F}$ implies $G \in \underline{F}$.

The formation of π-groups, of nilpotent groups, of supersoluble groups, of soluble groups are saturated; the formation of abelian groups is not saturated.

The following proposition is an easy consequence of the definition of formation:

Let G be a group and let $\underline{F}$ be a formation; then there exists a normal subgroup $G_{\underline{F}}$ of G such that for every normal subgroup N of G, $\frac{G}{N} \in \underline{F}$ if and only if $N \supseteq G_{\underline{F}}$. The subgroup $G_{\underline{F}}$ is said to be the $\underline{F}$-residual of G.

Let $\underline{F}$ be a formation. Then a subgroup H of a group G is called an $\underline{F}$-covering subgroup of G if $H \in \underline{F}$ and, for every subgroup S of G containing H and for every normal subgroup T of S such that $\frac{S}{T} \in \underline{F}$, one has $S = HT$.

W. Gaschütz [5] in 1963 proved the following theorem which generalizes the Hall and Carter theorems.

Theorem. Let G be a soluble group and let $\underline{F}$ be a saturated formation. Then:

a) G has some covering subgroup.

b) The $\underline{F}$-covering subgroups of G form a conjugacy class.

For the class $\underline{F}_\pi$ of all π-groups (π being a set of primes) the $\underline{F}$-covering subgroups are the Hall π-subgroups; for the class $\underline{N}$ of all nilpotent groups, the $\underline{N}$-covering subgroups are the Carter subgroups.

A generalization of saturated formations is provided by Schunck classes.

Recall that a group is said to be primitive if it contains a maximal subgroup H such that the intersection of all conjugates of H is 1.

A Schunck class is a class $\underline{F}$ of groups such that:

a) If $G \in \underline{F}$, then every epimorphic image of G is in $\underline{F}$;

b) If every primitive quotient of the group G is in $\underline{F}$, then $G \in \underline{F}$.

It is easy to see that every saturated formation is a Schunck class.

These classes were introduced by Schunck (a student of Gaschütz) in his "Diplomarbeit" published in 1967. Schunck [11] proved that the theorem of Gaschütz concerning $\underline{F}$-covering subgroups can be extended to the case in which $\underline{F}$ is a Schunck class.

A notion strictly related to that of an $\underline{F}$-covering subgroup is that of $\underline{F}$-projector.

Let $\underline{F}$ be a class of groups and G a group. An $\underline{F}$-projector of G is a subgroup H of G such that, for every normal subgroup N of G, $\frac{HN}{N}$ if $\underline{F}$-maximal in $\frac{G}{N}$ (i.e. the image of H is $\underline{F}$-maximal in every epimorphic image of G).

$\underline{F}$-covering subgroups do not coincide with $\underline{F}$-projectors. However, in 1969 Gaschütz [6] proved that if $\underline{F}$ is a Schunck class and G a soluble group, then every $\underline{F}$-projector of G is also an $\underline{F}$-covering subgroup and vice-versa.

The theory of formations was dualized by B. Fischer in his "Habilitationschrift" in 1966. In fact Fischer introduced the notion of Fitting class.

A <u>Fitting class</u> is a class $\underline{F}$ of groups such that:

1) If $G \in \underline{F}$, then every isomorphic image of G is in $\underline{F}$.
2) If $G \in \underline{F}$ and N is a normal subgroup of G, then $N \in \underline{F}$.
3) If N_1, N_2 are two normal subgroups of G and $N_1, N_2 \in \underline{F}$, then $N_1 N_2 \in \underline{F}$.

Let G be a group and let $\underline{F}$ be a Fitting class. There exists then a normal subgroup G_F of G such that, for every normal subgroup N of G, $N \in \underline{F}$ if and only if $N \supseteq G_F$. The subgroup G_F is called the <u>$\underline{F}$-radical</u> of G.

A subgroup H of a group G is called an <u>$\underline{F}$-injector</u> if, for every subnormal subgroup N of G, $H \cap N$ is $\underline{F}$-maximal in N.

In 1967, B. Fischer, W. Gaschütz and B. Hartley [4] dualized the theorems on projectors of saturated formations and Schunck classes in the following way:

<u>Theorem.</u> Let G be a soluble group and let $\underline{F}$ be a Fitting class. Then

a) G has some $\underline{F}$-injectors.
b) The $\underline{F}$-injectors of G form a conjugacy class.

Of course every $\underline{F}$-injector contains the $\underline{F}$-radical.

2. Normal Fitting classes

Let $\underline{H}$ be a Fitting class. A Fitting class $\underline{F} \subseteq \underline{H}$ is said to be $\underline{H}$-normal if, for every $G \in \underline{H}$, $G_{\underline{F}}$ is the unique $\underline{F}$-injector of G (i.e. $G_{\underline{F}}$ is $\underline{F}$-maximal). Let $\underline{S}$ be the class of soluble groups. A Fitting class is said to be normal if it is $\underline{S}$-normal.

In 1970 D. Blessenohl and W. Gaschütz [2] introduced the following method for constructing normal Fitting classes.

Let A be a (possibly infinite) abelian group and let Φ be a law which associates to every group $G \in \underline{S}$ a homomorphism Φ_G from G into A such that:

a) If $G, \overline{G} \in \underline{S}$ and χ is an isomorphism from G onto $\overline{G}$, then $\chi\Phi_{\overline{G}} = \Phi_G$.

b) If N is a normal subgroup of $G \in \underline{S}$, then Φ_N is the restriction of Φ_G to N.

c) $A = \langle \Phi_G(G) \mid G \in \underline{S} \rangle$.

Then the pair (Φ, A) is called a Fitting pair, and Φ is called a Fitting functor.

Blessenohl and Gaschütz proved that if (Φ, A) is a Fitting pair, then the class of groups $G \in \underline{S}$ such that $\Phi_G(G) = 1$ is a normal Fitting class. H. Lausch in 1973 [9] proved that every non-trivial normal Fitting class can be obtained in this way (for a convenient abelian group A).

Blessenohl and Gaschütz in [2] also gave two important examples of Fitting pairs. In the first example $A = \{\pm 1\}$ and a convenient representation χ of G as a permutation group is given such that for every $x \in G$, $\Phi_G(x) = 1$ if $\chi(x)$ is even, and $\Phi_G(x) = -1$ if $\chi(x)$ is odd. In the second example $\Phi_G(x)$ is the determinant of the matrix which is associated to x in an appropriate representation of G as a group of matrices.

A non-trivial Fitting class $\underline{F} \subseteq \underline{S}$ is normal if and only if, for every $G \in \underline{S}$, $G/G_{\underline{F}}$ is abelian (i.e. $G_{\underline{F}} \supseteq G'$). Consequently the intersection of all non-trivial normal Fitting classes is a non-trivial normal Fitting class, which is called the smallest normal Fitting class. It is a large class, which contains every nilpotent group and every group which is the derived group of some soluble group.

An important construction of a series of normal Fitting classes was obtained in 1977 by H. Laue, H. Lausch and G.R. Pain [8] using transfer. On the basis of a slight modification of this construction, T.R. Berger in 1981 characterized the smallest normal Fitting class. Let us expose briefly the results of Berger.

Let U be a group. For every soluble group G let $\Omega = \{H_1, \ldots, H_n\}$ be the set of all subnormal subgroups of G isomorphic to U. Let Φ_i be a given isomorphism of H_i onto U, $(i = 1, \ldots, n)$. For every $x \in G$, let $\overline{x}$ be the inner automorphism of G induced by x, and put $H_i^x = H_{ix}$ $(i=1,\ldots,r)$.

et K be the subgroup of AutU generated by (AutU)' and by all $\alpha\in$ AutU uch that $[U,\alpha]\subset U$. Put $g_G^U(x) = K\prod_{i=1}^{r}\Phi_i^{-1}\bar{x}|_{H_i}\Phi_{ix}$. Then g_G^U is a homomor-hism of G into the abelian group $B^U = \frac{\mathrm{Aut}U}{K}$, which does not depend on he choice of the automorphisms Φ_i; the class of all soluble groups G uch that, for every $x\in G$, $g_G^U(x) = 1$, is a normal Fitting class. Conse-uently (g^U,B^U) is a Fitting pair, which is called a Laue-Lausch-Pain itting pair since it is similar to the one considered by Laue, Lausch nd Pain.

Berger [1] proved the following important theorem:

<u>Theorem.</u> Let (f,A) be a Fitting pair, G a soluble group and x an lement of G such that $f(x) \neq 1$. Then there exists a subnormal subgroup of G and a Laue-Lausch-Pain Fitting pair (g^U,B^U) such that $g_G^U(x) \neq 1$. onsequently, if $\underline{M}$ is the smallest normal class, in order to find the -radical of a soluble group G, it is sufficient to find the intersec-ion of all the $\underline{F}$-radicals such that $\underline{F}$ is a normal Fitting class cor-esponding to the Fitting pair (g^U,B^U) for any subnormal subgroup U f G.

. <u>Constructive definition of normal Fitting classes. Normal classes.</u>

Formations, Schunck classes and Fitting classes are defined in a onstructive way, in the sense that if some group belongs to the class, ome other groups also belong to the same class. On the contrary, the efinition of normal Fitting class is not constructive. In order to tudy normal Fitting classes it is convenient to have a constructive efinition. In 1981 [12], I obtained some results in this direction.

A class $\underline{F}$ of soluble groups is said to enjoy the α-<u>property</u> if for very soluble group G such that:

a) G = HM with H an $\underline{F}$-maximal subgroup of G, M a maximal normal ubgroup of G, and $H\cap M$ a normal subgroup of G;

b) $[H,M] \nsubseteq H\cap M$;

c) for every normal subgroup N of G such that $H\cap M \subseteq N\subset M$, we have H,N] $\subseteq H\cap M$;

ne also has:

d) $M\in\underline{F}$.

I proved the following result:

<u>Theorem</u>. A Fitting class is normal if and only if it has the α-

property.

Sketch of proof. Let $\underline{F}$ be a normal Fitting class of soluble groups and let G be a soluble group satisfying a), b), c). If $M \subseteq G_F$, then $M \in \underline{F}$ because $\bar{M}$ is normal in G, and d) is satisfied. Otherwise let $M \not\subseteq G_F$. Then by a), b), c) it is easy to prove that $G_F \cap M = H \cap M \subset M$ and H is not normal in G, hence $H \neq G_F$. Consequently $p = |G : M| = |H : H\cap M| = |H : G_F \cap M| = |G : G_F| = |M : H\cap M|$. It follows that $|G : H\cap M| = |G : M||M : H\cap M| = p^2$, therefore $\frac{G}{H\cap M}$ is elementary abelian of order p^2 and H is normal in G, a contradiction.

Conversely, let $\underline{F}$ be a non-normal Fitting class of soluble groups; we shall prove that there exists a soluble group G with two subgroups H, M satisfying a), b), c), but not d). As $\underline{F}$ is not normal, there exist soluble groups G such that G_F is not $\underline{F}$-maximal. Let $\bar{G}$ be a group of minimal order among them, let $\bar{H}$ be an $\underline{F}$-injector of $\bar{G}$, and let $\bar{M}$ be a maximal normal subgroup of $\bar{G}$. Then it is easily seen that $\bar{G}_{\underline{F}} = \bar{M}_{\underline{F}} = H\cap\bar{M}$ and $G = H\bar{M}$. Moreover $[H, \bar{M}] \not\subseteq H\cap\bar{M}$, otherwise H is normal in $H\bar{M} = \bar{G}$ and $H \subseteq \bar{G}_{\underline{F}}$, a contradiction. Let $\underline{M}$ be the set of subnormal subgroups X of G such that $\bar{M}\cap H \subseteq X \subseteq \bar{M}$ and $N_{\bar{G}}(X) \supseteq H$. Then in $\underline{M}$ there exists a subgroup M such that $[H, M] \not\subseteq H\cap\bar{M} = H\cap M$ and, for every $N \in \underline{M}$ with $N \subset M$, $[H, N] \subseteq H\cap M$. Let $G = HM$. Then it is easy to prove that G, M, H satisfy a), b), c). But they do not satisfy d) because $H\cap M = \bar{M}_{\underline{F}}$ is $\underline{F}$-maximal in $\bar{M}$, hence in M, and therefore $M \notin \underline{F}$.

In order to appreciate the power of the α-property, it is interesting to consider the following generalization of normal Fitting classes, which I introduced in 1981 [13].

A class $\underline{F}$ of soluble groups is said to be <u>normal</u> if

a) For every $G \in \underline{F}$ every isomorphic image of G is also in $\underline{F}$;

b) For every $G \in \underline{F}$, every normal subgroup of G is also in $\underline{F}$;

c) For every $N_1, N_2 \in \underline{F}$, also $N_1 \times N_2 \in \underline{F}$;

d) $\underline{F}$ satisfies the α-property.

Of course, a class of soluble groups is a normal Fitting class if and only if it is a Fitting class and a normal class.

The definition of a normal class gives rise to the following problems:

1) Do there exist normal classes which are not Fitting classes?

2) Is the intersection of normal classes a normal class?

3) Characterize the minimal non-trivial normal classes.

No contributions to problems 1) and 2) have been given till now.

Partial contributions to problem 3) were given by A. Scarselli [10] and by myself [13].

In 1981, I proved that if $\underline{F}$ is a non-trivial normal class, then:

a) $\underline{F}$ contains all elementary abelian groups;

b) $\underline{F}$ contains all abelian 2-groups;

c) $\underline{F}$ contains the dihedral group of order 2^n $(n > 2)$;

d) $\underline{F}$ contains the generalized quaternion group of order 2^n $(n > 2)$;

e) $\underline{F}$ contains A_4;

f) $\underline{F}$ contains all groups of exponent p and order less than p^p, which have an abelian subgroup of index p.

In order to show the method used for obtaining these results, we give a sketch of the proofs of a), b), c), d).

Let $\underline{F}$ be a non-trivial normal class. Then there exists a soluble group $R \neq 1$ such that $R \in \underline{F}$. Let H be a minimal subnormal subgroup of R. Then $H \in \underline{F}$ and $|H| = q$, q being a prime. Let p be a prime with $p \neq q$, let n be the minimal positive integer such that $p^n \equiv 1 \pmod q$, and let M be an elementary abelian group of order p^n. Then M has an automorphism ω of order q. Let G be the semidirect product MH such that a generator of H induces ω on M. If H is not $\underline{F}$-maximal, then there exists a subgroup L of G such that $H \subset L$, $L \in \underline{F}$ and $|L|$ is divisible by p; then L contains a subnormal subgroup of order p belonging to $\underline{F}$. If H is $\underline{F}$-maximal in G, then, by the α-property, $M \in \underline{F}$; but M contains a subgroup of order p, normal in M and consequently belonging to $\underline{F}$. So in both cases $\underline{F}$ contains a group of order p for every prime p, and consequently it contains every elementary abelian group, and every abelian group of square-free exponent. Hence a) is proved.

Now let G be the semidirect product MS, M being the quaternion group, and S a cyclic group of order 3, permuting the three subgroups of order 4 of M. If Z is the center of M, let M = ZS. Then G = MH with $H \in \underline{F}$, and H maximal in G. If H is not $\underline{F}$-maximal, then $G \in \underline{F}$ so $M \in \underline{F}$; if H is $\underline{F}$-maximal, then by the α-property $M \in \underline{F}$. So the quaternion group is in $\underline{F}$. Consequently also the cyclic group of order 4 belongs to $\underline{F}$.

Let G be a dihedral group of order 2^{n+1} $(n \geq 3)$. Then G = MH, where M is a dihedral subgroup of order 2^{n-1}, and H is a non-normal dihedral subgroup of order 2^{n-1} when $n > 3$ and an elementary abelian subgroup of order 4 when $n = 3$. Moreover H is contained in a unique subgroup K of order 2^n, and K is also dihedral. Suppose $H \in \underline{F}$. If H is not $\underline{F}$-maximal, then $K \in \underline{F}$. If H is $\underline{F}$-maximal, then by the α-property $M \in \underline{F}$. In both cases, the dihedral group of order 2^n belongs to $\underline{F}$. As the elementary abelian group of order 4 is in $\underline{F}$, by induction we obtain c). But the dihedral group of order 2^n contains a cyclic normal subgroup of order 2^{n-1}; so every cyclic 2-group and every abelian 2-group belong to $\underline{F}$, and b) is proved. In a similar way we can prove d).

In 1983 A. Scarselli proved that every non-trivial normal class contains:

a) every abelian group;

b) every group of order p^n and exponent exactly p^{n-1} (p being an odd prime).

References

1. T.R. Berger, The smallest normal Fitting class revealed, Proc. Lond. Math. Soc., III Ser., 42 (1981), 59-86.
2. D. Blessenohl, W. Gaschütz, Ueber normale Schunck- und Fittingklassen, Math. Z. 118 (1970), 1-8.
3. R. Carter, Nilpotent self-normalizing subgroups of soluble groups, Math. Z. 75 (1961), 136-139.
4. B. Fischer, W. Gaschütz, B. Hartley, Injektoren endlicher auflösbarer Gruppen, Math. Z. 102 (1967), 337-339.
5. W. Gaschütz, Zur Theorie der endlichen auflösbarer Gruppen, Math. Z. 80 (1963), 300-305.
6. W. Gaschütz, Selected topics in the theory of soluble groups, Lectures in the 9th Summer Res. Inst., Australian Math. Soc., Canberra 1969.
7. P. Hall, A note on soluble groups, J. Lond. Math. Soc. 3 (1928), 98-105.
8. H. Laue, H. Lausch, R. Pain, Verlagerung und normale Fittingklassen endlicher Gruppen, Math. Z. 154 (1977), 257-260.
9. H. Lausch, On normal Fitting classes, Math. Z. 130 (1973), 67-72.
10. A. Scarselli, Sulla intersezione delle classi normali, Atti Acc. Naz. Lincei, VIII Ser., Rend., Cl. Sci. Fis. Mat. Nat. 74 (1983), 211-215.
11. H. Schunck, H-Untergruppen in endlicher auflösbarer Gruppen, Math. Z. 97 5(1967), 326-330.
12. G. Zappa, Un'osservazione sulle classi di Fitting normali, Atti Acc. Naz. Lincei, VIII Ser., Rend., Cl. Sci. Fis. Mat. Nat. 70 (1981), 1-5.
13. G. Zappa, On the normal classes of finite soluble groups, Studia Sci. Math. Hung. 16 (1981), 175-179.

"This paper is in final form and no version of it will be submitted for publication elsewhere"

ON THE NILPOTENCY OF NIL ALGEBRAS

E.I. Zel'manov
Institute of Mathematics
Siberian Branch of the Academy of Sciences of the USSR
630090 Novosibirsk, USSR

§1. Local nilpotency of nil algebras.

In 1941 A.G. Kurosh formulated two problems for nil algebras which were similar to Burnside's problems. We begin by recalling W. Burnside's famous problems for periodic groups.

The General Burnside Problem (GBP) asks: Is it true that every finitely generated periodic group is finite?

The Ordinary Burnside Problem (known simply as THE Burnside Problem, BP) asks: Is it true that every finitely generated group of bounded exponent is finite?

The Restricted Burnside Problem (RBP) asks: Is it true that there is a bound (depending on the number of generators and the exponent) for the order of all finitely generated groups of bounded exponent, which ARE finite?

Using analogous notation, the Kurosch problems can now be stated as follows, cf. [1]:

The General Kurosch Problem (GKP) asks: Is every finitely generated nil algebra nilpotent (equivalently, is every nil algebra locally nilpotent)?

The Ordinary Kurosch Problem (KP) asks: Is every finitely generated nil algebra of bounded degree nilpotent (equivalently, is every nil algebra of bounded degree locally nilpotent)?

We shall consider the Kurosh problems in three big classes of algebras: (1) associative algebras, (2) Lie algebras, (3) nonassociative (alternative and Jordan) algebras.

1.1. The Kurosh problem for associative algebras.

In 1964 E.S. Golod and I.R. Shafarevich [2] gave an example of a finitely generated nil algebra over an arbitrary field which is not nilpotent. Moreover, they used it to construct the first counter-example to the GBP, as follows.

Let A be a nil algebra over the field $\mathbb{Z}_p$, $|\mathbb{Z}_p| = p \geq 2$. Consider the adjoint group $G(A) = \{1+a,\ a \in A\}$, where 1 is a formal unit. If $a^{p^n}=0$ then $(1+a)^{p^n} = 1 + a^{p^n} = 1$. Hence G(A) is a periodic group. Suppose that A is generated by the elements $a_1,\dots a_m$ and the subgroup H of G(A) is generated by the elements $1+a_i$, $1 \leq i \leq m$. It is easy to see that the finiteness of H implies the finiteness (and nilpotency) of A. Thus if A is not nilpotent then H is a counter-example to the General Burnside Problem.

We remark that in the last 20 years there has appeared a considerable array of infinite finitely generated periodic groups (Alyoshin, Suschansky, Grigorchuk) not to mention the Ol'shansky Monsters, which are counter examples to the GBP. However thus far the example due to E.S. Golod and I.R. Shafarevich remains the only counter example to the General Kurosh Problem (GKP).

For the (ordinary) Kurosch Problem (KP) we have a quite different situation. Unlike Group Theory, where even the ordinary Burnside Problem has a negative answer (P.S. Novikov and S.I. Adian [3]), the Kurosh Problem has only positive answers in all important classes of algebras considered so far. The first result in this series was due to J. Levitzki [4].

Theorem (J. Levitzki) An associative nil algebra which satisfies some polynomial identity (PI) is locally nilpotent.

Levitzki's proof consists of two lemmas.

Lemma 1.1. Any associative algebra A contains a largest locally nilpotent ideal Loc(A) whose quotient algebra A/Loc(A) does not contain any nonzero locally nilpotent ideal.

Lemma 1.2. Any nonzero associative algebra which satisfies some PI contains a nonzero element of 2nd order, that is an element $a \in A$ such that $a^2 = aAa = 0$.

Clearly any 2nd order element generates a nilpotent ideal and so

ies in Loc(A). Now if $\bar{A} = A/Loc(A) \neq 0$ then by Lemma 1.2 $\bar{A}$ contains a nonzero 2nd order element which contradicts Lemma 1.1.

We indicated this scheme of proof because some of its features such as "locally nilpotent radical + elements of 2nd order" appear in other big classes of algebras: in nonassociative algebras and even (though on quite a different technical level) in Lie algebras (!).

A.I. Shirshov [5] suggested another purely combinatorial direct approach to the Kurosh Problem.

A.I. Shirshov's Height Theorem. Let A be an algebra over a ring of scalars Φ, satisfying a polynomial identity of degree d. Suppose A is generated by a set of elements $a_1,\ldots,a_m$. Then there exists an integer function H(m,d) such that any monomial a in $\{a_i\}$ is a Φ-linear combination of monomials $v_1^{n_1}\ldots v_h^{n_h}$ (of the same degree as a) where $h < H(m,d)$ and each v_i has degree $<d$ in $\{a_i\}$.

Corollary. If every monomial in $\{a_i\}$ of degree $<d$ is nilpotent of degree s, then $A^{(s-1)dH(m,d)+1} = 0$.

It is very important to observe that the nilpotency assumption here is imposed not on every element of A, but only on the monomials in the generators (even on a finite collection of them).

1.2. The Burnside-like Problems for Lie algebras.

For any element a of a Lie algebra L the equality $a^2 = 0$ holds. Hence the notion of nil element of a Lie algebra needs modification. Denote by ad(a) the operator $ad(a):L\ni x \to [x,a]$. It is natural to call an element $a\in L$ nilpotent if the operator ad(a) is nilpotent. Respectively L is a nil algebra if each one of its elements is nilpotent. With these definitions both the general and the restricted Burnside-like problems become meaningful for the class of Lie algebras. As for the first one, the negative solution is immediately provided by the example of E.S. Golod and I.R. Shafarevich. Indeed, it suffices to consider an associative nil algebra A which is not locally nilpotent and then take its commutator Lie algebra (A,[,]). The situation with the restricted problem for Lie algebras is much more difficult and intriguing. To appreciate this let us consider its connections to the Restricted Burnside Problem (RBP).

Consider the free Burnside group B(m,n) - the quotient group of the free group $F\langle x_1,\dots,x_m\rangle$ on m generators by its normal subgroup generated by $\{x^n,\ x\in F\langle x_1,\dots,x_m\rangle\}$. It is known (cf. P.S. Novikov, S.I. Adian [3]) that B(m,n) is infinite. The RBP can be formulated as follows:

Is it true that B(m,n) has only finitely many nonisomorphic finite homomorphic images?

Assume n = p to be a prime integer. Recall that a group with the identity $x^p = 1$ is called an elementary p-group and that every finite p-group is nilpotent. Clearly, in order to estimate the order of a finite m-generated elementary p-group by a function of m and p it suffices to estimate its nilpotency degree by some function of this kind.

Let G be a finite elementary p-group generated by the elements $x_1,\dots,x_m$. Consider its lower central series

$$G = G_0 > G_1 > G_2 > \dots > G_{s+1} = \langle 1\rangle,\ G_{i+1} = [G_1, G],\ 0 \leq i \leq s.$$

The factor G_i/G_{i+1} may be viewed as a vector space over the field $\mathbb{Z}_p$, $|\mathbb{Z}_p| = p$. Let us consider the direct sum

$$L = L(G) = \bigoplus_{i=0}^{s} G_i/G_{i+1}$$

and define the Lie brackets via

$$[a_iG_{i+1}, b_jG_{j+1}] = (a_i, b_j)G_{i+j+1}.$$

where $a_i\in G_i$, $b_j\in G_j$, and (,) denotes group commutators. It is easy to see that L is a finite dimensional Lie algebra which has the same nilpotency degree as G. W.Magnus [6] proved that L satisfies the identity $ad(x)^{p-1} = 0$. Thus for the proof of RBP (for n=p) it remained to settle in affirmative the following Kurosch-like Problem for Lie algebras:

Is it true that any m-generated Lie algebra over the field $\mathbb{Z}_p$, which satisfies the identity $ad(x)^{p-1} = 0$ is nilpotent?

This task was successfully accomplished by A.I. Kostrikin in 1958 (cf. [7], [8]). A detailed exposition of A.I. Kostrikin's proof and some related ideas is given in [9]. We shall make only a few comments on it. The proof is based on the so called "Locally nilpotent radical + elements of 2nd order" approach. Let L be an algebra over a field of characteristic p (or zero) which satisfies the identity $ad(x)^n = 0$ $(n < p)$. Roughly speaking, the proof consists of three nonequal parts:

1) The algebra L contains a largest locally nilpotent ideal Loc(L) such that Loc(L/Loc(L)) = 0.

2) If L ≠ 0 then L contains a nonzero element of 2nd order (that is such an element a that $ad(a)^2 = 0$).

3) Any element of 2nd order of L lies in Loc(L).

Unlike the case of associative algebras where the third part is prac-ically for free, for Lie algebras it is indeed the most difficult com-onent of the proof.

It would be interesting to find an explicit estimate N(m,p) for he nilpotency degree of m-generated Lie algebras over the field $\mathbb{Z}_p$ hich satisfy the identity $ad(x)^n = 0$, $n < p$.

Usually proofs based on the locally nilpotent radical (such as the roof of J. Levitzki or the proof of A.I. Kostrikin) are called ineffec-ive. This is not quite correct. The locally nilpotent radical is an ffective tool from the point of view of Recursive Function Theory, as hown by the following obvious proposition.

Proposition. Let $\mathfrak{m}$ be a variety of linear algebras over a field Φ uch that for any ideal I in some algebra $A \in \mathfrak{m}$, its cube I^3 is also an deal in A. Then the following assertions are equivalent:

1) Any algebra $A \in \mathfrak{m}$ contains a largest locally nilpotent ideal oc(A) such that Loc(A/Loc(A)) = 0.

2) Let $A \in \mathfrak{m}$ be an m-generated algebra, and let I be an ideal of A uch that $A^k \subseteq I$. Then the algebra I is generated by $\leq f(m,k)$ elements.

The function f(m,k) depends on $\mathfrak{m}$ and usually is a primitive recur-ive function. Now it remains to replace the "ineffective" Assertion 1 n such a proof by its "effective" version 2. Thus the function N(m,p) ay be estimated from above by a primitive recursive function. However, lthough fine from the Recursive Function Theory point of view, this stimate is absolutely unreasonable from the point of view of Common ense.

Problem 1. Find a reasonable estimate for N(m,p).

S.I. Adian and N.N. Repin [10] proved that for sufficiently big rimes p, one has $N(2,p) \geq 2^{p/15}$.

The author feels that before attacking Problem 1 it might be na-ural to turn to associative nil algebras since the latter seem to be uch more agreable objects than their Lie analogs, especially as far as urosch-like Problems are concerned.

Problem 2. Find a reasonable estimate for the nilpotency degree $_{ass}(m,n)$ of the associative m-generated ring which satisfies the dentity $x^n = 0$.

If our associative nil ring has no additive $n!$ -torsion, then $N(m,n) \leq n^2$ as shown by Yu.P. Razmyslov [11]. However, bearing Problem 1 in mind, we are mostly interested in rings without such restrictions on the additive structure.

The BRP for arbitrary n, as well as the Kurosch-like Problem for nil Lie algebras, remains open. However unlike the prime exponent case, even for $n = p^s$ (p prime), RBP doesn't immediately follow from a positive solution of the Kurosch-like Problem for Lie algebras. Indeed, let G be a group of exponent p^s and let $L(G) = \bigoplus_{i=1}^{\infty} G_i/G_{i+1}$ be its adjoint Lie algebra. Then L(G) (generally speaking) is not known to satisfy any identity $ad(x)^k = 0$. However:

1) there exists an integer $s > 1$ such that for any homogeneous element $a = a_i G_{i+1} \in L(G)$, $a_i \in G_i$, the equality $ad(a)^s = 0$ holds (H. Zassenhaus [12], O. Grün [13], I.N. Sanov [14]), and

2) L(G) satisfies some polynomial identity (G. Higman [15]).

Thus for the positive solution of RBP for $n = p^s$ if is sufficient to do for Lie algebras what has been done by A.I. Shirshov for associative algebras.

<u>Problem 3.</u> Suppose that a Lie algebra L over $\mathbb{Z}_p$ is generated by a subset M (not a subspace!) which is closed with respect to commutation. Assume further that (1) L satisfies some polynomial identity, and (2) there exists an integer $s > 1$ such that for any element $a \in M$ the equality $ad(a)^s = 0$ holds.

Is it true that L is locally nilpotent?

One might notice that L(G) is actually an algebra over $\mathbb{Z}/(p^s)$, not over $\mathbb{Z}/(p)$, however this is a minor obstacle.

If we agree that the final classification of finite simple groups has been established, then to solve the RBP for arbitrary n it would remain to solve it for $n = p^s$, where p is prime, and the latter problem has been reduced to the Problem 3 on Lie algebras.

1.3. Nonassociative algebras

In this chapter we consider two important classes of nonassociative algebras: the variety of alternative algebras and the variety of Jordan algebras. Recall that an algebra A is called alternative if it satisfies the identities

(1) $x^2 \cdot y = x(xy)$

(2) $y.x^2 = (yx)x$

(which is equivalent to the condition that any 2-generated subalgebra of A is associative (E. Artin)). The basic examples of such algebras are the associative algebras and the octonions.

We call an algebra J over a ring of scalars $\Phi \ni 1/2$ a Jordan algebra if it satisfies the identities

(1) $xy = yx$

(2) $x^2(yx) = (x^2y)x$,

Jordan algebras first appeared as the algebras of observables in Quantum Mechanics; however their further development has been mostly due to their close connections with Lie algebras and groups, analysis, geometry etc. The basic examples are: 1) the Hermitian elements of an associative algebra with respect to some involution, 2) the Jordan algebra of a symmetric bilinear form, 3) the Hermitian 3×3 matrices over the octonions with respect to the standard involution.

The aim of the structure theories in the classes of alternative algebras and Jordan algebras is to reduce them back to the basic examples. It is amazing that the major obstacle in doing this (and an unavoidable one) was the solution of the Kurosh Problem in these classes. The pioneering work in this context was done in the 50's by A.I. Shirshov. He generalized his combinatorial method to nonassociative algebras and proved that any alternative (resp., special Jordan) nil algebra with PI is locally nilpotent. The restriction of speciality for Jordan algebras was dropped by the author, however again by the "locally nilpotent radical + elements of the 2nd order" method, with the strong impact of A.I. Kostrikin's work [7]. Note that the locally nilpotent radical in Jordan algebras was studied by K.A.Ževlakov [16].

§2. Global Nilpotency of nil algebras

With the positive solutions of the Kurosch-like Problems in various classes of algebras there naturally arose the question whether these results were essentially local. In other words: is it true that an algebra with $x^n = 0$ is not only locally but also globally nilpotent? Rather soon it became clear that the answer depended on the additive structure of the algebra. Indeed, there are nonnilpotent $\mathbb{Z}_p$-algebras which satisfy the identity $x^p = 0$: consider for example the algebra of truncated polynomials in a countable set of variables $\mathbb{Z}_p[x_i, i \geq 1]/(x_i^p, i \geq 1)$. However Ya.S. Dubnov, V.K. Ivanov ([17], 1943)* and independently M.

Nagata ([18], 1952) and G. Higman ([19], 1956) proved that if an associative ring has no elements of additive order $\leq n$ and satisfies the identity $x^n=0$, then it is nilpotent. As for the nilpotency degree $N_{ass}(n)$, Yu.P. Razmyslov [11] proved that $N_{ass}(n) < n^2$, while E.N. Kuz'min [20] proved that $N_{ass}(n) \geq \frac{n(n+1)}{2}$, and conjectured equality holds.

K.A. Ževlakov extended the theorem stated above to the class of alternative algebras with $x^n = 0$ [21].

For Lie rings the problem reads as follows: is it true that a Lie ring without nonzero elements of additive order $\leq n$, and which satisfies the identity $ad(x)^n = 0$, is nilpotent? (cf. [7]). If this problem had a positive solution then the estimate obtained for the nilpotency degree would also serve as a uniform bound for the nilpotency degree of any nilpotent p-group, without regard to the number of generators. However in 1971 Yu.P. Razmyslov [23] constructed a nonnilpotent Lie algebra over $\mathbb{Z}_p$ which satisfies the identity $ad(x)^{p-2} = 0$ for any prime integer $p \geq 5$. There remained the conjecture (cf. [22]) that $ad(x)^n = 0$ still imples nilpotency under more severe restrictions on the characteristic, for instance, when the characteristic is zero.

<u>Problem A</u> (cf. A.I. Kostrikin [22]). Is it true that a Lie algebra over a field of zero characteristic which satisfies $ad(x)^n = 0$ is nilpotent?

For Jordan algebras with $x^n = 0$ one should speak not of nilpotency but of solvability because there is an array of solvable nonnilpotent Jordan algebras (cf. G.V. Dorofeev, [24]).

<u>Problem B</u> (A.I. Shirshov, [25]). Is it true that a Jordan algebra over a field of zero characteristic which satisfies $x^n = 0$ is solvable?

In what follows all algebras are considered over a field Φ of zero characteristic.

In the most general setup, the problem of Global Nilpotency of linear algebras may be formulated as follows:

Let m be a variety of linear algebras, A_m its free algebra of countable rank, $f(x_1,\dots,x_n)\in A_m$. Let us consider the set of all values $f(A)$ of f in A and the smallest ideal $I_f(A) = Id_A(f(A))$ in A which contains $f(A)$; $I_f(A)$ is the set of values assumed on A by the T-ideal I_f in A_m generated by f, so $I_f(A)$ is nilpotent for all A precisely when I_f is nilpotent in A_m.

* This work was totally overlooked by specialists in Ring Theory. Its discovery for this group of algebraists is due to E. Formanek.

Global Nilpotency Problem. Is it true that I_f is nilpotent?

Recall that an ideal I of a nonassociative algebra A is called nilpotent (of degree m) if any product of elements of A which includes at least m factors from I vanishes.

This rather general setup includes one more interesting problem, which at first sight has nothing to do with Kurosch-like Problems. Denote by Alt[X] the free alternative algebra of countable rank, I.P. Shestakov [26] and M. Slater [27] proved that its radical J(Alt[X]) (quasiregular- = nil- = locally nilpotent- = Baer-radical) consists of those elements which vanish identically in all associative algebras and octonions. If it were our aim to consider the variety generated just by these two basic classes of algebras we should find that it is defined by the identities ($f=0$, $f\in J(\mathrm{Alt}[X])$. Thus J(Alt[X]) is the price we pay for the conciseness of the definition of alternative algebras.

Problem C (K.A. Ževlakov). Is it true that J(Alt[X]) is nilpotent?

A.V. Il'tyakov proved that for a certain finite collection of elements $f_1,\ldots,f_s\in J(\mathrm{Alt}[X])$, it holds that $J(\mathrm{Alt}[X]) = I_{f_1}+\ldots+I_{f_s}$. Hence for the solution of Problem C if suffices to prove the nilpotency of the ideal I_f, $f\in J(\mathrm{Alt}[X])$ which is clearly a Global Nilpotency Problem.

Our aim now is to explain the approach to General Nilpotency Problems through superalgebras. This approach proved to be helpful for the solutions of problems A, B, C (the latter one in joint work by I.P. Shestakov and the author).

Let $X = \{x_i,\ i \geq 1\}$ be a set of free generators of A_m, and set $\Lambda = \{x_{i_1},\ldots,x_{i_m}\}\subset X$, $i_1<\ldots<i_m$. Any permutation $\sigma\in S_m$ induces the automorphism $\hat{\sigma}$ of A : $x_{i_k}\to x_{i_{\sigma(k)}}$, $x_j\to x_j$ for $x_j\notin\Lambda$. Denote by $H_\Lambda = \sum_{\sigma\in S_m}\hat{\sigma}$ the symmetrization in Λ and by $S_\Lambda = \sum_{\sigma\in S_m}(-1)^\sigma\hat{\sigma}$ the skew-symmetrization in Λ.

The following lemma is an obvious corollary of the Young Diagram Method [28].

Lemma 2.1. Let $h(\ldots x_1\ldots x_m\ldots)$ be a nonzero element of A_m which is multilinear in $x_1,\ldots,x_m$. There exist a subset $\Lambda = \{x_{i_1},\ldots,x_{i_t}\}\subset \{x_1,\ldots,x_m\}$, $t\geq[\sqrt{m}]$ and a permutation $\tau\in S_m$ such that either $H_\Lambda h^\tau\neq 0$ or $S_\Lambda h^\tau\neq 0$.

Suppose that

$$\ldots f(a_{11},\ldots,a_{1n})\ldots f(a_{d1},\ldots,a_{dn})\ldots \neq 0$$

with some brackets, where the dots indicate products (and parentheses) with other variables.

The product is multilinear in $a_{11},\ldots,a_{d1}$. From Lemma 2.1 it follows that for some elements b_{ij}, $1 \leq i \leq t = [\sqrt{d}]$, $1 \leq j \leq n$, either

$$\sum_{\sigma \in S_t} \ldots f(b_{\sigma(1)1},b_{12},\ldots,b_{1n})\ldots f(b_{\sigma(t)1},\ldots,b_{tn})\ldots \neq 0 \text{ or}$$

$$\sum_{\sigma \in S_t} (-1)^{\sigma}\ldots f(b_{\sigma(1)1},b_{12},\ldots,b_{1n})\ldots f(b_{\sigma(t)1},\ldots,b_{tn})\ldots \neq 0$$

This nonzero (skew)symmetrization is multilinear in $b_{12},\ldots,b_{t2}$ and again we may apply Lemma 2.1. After n applications of Lemma 2.1 we get a system of elements $c_{ij} \in A_m$, $1 \leq i \leq m$, $1 \leq j \leq n$, such that

$$Q_1\ldots Q_n(\ldots f(c_{11},\ldots,c_{1n})\ldots f(c_{m1},\ldots,c_{mn})\ldots) \neq 0$$

where Q_j is either a symmetrization or a skew-symmetrization in $c_{1j},\ldots,c_{mj}$.

If Q_j is a symmetrization then replace each of the elements $c_{1j},\ldots,c_{mj}$ by a free generator x_j which has not been used previously. The resulting expression is evidently nonzero. But what is one supposed to do when Q_j is a skew-symmetrization? In this situation A.R. Kemer made the move which in any chess notation would deserve the !!-mark.

Kemer's Super-Trick. If the expression $h(c_1,\ldots,c_k,\ldots)$ is multilinear and skew-symmetric in $c_1,\ldots,c_k$ then replace each of the elements $c_1,\ldots,c_k$ by the sum $c = \sum_{i=1}^{k} c_i \otimes e_i$, where $e_1,\ldots,e_k$ are anticommuting Grassmann variables from the Grassmann algebra G on an infinite dimensional vector space. Then the resulting expression is nonzero, for

$$h(\Sigma c_i \otimes e_i,\ldots,\Sigma c_i \otimes e_i,\ldots) = \Sigma(-1)^{\sigma} h(c_{\sigma(1)},\ldots,c_{\sigma(k)},\ldots)\otimes e_1\ldots e_k =$$
$$= k!h(c_1,\ldots,c_k,\ldots)\otimes e_1\ldots e_k \neq 0$$

This Trick brings superalgebras into consideration, since $A \otimes G$ is a superalgebra.

Assume that $Q_1,\ldots,Q_r$ are skew-symmetrizations while $Q_{r+1},\ldots,Q_n$ are symmetrizations, and let $\{e_{ij}, 1 \leq i, 1 \leq j \leq r\}$ be r commuting systems of Grassmann variables, that is

$e_{ij}e_{kj} + e_{kj}e_{ij} = 0$, $e_{ij}e_{pq} = e_{pq}e_{ij}$ for $q \neq j$; $G_j = \langle e_{ij}, i \geq 1\rangle$ is the j-th Grassmann algebra.

Consider the element

$\hat{f} = f(\sum_i c_{i1} \otimes e_{i1}, \sum_i c_{i2} \otimes e_{i2}, \ldots, \sum_i c_{ir} \otimes e_{ir}, x_{r+1}, \ldots, x_n) \in$

$\in A_m \otimes G_1 \otimes \ldots \otimes G_r$.

Then $\ldots\hat{f}\ldots\hat{f}\ldots\hat{f}\ldots \neq 0$ since its component $A_m \otimes e_{11} \ldots e_{r1} \ldots e_{1r} \ldots e_{nr}$ is a nonzero multiple of the original $Q_1 \ldots Q_n(\ldots f \ldots f \ldots)$.

Now to verify that the ideal I_f is nilpotent it is sufficient to prove that $\hat{f}$ generates a nilpotent ideal in $A_m \otimes G_1 \otimes \ldots \otimes G_r$. However the nilpotency degree of this ideal should not depend on the number of summands in $\sum_i c_{ij} \otimes e_{ij}$, otherwise our argument ends up with nothing. Thus we should substitute the infinite series $\sum_{i=1}^{\infty} c_{ij} \otimes e_{ij}$ into f and prove that the resulting element generates a nilpotent ideal. Now it is time to turn to the appropriate formalism.

Denote by $\mathbb{Z}_2^\infty$ the direct sum of countably many copies of $\mathbb{Z}_2$. By a chromatic superalgebra we mean any $\mathbb{Z}_2^\infty$-graded algebra $A = \oplus A_\alpha$, $\alpha \in \mathbb{Z}_2^\infty$.

Example. The Grassmann algebra G is equipped with the natural $\mathbb{Z}_2$-grading. Denote by $G^{\otimes n}$ its n-th tensor power, $G \subset G^{\otimes 2} \subset \ldots \subset G^{\otimes n} \subset \ldots$ Then

$G^\infty = \bigcup_{n=1}^{\infty} G^{\otimes n}$ is the natural example of a chromatic superalgebra,

$G^\infty = \cup G_\alpha^\infty$, $\alpha \in \mathbb{Z}_2^\infty$.

For an arbitrary chromatic superalgebra $A = \oplus A_\alpha$ consider the tensor product $A \otimes G^\infty$ and its subalgebra

$$G(A) = \oplus A_\alpha \otimes G_\alpha^\infty, \quad \alpha \in \mathbb{Z}_2^\infty.$$

We call G(A) the Grassmann envelope of the superalgebra A.

Definition. Let m be a variety of linear algebras. We say that a chromatic superalgebra A is an m-superalgebra if $G(A) \in m$.

The chromatic m-superalgebras form a variety in the sense of [29]. Consider the free chromatic m-superalgebra $S = \oplus S_\alpha$, with $x_{\alpha i}$, $i \geq 1$, being its free generators of weight α, $\alpha \in \mathbb{Z}_2^\infty$. It was shown above that to verify the nilpotency of the ideal I_f it is sufficient to prove that the element $f(x_{\alpha_1 1}, \ldots, x_{\alpha_r 1}, x_{01}, \ldots, x_{0n})$, where $\alpha_i = (0 \ldots \underset{i}{1}\ 0 \ldots) \in \mathbb{Z}_2^\infty$, generates a nilpotent ideal in S. In this connection there arises the following problem.

Prime Chromatic Superalgebras Problem. Let $A = \bigoplus_{\alpha} A_{\alpha}$ be a prime chromatic m-superalgebra. Is if true that $f = 0$ holds identically in $G(A)$?

With the help of the so called discoloration functor this problem may be reduced to the similar problem on the ordinary (not chromatic) superalgebras.

Prime Superalgebras Problem. Let $A = A_0 + A_1$ be a prime m-superalgebra. Is it true that $f = 0$ holds identically in $G(A) = A_0 \otimes G_0 + A_1 \otimes G_1$?

This problem is the core of our approach to Global Nilpotency. First the Global Nilpotency Problem is supposed to be reduced to the Prime Superalgebras Problem. We don't claim that this reduction is always possible. However, by essentially the same argument we succeeded in doing it for Problems A, B, C. Next we attack the Prime Superalgebras Problem which is local by its nature.

Let m be the variety of Lie algebras which satisfy the identity $ad(x)^n = 0$ or the variety of Jordan algebras which satisfy the identity $x^n = 0$. The positive solution of the Prime Superalgebras Problem for m results from the following theorem.

Theorem ([30], [31]). There are no nonzero prime m-superalgebras.

For Lie algebras the proof is essentially based on A.I. Kostrikin's theorem from [7] (for the detailed exposition cf. also [9]).

Theorem (A.I. Kostrikin). There are no nonzero prime Lie algebras in m.

As for Problem C and the related Prime Superalgebras Problem, I.P. Shestakov and the author obtained a complete classification of the prime alternative superalgebras of characteristic $\neq 3$.

Theorem (I.P. Shestakov, E.I. Zel'manov). Any prime alternative superalgebra $A = A_0 + A_1$ over a field of characteristic $\neq 3$ is either associative or $A_1 = (0)$ while A_0 is a Cayley-Dickson ring.

Finally, we formulate the theorems on Global Nilpotency which follow from these assertions. The characteristic is assumed to be zero.

Theorem [30]. Any Lie algebra which satisfies the identity $ad(x)^n=0$ is nilpotent.

Theorem [31]. Any Jordan algebra which satisfies the identity $x^n=0$ is solvable.

Theorem (I.P. Shestakov, E.I. Zel'manov). J(Alt[X]) is nilpotent.

References

1. A.G. Kurosh. Problems in ring theory which are related to the Burnside problem on periodic groups (in Russian). Izv. Akad. Nauk SSSR, Ser. Mat., 5 (1941), 233-240.
2. E.S. Golod. On nil-algebras and finitely approximable p-groups (in Russian). Izv. Akad. Nauk SSSR, Ser. Mat., 28 (1964), 273-276.
3. P.S. Novikov, S.I. Adian. On infinite periodic groups. I, II, III (in Russian). Izv. Akad. Nauk SSSR, Ser. Mat., 32 (1968), 212-244, 251-524, 709-731; (English translation: Math. USSR, Izv., 2 (1968), 209-236, 241-480, 665-685 (1969)).
4. J. Levitzki. On the radical of a general ring. Bull. Am. Math. Soc., 49 (1943), 462-466.
5. A.I. Shirshov. On rings with identical relations (in Russian). Mat. Sb., 43 (85), (1957), 277-283.
6. W. Magnus, A connection between the Baker-Hausdorff formula and a problem of Burnside. Ann. Math., 52(1950), 111-126; 57 (1953), 606.
7. A.I. Kostrikin. On the Burnside problem (in Russian). Izv. Akad. Nauk SSSR, Ser. Mat., 23 (1959), 3-34.
8. A.I. Kostrikin. Sandwiches in Lie algebras (in Russian). Mat. Sb., 110 (152), (1979), 3-12; (English translation: Math. USSR, Sb., 38 (1981), 1-9).
9. A.I. Kostrikin. Around Burnside (in Russian). "Nauka"; Moscow, 1986.
10. S.I. Adian, N.N. Repin. Exponential lower estimate for the degree of nilpotency of Engel Lie algebras (in Russian). Mat. Zametki, 39 (1986), 444-452; (English translation: Math. Notes 39 (1986), 244-249).
11. Yu.P. Razmyslov. Trace identities of full matrix algebras over a field of characteristic zero (in Russian). Izv. Akad. Nauk SSSR, Ser. Mat., 38 (1974), 723-756. (English translation: Math. USSR, Izv., 8 (1974), 727-760 (1975)).
12. H. Zassenhaus. Über Liesche Ringe mit Primzahlcharacteristic. Abh. Math. Semin. Univ. Hamb., 13 (1939), 1-100.
13. O. Grün. Zusammenhang zwischen Potenzbildung und Kommutatorbildung. J. Reine Angew. Math., 182 (1940), 158-177.
14. I.N. Sanov. On a system of relations in periodic groups whose exponent is a power of a prime integer (in Russian). Izv. Akad. Nauk SSSR, Ser., Mat., 15 (1951), 477-502.
15. G. Higman. Lie ring methods in the theory of finite nilpotent groups. Proc. Intern. Congr. Math., Edinburgh, 1958, 307-312.

16. K.A.Ževlakov. Solvability and nilpotency of Jordan rings (in Russian). Algebra Logika, 5 (1966), 37-58.

17. Ya.S. Dubnov, V.K. Ivanov. On the reduction of degrees of affinor polynomials (in Russian). Dokl. Akad. Nauk SSSR, 41 (1943), 99-102.

18. M. Nagata. On the nilpotency of nil algebras. J. Math. Soc. Japan, 4 (1952), 296-301.

19. G. Higman. On a conjecture of Nagata. Math. Proc. Camb. Philos. Soc., 52 (1956), 1-4.

20. E.N. Kuz'min. On a theorem of Nagata-Higman (in Russian). In: Mathematical Structures. Computational mathematics. Mathematical modelling. Bulgarian Academy of Sciences, Sofia, 1975, pp. 101-107.

21. K.A.Ževlakov, A.M. Slin'ko, I.P. Shestakov, A.I. Shirshov. Rings that are nearly associative (in Russian). "Nauka"; Moscow, (English translation: Academic Press; New York, London, 1982).

22. A.I. Kostrikin. Lie algebras and finite groups (in Russian). Tr. Mat. Inst. Steklova, 168 (1984), 132-154; (English translation: Proc. Steklov Inst. Math., 168 (1986), 137-159).

23. Yu.P. Razmyslov. On the algebras satisfying the Engel condition (in Russian). Algebra Logika, 10 (1971), 33-44; (English translation: Algebra Logic, 10 (1971), 21-29 (1972)).

24. G.V. Dorofeev. An example of a solvable alternative ring which is not nilpotent (in Russian). Usp. Mat. Nauk, 15 (1960), No. 3, 147-150.

25. Dnestrovskaya tetrad': Open problems in ring and module theory (in Russian). Novosibirsk, 1976.

26. I.P. Shestakov. Radicals and nilpotent elements of free alternative algebras (in Russian). Algebra Logika, 14 (1975), 354-365; (English translation: Algebra Logic, 14 (1975), 219-226 (1976)).

27. M. Slater. Prime alternative rings. I, II, III. J. Algebra, 15 (1970), 229-243, 244-251; 21 (1972), 394-409).

28. H. Weyl. The classical groups, their invariants and representations. Princeton Univ. Press, Princeton, N.J., 1946.

29. A.I. Mal'cev. Algebraic systems (in Russian). "Nauka"; Moscow, 1970; (English translation: Die Grundlehren der mathematischen Wissenschaften, Band 192, Springer-Verlag; Berlin, Heidelberg, New York, 1973).

30. E.I. Zel'manov. On Lie algebras with the Engel property (in Russian)- Dokl. Akad. Nauk SSSR, 292 (1987), 265-268; (English translation: to appear in Soviet Math. Doklady).

31. E.I. Zel'manov. On Jordan nil algebras of bounded degree (in Russian). Varieties of algebraic systems. Preprint No. 647. Computer Center. Novosibirsk, 1986.

Note added in proof. The author has shown that RBP does follow from a positive solution of the Kurosch-like Problem for Lie algebras (to appear in Mat, Sb., No 1, 1989), but in a rather roundabout way, and not at all immediately: cf. the discussion preceding Problem 3 in Section 1.

"This paper is in final form and no version of it will be submitted for publication elsewhere"

Lecture Notes aim to report new developments - quickly, informally and at a high level. The following describes criteria and procedures which apply to proceedings volumes. The editors of a volume are strongly advised to inform contributors about these points at an early stage.

§1. One (or more) expert participant(s) of the meeting should act as the responsible editor(s) of the proceedings. They select the papers which are suitable (cf. §§ 2, 3) for inclusion in the proceedings, and have them individually refereed (as for a journal). It should not be assumed that the published proceedings must reflect conference events faithfully and in their entirety. Contributions to the meeting which are not included in the proceedings can be listed by title. The series editors will normally not interfere with the editing of a particular proceedings volume - except in fairly obvious cases, or on technical matters, such as described in §§ 2, 3. The names of the responsible editors appear on the title page of the volume.

§2. The proceedings should be reasonably homogeneous (concerned with a limited area). For instance, the proceedings of a congress on "Analysis" or "Mathematics in Wonderland" would normally not be sufficiently homogeneous.

One or two longer survey articles on recent developments in the field are often very useful additions to such proceedings - even if they do not correspond to actual lectures at the congress. An extensive introduction on the subject of the congress would be desirable.

§3. The contributions should be of a high mathematical standard and of current interest. Research articles should present new material and not duplicate other papers already published or due to be published. They should contain sufficient information and motivation and they should present proofs, or at least outlines of such, in sufficient detail to enable an expert to complete them. Thus resumes and mere announcements of papers appearing elsewhere cannot be included, although more detailed versions of a contribution may well be published in other places later.

Surveys, if included, should cover a sufficiently broad topic, and should in general not simply review the author's own recent research. In the case of surveys, exceptionally, proofs of results may not be necessary.

"Mathematical Reviews" and "Zentralblatt für Mathematik" require that papers in proceedings volumes carry an explicit statement that they are in final form and that no similar paper has been or is being submitted elsewhere, if these papers are to be considered for a review. Normally, papers that satisfy the criteria of the Lecture Notes in Mathematics series also satisfy this

.../...

requirement, but we would strongly recommend that the contributing authors be asked to give this guarantee explicitly at the beginning or end of their paper. There will occasionally be cases where this does not apply but where, for special reasons, the paper is still acceptable for LNM.

§4. Proceedings should appear soon after the meeeting. The publisher should, therefore, receive the complete manuscript within nine months of the date of the meeting at the latest.

§5. Plans or proposals for proceedings volumes should be sent to one of the editors of the series or to Springer-Verlag Heidelberg. They should give sufficient information on the conference or symposium, and on the proposed proceedings. In particular, they should contain a list of the expected contributions with their prospective length. Abstracts or early versions (drafts) of some of the contributions are very helpful.

§6. Lecture Notes are printed by photo-offset from camera-ready typed copy provided by the editors. For this purpose Springer-Verlag provides editors with technical instructions for the preparation of manuscripts and these should be distributed to all contributing authors. Springer-Verlag can also, on request, supply stationery on which the prescribed typing area is outlined. Some homogeneity in the presentation of the contributions is desirable.

Careful preparation of manuscripts will help keep production time short and ensure a satisfactory appearance of the finished book. The actual production of a Lecture Notes volume normally takes 6 -8 weeks.

Manuscripts should be at least 100 pages long. The final version should include a table of contents and as far as applicable a subject index.

§7. Editors receive a total of 50 free copies of their volume for distribution to the contributing authors, but no royalties. (Unfortunately, no reprints of individual contributions can be supplied.) They are entitled to purchase further copies of their book for their personal use at a discount of 33.3 %, other Springer mathematics books at a discount of 20 % directly from Springer-Verlag. Contributing authors may purchase the volume in which their article appears at a discount of 33.3 %.

Commitment to publish is made by letter of intent rather than by signing a formal contract. Springer-Verlag secures the copyright for each volume.

LECTURE NOTES

ESSENTIALS FOR THE PREPARATION OF CAMERA-READY MANUSCRIPTS

Springer
Springer-Verlag
Berlin Heidelberg New York
London Paris Tokyo Hong Kong

The preparation of manuscripts which are to be reproduced by photo-offset require special care. Manuscripts which are submitted in technically unsuitable form will be returned to the author for retyping. There is normally no possibility of carrying out further corrections after a manuscript is given to production. Hence it is crucial that the following instructions be adhered to closely. If in doubt, please send us 1 - 2 sample pages for examination.

General. The characters must be uniformly black both within a single character and down the page. Original manuscripts are required: photocopies are acceptable only if they are sharp and without smudges.

On request, Springer-Verlag will supply special paper with the text area outlined. The standard TEXT AREA (OUTPUT SIZE if you are using a 14 point font) is 18 x 26.5 cm (7.5 x 11 inches). This will be scale-reduced to 75% in the printing process. If you are using computer typesetting, please see also the following page.

Make sure the TEXT AREA IS COMPLETELY FILLED. Set the margins so that they precisely match the outline and type right from the top to the bottom line. (Note that the page number will lie outside this area). Lines of text should not end more than three spaces inside or outside the right margin (see example on page 4).

Type on one side of the paper only.

Spacing and Headings (Monographs). Use ONE-AND-A-HALF line spacing in the text. Please leave sufficient space for the title to stand out clearly and do NOT use a new page for the beginning of subdivisons of chapters. Leave THREE LINES blank above and TWO below headings of such subdivisions.

Spacing and Headings (Proceedings). Use ONE-AND-A-HALF line spacing in the text. Do not use a new page for the beginning of subdivisons of a single paper. Leave THREE LINES blank above and TWO below headings of such subdivisions. Make sure headings of equal importance are in the same form.

The first page of each contribution should be prepared in the same way. The title should stand out clearly. We therefore recommend that the editor prepare a sample page and pass it on to the authors together with these instructions. Please take the following as an example. Begin heading 2 cm below upper edge of text area.

MATHEMATICAL STRUCTURE IN QUANTUM FIELD THEORY

John E. Robert
Mathematisches Institut, Universität Heidelberg
Im Neuenheimer Feld 288, D-6900 Heidelberg

Please leave THREE LINES blank below heading and address of the author, then continue with the actual text on the same page.

Footnotes. These should preferable be avoided. If necessary, type them in SINGLE LINE SPACING to finish exactly on the outline, and separate them from the preceding main text by a line.

Symbols. Anything which cannot be typed may be entered by hand in BLACK AND ONLY BLACK ink. (A fine-tipped rapidograph is suitable for this purpose; a good black ball-point will do, but a pencil will not). Do not draw straight lines by hand without a ruler (not even in fractions).

Literature References. These should be placed at the end of each paper or chapter, or at the end of the work, as desired. Type them with single line spacing and start each reference on a new line. Follow "Zentralblatt für Mathematik"/"Mathematical Reviews" for abbreviated titles of mathematical journals and "Bibliographic Guide for Editors and Authors (BGEA)" for chemical, biological, and physics journals. Please ensure that all references are COMPLETE and ACCURATE.

IMPORTANT

Pagination. For typescript, number pages in the upper right-hand corner in LIGHT BLUE OR GREEN PENCIL ONLY. The printers will insert the final page numbers. For computer type, you may insert page numbers (1 cm above outer edge of text area).

It is safer to number pages AFTER the text has been typed and corrected. Page 1 (Arabic) should be THE FIRST PAGE OF THE ACTUAL TEXT. The Roman pagination (table of contents, preface, abstract, acknowledgements, brief introductions, etc.) will be done by Springer-Verlag.

If including running heads, these should be aligned with the inside edge of the text area while the page number is aligned with the outside edge noting that right-hand pages are odd-numbered. Running heads and page numbers appear on the same line. Normally, the running head on the left-hand page is the chapter heading and that on the right-hand page is the section heading. Running heads should not be included in proceedings contributions unless this is being done consistently by all authors.

Corrections. When corrections have to be made, cut the new text to fit and paste it over the old. White correction fluid may also be used.

Never make corrections or insertions in the text by hand.

If the typescript has to be marked for any reason, e.g. for provisional page numbers or to mark corrections for the typist, this can be done VERY FAINTLY with BLUE or GREEN PENCIL but NO OTHER COLOR: these colors do not appear after reproduction.

COMPUTER-TYPESETTING. Further, to the above instructions, please note with respect to your printout that
- the characters should be sharp and sufficiently black;
- it is not strictly necessary to use Springer's special typing paper. Any white paper of reasonable quality is acceptable.

If you are using a significantly different font size, you should modify the output size correspondingly, keeping length to breadth ratio 1 : 0.68, so that scaling down to 10 point font size, yields a text area of 13.5 x 20 cm (5 3/8 x 8 in), e.g.

Differential equations.: use output size 13.5 x 20 cm.

Differential equations.: use output size 16 x 23.5 cm.

Differential equations.: use output size 18 x 26.5 cm.

Interline spacing: 5.5 mm base-to-base for 14 point characters (standard format of 18 x 26.5 cm).
If in any doubt, please send us 1 - 2 sample pages for examination. We will be glad to give advice.

Vol. 1201: Curvature and Topology of Riemannian Manifolds. Proceedings, 1985. Edited by K. Shiohama, T. Sakai and T. Sunada. VII, 336 pages. 1986.

Vol. 1202: A. Dür, Möbius Functions, Incidence Algebras and Power Series Representations. XI, 134 pages. 1986.

Vol. 1203: Stochastic Processes and Their Applications. Proceedings, 1985. Edited by K. Itô and T. Hida. VI, 222 pages. 1986.

Vol. 1204: Séminaire de Probabilités XX, 1984/85. Proceedings. Edité par J. Azéma et M. Yor. V, 639 pages. 1986.

Vol. 1205: B.Z. Moroz, Analytic Arithmetic in Algebraic Number Fields. VII, 177 pages. 1986.

Vol. 1206: Probability and Analysis, Varenna (Como) 1985. Seminar. Edited by G. Letta and M. Pratelli. VIII, 280 pages. 1986.

Vol. 1207: P.H. Bérard, Spectral Geometry: Direct and Inverse Problems. With an Appendix by G. Besson. XIII, 272 pages. 1986.

Vol. 1208: S. Kaijser, J.W. Pelletier, Interpolation Functors and Duality. IV, 167 pages. 1986.

Vol. 1209: Differential Geometry, Peñíscola 1985. Proceedings. Edited by A.M. Naveira, A. Ferrández and F. Mascaró. VIII, 306 pages. 1986.

Vol. 1210: Probability Measures on Groups VIII. Proceedings, 1985. Edited by H. Heyer. X, 386 pages. 1986.

Vol. 1211: M.B. Sevryuk, Reversible Systems. V, 319 pages. 1986.

Vol. 1212: Stochastic Spatial Processes. Proceedings, 1984. Edited by P. Tautu. VIII, 311 pages. 1986.

Vol. 1213: L.G. Lewis, Jr., J.P. May, M. Steinberger, Equivariant Stable Homotopy Theory. IX, 538 pages. 1986.

Vol. 1214: Global Analysis – Studies and Applications II. Edited by Yu.G. Borisovich and Yu.E. Gliklikh. V, 275 pages. 1986.

Vol. 1215: Lectures in Probability and Statistics. Edited by G. del Pino and R. Rebolledo. V, 491 pages. 1986.

Vol. 1216: J. Kogan, Bifurcation of Extremals in Optimal Control. VIII, 106 pages. 1986.

Vol. 1217: Transformation Groups. Proceedings, 1985. Edited by S. Jackowski and K. Pawalowski. X, 396 pages. 1986.

Vol. 1218: Schrödinger Operators, Aarhus 1985. Seminar. Edited by E. Balslev. V, 222 pages. 1986.

Vol. 1219: R. Weissauer, Stabile Modulformen und Eisensteinreihen. III, 147 Seiten. 1986.

Vol. 1220: Séminaire d'Algèbre Paul Dubreil et Marie-Paule Malliavin. Proceedings, 1985. Edité par M.-P. Malliavin. IV, 200 pages. 1986.

Vol. 1221: Probability and Banach Spaces. Proceedings, 1985. Edited by J. Bastero and M. San Miguel. XI, 222 pages. 1986.

Vol. 1222: A. Katok, J.-M. Strelcyn, with the collaboration of F. Ledrappier and F. Przytycki, Invariant Manifolds, Entropy and Billiards; Smooth Maps with Singularities. VIII, 283 pages. 1986.

Vol. 1223: Differential Equations in Banach Spaces. Proceedings, 1985. Edited by A. Favini and E. Obrecht. VIII, 299 pages. 1986.

Vol. 1224: Nonlinear Diffusion Problems, Montecatini Terme 1985. Seminar. Edited by A. Fasano and M. Primicerio. VIII, 188 pages. 1986.

Vol. 1225: Inverse Problems, Montecatini Terme 1986. Seminar. Edited by G. Talenti. VIII, 204 pages. 1986.

Vol. 1226: A. Buium, Differential Function Fields and Moduli of Algebraic Varieties. IX, 146 pages. 1986.

Vol. 1227: H. Helson, The Spectral Theorem. VI, 104 pages. 1986.

Vol. 1228: Multigrid Methods II. Proceedings, 1985. Edited by W. Hackbusch and U. Trottenberg. VI, 336 pages. 1986.

Vol. 1229: O. Bratteli, Derivations, Dissipations and Group Actions on C*-algebras. IV, 277 pages. 1986.

Vol. 1230: Numerical Analysis. Proceedings, 1984. Edited by J.-P. Hennart. X, 234 pages. 1986.

Vol. 1231: E.-U. Gekeler, Drinfeld Modular Curves. XIV, 107 pages. 1986.

Vol. 1232: P.C. Schuur, Asymptotic Analysis of Soliton Problems. VIII, 180 pages. 1986.

Vol. 1233: Stability Problems for Stochastic Models. Proceedings, 1985. Edited by V.V. Kalashnikov, B. Penkov and V.M. Zolotarev. VI, 223 pages. 1986.

Vol. 1234: Combinatoire énumérative. Proceedings, 1985. Edité par G. Labelle et P. Leroux. XIV, 387 pages. 1986.

Vol. 1235: Séminaire de Théorie du Potentiel, Paris, No. 8. Directeurs: M. Brelot, G. Choquet et J. Deny. Rédacteurs: F. Hirsch et G. Mokobodzki. III, 209 pages. 1987.

Vol. 1236: Stochastic Partial Differential Equations and Applications. Proceedings, 1985. Edited by G. Da Prato and L. Tubaro. V, 257 pages. 1987.

Vol. 1237: Rational Approximation and its Applications in Mathematics and Physics. Proceedings, 1985. Edited by J. Gilewicz, M. Pindor and W. Siemaszko. XII, 350 pages. 1987.

Vol. 1238: M. Holz, K.-P. Podewski and K. Steffens, Injective Choice Functions. VI, 183 pages. 1987.

Vol. 1239: P. Vojta, Diophantine Approximations and Value Distribution Theory. X, 132 pages. 1987.

Vol. 1240: Number Theory, New York 1984–85. Seminar. Edited by D.V. Chudnovsky, G.V. Chudnovsky, H. Cohn and M.B. Nathanson. V, 324 pages. 1987.

Vol. 1241: L. Gårding, Singularities in Linear Wave Propagation. III, 125 pages. 1987.

Vol. 1242: Functional Analysis II, with Contributions by J. Hoffmann-Jørgensen et al. Edited by S. Kurepa, H. Kraljević and D. Butković. VII, 432 pages. 1987.

Vol. 1243: Non Commutative Harmonic Analysis and Lie Groups. Proceedings, 1985. Edited by J. Carmona, P. Delorme and M. Vergne. V, 309 pages. 1987.

Vol. 1244: W. Müller, Manifolds with Cusps of Rank One. XI, 158 pages. 1987.

Vol. 1245: S. Rallis, L-Functions and the Oscillator Representation. XVI, 239 pages. 1987.

Vol. 1246: Hodge Theory. Proceedings, 1985. Edited by E. Cattani, F. Guillén, A. Kaplan and F. Puerta. VII, 175 pages. 1987.

Vol. 1247: Séminaire de Probabilités XXI. Proceedings. Edité par J. Azéma, P.A. Meyer et M. Yor. IV, 579 pages. 1987.

Vol. 1248: Nonlinear Semigroups, Partial Differential Equations and Attractors. Proceedings, 1985. Edited by T.L. Gill and W.W. Zachary. IX, 185 pages. 1987.

Vol. 1249: I. van den Berg, Nonstandard Asymptotic Analysis. IX, 187 pages. 1987.

Vol. 1250: Stochastic Processes – Mathematics and Physics II. Proceedings 1985. Edited by S. Albeverio, Ph. Blanchard and L. Streit. VI, 359 pages. 1987.

Vol. 1251: Differential Geometric Methods in Mathematical Physics. Proceedings, 1985. Edited by P.L. García and A. Pérez-Rendón. VII, 300 pages. 1987.

Vol. 1252: T. Kaise, Représentations de Weil et GL_2 Algèbres de division et GL_n. VII, 203 pages. 1987.

Vol. 1253: J. Fischer, An Approach to the Selberg Trace Formula via the Selberg Zeta-Function. III, 184 pages. 1987.

Vol. 1254: S. Gelbart, I. Piatetski-Shapiro, S. Rallis. Explicit Constructions of Automorphic L-Functions. VI, 152 pages. 1987.

Vol. 1255: Differential Geometry and Differential Equations. Proceedings, 1985. Edited by C. Gu, M. Berger and R.L. Bryant. XII, 243 pages. 1987.

Vol. 1256: Pseudo-Differential Operators. Proceedings, 1986. Edited by H.O. Cordes, B. Gramsch and H. Widom. X, 479 pages. 1987.

Vol. 1257: X. Wang, On the C*-Algebras of Foliations in the Plane. V, 165 pages. 1987.

Vol. 1258: J. Weidmann, Spectral Theory of Ordinary Differential Operators. VI, 303 pages. 1987.

Vol. 1259: F. Cano Torres, Desingularization Strategies for Three-Dimensional Vector Fields. IX, 189 pages. 1987.

Vol. 1260: N.H. Pavel, Nonlinear Evolution Operators and Semi-groups. VI, 285 pages. 1987.

Vol. 1261: H. Abels, Finite Presentability of S-Arithmetic Groups. Compact Presentability of Solvable Groups. VI, 178 pages. 1987.

Vol. 1262: E. Hlawka (Hrsg.), Zahlentheoretische Analysis II. Seminar, 1984–86. V, 158 Seiten. 1987.

Vol. 1263: V.L. Hansen (Ed.), Differential Geometry. Proceedings, 1985. XI, 288 pages. 1987.

Vol. 1264: Wu Wen-tsün, Rational Homotopy Type. VIII, 219 pages. 1987.

Vol. 1265: W. Van Assche, Asymptotics for Orthogonal Polynomials. VI, 201 pages. 1987.

Vol. 1266: F. Ghione, C. Peskine, E. Sernesi (Eds.), Space Curves. Proceedings, 1985. VI, 272 pages. 1987.

Vol. 1267: J. Lindenstrauss, V.D. Milman (Eds.), Geometrical Aspects of Functional Analysis. Seminar. VII, 212 pages. 1987.

Vol. 1268: S.G. Krantz (Ed.), Complex Analysis. Seminar, 1986. VII, 195 pages. 1987.

Vol. 1269: M. Shiota, Nash Manifolds. VI, 223 pages. 1987.

Vol. 1270: C. Carasso, P.-A. Raviart, D. Serre (Eds.), Nonlinear Hyperbolic Problems. Proceedings, 1986. XV, 341 pages. 1987.

Vol. 1271: A.M. Cohen, W.H. Hesselink, W.L.J. van der Kallen, J.R. Strooker (Eds.), Algebraic Groups Utrecht 1986. Proceedings. XII, 284 pages. 1987.

Vol. 1272: M.S. Livšic, L.L. Waksman, Commuting Nonselfadjoint Operators in Hilbert Space. III, 115 pages. 1987.

Vol. 1273: G.-M. Greuel, G. Trautmann (Eds.), Singularities, Representation of Algebras, and Vector Bundles. Proceedings, 1985. XIV, 383 pages. 1987.

Vol. 1274: N. C. Phillips, Equivariant K-Theory and Freeness of Group Actions on C*-Algebras. VIII, 371 pages. 1987.

Vol. 1275: C.A. Berenstein (Ed.), Complex Analysis I. Proceedings, 1985–86. XV, 331 pages. 1987.

Vol. 1276: C.A. Berenstein (Ed.), Complex Analysis II. Proceedings, 1985–86. IX, 320 pages. 1987.

Vol. 1277: C.A. Berenstein (Ed.), Complex Analysis III. Proceedings, 1985–86. X, 350 pages. 1987.

Vol. 1278: S.S. Koh (Ed.), Invariant Theory. Proceedings, 1985. V, 102 pages. 1987.

Vol. 1279: D. Ieşan, Saint-Venant's Problem. VIII, 162 Seiten. 1987.

Vol. 1280: E. Neher, Jordan Triple Systems by the Grid Approach. XII, 193 pages. 1987.

Vol. 1281: O.H. Kegel, F. Menegazzo, G. Zacher (Eds.), Group Theory. Proceedings, 1986. VII, 179 pages. 1987.

Vol. 1282: D.E. Handelman, Positive Polynomials, Convex Integral Polytopes, and a Random Walk Problem. XI, 136 pages. 1987.

Vol. 1283: S. Mardešić, J. Segal (Eds.), Geometric Topology and Shape Theory. Proceedings, 1986. V, 261 pages. 1987.

Vol. 1284: B.H. Matzat, Konstruktive Galoistheorie. X, 286 pages. 1987.

Vol. 1285: I.W. Knowles, Y. Saitō (Eds.), Differential Equations and Mathematical Physics. Proceedings, 1986. XVI, 499 pages. 1987.

Vol. 1286: H.R. Miller, D.C. Ravenel (Eds.), Algebraic Topology. Proceedings, 1986. VII, 341 pages. 1987.

Vol. 1287: E.B. Saff (Ed.), Approximation Theory, Tampa. Proceedings, 1985–1986. V, 228 pages. 1987.

Vol. 1288: Yu. L. Rodin, Generalized Analytic Functions on Riemann Surfaces. V, 128 pages, 1987.

Vol. 1289: Yu. I. Manin (Ed.), K-Theory, Arithmetic and Geometry. Seminar, 1984–1986. V, 399 pages. 1987.

Vol. 1290: G. Wüstholz (Ed.), Diophantine Approximation and Transcendence Theory. Seminar, 1985. V, 243 pages. 1987.

Vol. 1291: C. Mœglin, M.-F. Vignéras, J.-L. Waldspurger, Correspondances de Howe sur un Corps p-adique. VII, 163 pages. 1987.

Vol. 1292: J.T. Baldwin (Ed.), Classification Theory. Proceedings, 1985. VI, 500 pages. 1987.

Vol. 1293: W. Ebeling, The Monodromy Groups of Isolated Singularities of Complete Intersections. XIV, 153 pages. 1987.

Vol. 1294: M. Queffélec, Substitution Dynamical Systems – Spectral Analysis. XIII, 240 pages. 1987.

Vol. 1295: P. Lelong, P. Dolbeault, H. Skoda (Réd.), Séminaire d'Analyse P. Lelong – P. Dolbeault – H. Skoda. Seminar, 1985/1986. VII, 283 pages. 1987.

Vol. 1296: M.-P. Malliavin (Ed.), Séminaire d'Algèbre Paul Dubreil et Marie-Paule Malliavin. Proceedings, 1986. IV, 324 pages. 1987.

Vol. 1297: Zhu Y.-l., Guo B.-y. (Eds.), Numerical Methods for Partial Differential Equations. Proceedings. XI, 244 pages. 1987.

Vol. 1298: J. Aguadé, R. Kane (Eds.), Algebraic Topology, Barcelona 1986. Proceedings. X, 255 pages. 1987.

Vol. 1299: S. Watanabe, Yu.V. Prokhorov (Eds.), Probability Theory and Mathematical Statistics. Proceedings, 1986. VIII, 589 pages. 1988.

Vol. 1300: G.B. Seligman, Constructions of Lie Algebras and their Modules. VI, 190 pages. 1988.

Vol. 1301: N. Schappacher, Periods of Hecke Characters. XV, 160 pages. 1988.

Vol. 1302: M. Cwikel, J. Peetre, Y. Sagher, H. Wallin (Eds.), Function Spaces and Applications. Proceedings, 1986. VI, 445 pages. 1988.

Vol. 1303: L. Accardi, W. von Waldenfels (Eds.), Quantum Probability and Applications III. Proceedings, 1987. VI, 373 pages. 1988.

Vol. 1304: F.Q. Gouvêa, Arithmetic of p-adic Modular Forms. VIII, 121 pages. 1988.

Vol. 1305: D.S. Lubinsky, E.B. Saff, Strong Asymptotics for Extremal Polynomials Associated with Weights on ℝ. VII, 153 pages. 1988.

Vol. 1306: S.S. Chern (Ed.), Partial Differential Equations. Proceedings, 1986. VI, 294 pages. 1988.

Vol. 1307: T. Murai, A Real Variable Method for the Cauchy Transform and Analytic Capacity. VIII, 133 pages. 1988.

Vol. 1308: P. Imkeller, Two-Parameter Martingales and Their Quadratic Variation. IV, 177 pages. 1988.

Vol. 1309: B. Fiedler, Global Bifurcation of Periodic Solutions with Symmetry. VIII, 144 pages. 1988.

Vol. 1310: O.A. Laudal, G. Pfister, Local Moduli and Singularities. V, 117 pages. 1988.

Vol. 1311: A. Holme, R. Speiser (Eds.), Algebraic Geometry, Sundance 1986. Proceedings. VI, 320 pages. 1988.

Vol. 1312: N.A. Shirokov, Analytic Functions Smooth up to the Boundary. III, 213 pages. 1988.

Vol. 1313: F. Colonius, Optimal Periodic Control. VI, 177 pages. 1988.

Vol. 1314: A. Futaki, Kähler-Einstein Metrics and Integral Invariants. IV, 140 pages. 1988.

Vol. 1315: R.A. McCoy, I. Ntantu, Topological Properties of Spaces of Continuous Functions. IV, 124 pages. 1988.

Vol. 1316: H. Korezlioglu, A.S. Ustunel (Eds.), Stochastic Analysis and Related Topics. Proceedings, 1986. V, 371 pages. 1988.

Vol. 1317: J. Lindenstrauss, V.D. Milman (Eds.), Geometric Aspects of Functional Analysis. Seminar, 1986–87. VII, 289 pages. 1988.

Vol. 1318: Y. Felix (Ed.), Algebraic Topology – Rational Homotopy. Proceedings, 1986. VIII, 245 pages. 1988.

Vol. 1319: M. Vuorinen, Conformal Geometry and Quasiregular Mappings. XIX, 209 pages. 1988.